Erling B. Andersen

The Statistical Analysis of Categorical Data

With 41 Figures

Springer-Verlag Berlin Heidelberg New York
London Paris Tokyo Hong Kong

Professor Dr. Erling B. Andersen
Department of Statistics
University of Copenhagen
Studiestræde 6
DK-1455 Copenhagen K
Denmark

ISBN 3-540-52139-9 Springer-Verlag Berlin Heidelberg New York Tokyo
ISBN 0-387-52139-9 Springer-Verlag New York Berlin Heidelberg Tokyo

CIP-Titelaufnahme der Deutschen Bibliothek
Andersen, Erling B.:
The statistical analysis of categorical data / Erling B. Andersen.
– Berlin; Heidelberg; New York; London; Paris; Tokyo;
Hong Kong : Springer, 1990
ISBN 3-540-52139-9 (Berlin ...)
ISBN 0-387-52139-9 (New York ...)

Printing: Weihert-Druck GmbH, Darmstadt
Bookbinding: T. Gansert GmbH, Weinheim-Sulzbach
2142/7130-543210

Preface

The aim of this book is to give an up to date account of the most commonly uses statistical models for categorical data. The emphasis is on the connection between theory and applications to real data sets. The book only covers models for categorical data. Various models for mixed continuous and categorical data are thus excluded.

The book is written as a textbook, although many methods and results are quite recent. This should imply, that the book can be used for a graduate course in categorical data analysis. With this aim in mind chapters 3 to 12 are concluded with a set of exercises. In many cases, the data sets are those data sets, which were not included in the examples of the book, although they at one point in time were regarded as potential candidates for an example.

A certain amount of general knowledge of statistical theory is necessary to fully benefit from the book. A summary of the basic statistical concepts deemed necessary prerequisites is given in chapter 2.

The mathematical level is only moderately high, but the account in chapter 3 of basic properties of exponential families and the parametric multinomial distribution is made as mathematical precise as possible without going into mathematical details and leaving out most proofs.

The treatment of statistical methods for categorical data in chapters 4 to 12 is based on development of models and on derivation of parameters estimates, test quantities and diagnostics for model departures. All the introduced methods are illustrated by data sets almost exclusively from Danish sources. If at all possible, the data source is given.

Almost all statistical computations require the use of a personal or main frame computer. A desk calculator will only in few cases suffice. As a general rule the methods in chapters 4 to 7 are covered by standard statistical software packages like SAS, BMDP, SPSS or GENSTAT. This is not the case for the methods in chapters 8 to 12. Søren V.

Andersen and the author have developed a software package for personal computers, called CATANA, which cover all models in chapters 8 to 12. This package is necessary in order to check the calculations in the examples or to work through the exercises. Information on how to obtain a diskette with CATANA, which will be released in early 1990, can be obtained by writing to the author.

A fair share of the examples and exercises are based on the Danish Welfare Study and I wish to thank the director of this study professor Erik J. Hansen, who through the Danish Data Archive put the data file from the Welfare Study to my disposal, and has been extremely helpful with extra information on the data.

Part of the book was written during visits to the United States and France. I wish to thank first of all Leo Goodman, but also Peter Bickel, Terry Speed, Jan de Leeuw, Shelby Haberman, Peter McCullogh, Darrell Bock, Clifford Clogg, Paul Holland, Robert Mislevy and Murray Aitkin in the United States and Yves Escoufier, Henri Caussinus and Paul Falguerolles in France for stimulating discussions. Many other persons have contributed to the book through discussions and criticism. It is impossible to name all, but the help of Svend Kreiner, Nils Kousgaard and Anders Milhøj is appreciated.

I also wish to thank the Danish Social Science Research Council, who financed my visits to the United States and France.

The book would never have been a reality without the care and enthusiasm with which my secretary Mirtha Cereceda typed and retyped the manuscript many times. I owe her my most sincere thanks for a very competent job.

Finally a special thank you to the many students who suffered through courses based on early drafts of the book.

Copenhagen, October 1989
Erling B. Andersen

Contents

1. Categorical Data

This book is about categorical data, i.e. data which can only take a finite or countable number of values. Typical situations, which give rise to a statistical analysis of categorical data are the following:

Consider first a population of individuals. For each of these a variable can be measured for which the possible values are the numbers from 1 to m. The variable may for example be the social class, the individual belongs to, with the possible social classes numbered 1 to 5. From the population a sample of n individuals is selected at random and for each of the sampled individuals the social class, he or she belongs to, is observed. The data then consists of the counts $x_1,...x_5$ of number of individuals observed in each social class. Based on this data set one statistical problem is to estimate the percentages of individuals in the total population, which belong to the five social classes. Another would be to test hypotheses concerning the distribution of the population on social classes.

Consider secondly the counting of traffic accidents. One may e.g. be interested in the increase or decrease of accidents following the enforcement of a given safety measure for the traffic, like speed limitations. A data set could then consist of the number of traffic accidents $x_1,...,x_{12}$ for the months of a particular year. If a safety measure has been introduced during the year, the statistical problem would be to check if the data bear evidence of a decrease in number of accidents, which cannot be ascribed to random fluctuations in the traffic counts.

Consider thirdly an economic variable, which takes a wide range of values on the real line, like income, but where the observed values for practical reasons are only registered in intervals, i.e. it is only observed which income interval a given income belongs to. The data set then consists of the numbers of incomes $x_1,...,x_m$ in each of the income intervals $(0,t_1], (t_2,t_3],...,(t_{m-1},+\infty)$.

The three kinds of data considered above are very different in nature and require, therefore, different statistical models. There are, however, a number of common features for categorical data. Such a basic structure allows for a unified treatment.

The basic statistical distribution for categorical data is the multinomial distribution. It describes the distribution of a random sample from a given large population over the categories of a variable measurable for each individual in the population. But also other distributions play important roles.

The statistical models and methods of subsequent chapters can be exemplified by five concrete sets of data.

Example 1.1.

Table 1.1 show the number of persons killed in the traffic in Denmark 1970 to 1980. The number of persons killed during a given year is a categorical random variable for which the possible values are all non–negative integers. Based on the data in table 1.1 it is possible to evaluate the extent to which the risk of being killed in the traffic has changed over the years 1970 to 1980. Since considerable speed limitations were introduced in Denmark in 1973 a more concrete question is whether these speed limits have caused a decrease in the risk of being killed in the traffic. In order to answer this question, a statistical model must be formulated for the data in table 1.1, i.e. a probability distribution must be specified for the number of killed persons in a given year. The statistical model should

Tabel 1.1. Number of persons killed in the traffic in Denmark 1970 to 1980.

Year	Number of killed persons
1970	1208
1971	1213
1972	1116
1973	1132
1974	766
1975	857
1976	857
1977	828
1978	849
1979	730
1980	690

Source: Road traffic accidents 1981. Publication 1982:8.
Statistics Denmark. Table 1.1.

include the risk of being killed as a parameter, to be estimated from the data. A comparison of the estimated risks for the years up to and including 1973 and the corresponding estimated risks for the years after 1974 can then tell the statistician if the drop in number of killed persons from 1973 and 1974 is a clear indicator of a corresponding drop in the general risk of being killed in the traffic. △.

Example 1.2.

As part of a large scale investigation of job satisfaction in 1968, a random sample of blue collar workers in Denmark were interviewed. The main questions were a number of questions concerning aspects of the workers job satisfactions. Based on the answers the workers were categorized as having high or low job satisfaction. At the same time the workers supervisors were asked similar questions leading to a categorization of the supervisors as those with high and low job satisfaction. In addition the quality of the factory management was classified as good or bad based on an external evaluation. Thus three categorical variables are measurable for each worker:

A: The workers own job satisfaction.

B. The job satisfaction of the workers supervisor.

C: The quality of the management.

Table 1.2 show the sample cross–classified according to these three categorical variables. Such a table is called a **contingency table**.

Based on the data in table 1.2 it is possible to study the way the job satisfaction of a worker depends on his work environment, exemplified by the job satisfaction of the supervisor and the quality of the management. A very primitive analysis based on comparisons of relative frequencies will indicate the type of statistical conclusions, which can be reached. Consider tables 1.3 and 1.4, where the percentage of workers with low and high job satisfaction are shown for various parts of the sample. In table 1.3 the percentages are shown for the two levels of job satisfaction of the supervisor. In table 1.4 the percentage of workers with low and high job satisfaction is in addition subdivided according to the quality of the management.

Table 1.2. A sample of 715 blue collar workers cross–classified according to three categorical variables: Own job satisfaction, supervisors job satisfaction and quality of management.

Quality of management	Supervisors job satisfaction	Own job satisfaction Low	High
Bad	Low	103	87
	High	32	42
Good	Low	59	109
	High	78	205

Source: Petersen (1968). Table M/7.

Table 1.3. Percentage of workers with high and low job satisfaction for the two levels of the supervisors job satisfaction.

Supervisors job satisfaction	Own job satisfaction Low	High	Total
Low	43	55	100
High	31	66	100

Tables 1.3 and 1.4 show that the difference between the percentage of workers with high and low job satisfaction is smaller if the two levels of the supervisors job satisfaction are studied independently as in table 1.4. The quality of the management thus seems to influence the job satisfaction of workers more than the job satisfaction of the supervisor.

△·

Table 1.4. Percentage of workers with high and low job satisfaction for the two levels of quality of management jointly with the two levels of the supervisors job satisfaction.

Quality of management	Supervisors job satisfaction	Own job satisfaction Low	High	Total
Bad	Low	54	46	100
	High	43	56	100
Good	Low	35	65	100
	High	28	72	100

Example 1.3

The Danish National Institute for Social Science Research interviewed in 1974 a random sample of 5160 Danes between 20 and 69 years old in order to investigate the general welfare in Denmark. In table 1.5 the distribution of this sample with respect to five age groups is shown.

Each of the sampled persons represents an observation of a categorical variable with the five age groups as categories. If the sample is drawn completely at random, the multinomial distribution describes the observed distribution over age groups. The parameters of this distribution are the probabilities that a randomly drawn person belong to each of the age groups. These probabilities are according to Laplace's law equal to the frequencies of persons in the total Danish population in 1974 in the various age groups. These frequencies are shown as the last column in table 1.3.

The statistical problem arising from table 1.5 is whether the sample is representative of the population. In more concrete terms the statistical problem is to determine if the observed distribution is consistent with the teoretical distribution represented by the frequencies over the age groups in the total population. An equivalent formulation of this problem is to ask , if column two in table 1.5 is equal to column three apart from random errors. If the answer is affirmative one may claim that the sample is representative as regards age. △.

Table 1.5. A random sample of persons in Denmark in 1974 distributed according to age, and the age group distribution of the total population in Denmark in 1974.

Age groups – years –	The sample – % –	The population – % –
20–29	24.3	24.6
30–39	24.3	23.0
40–49	17.5	17.8
50–59	17.7	18.4
60–69	16.1	16.2
Total	99.9	100.00
Number of persons	5166	3124455

Source: Hansen (1978). Table 4.10.

Example 1.4.

Many of the methods in this book are concerned with contingency tables, where two or more categorical variables are cross–classified for a sample of persons. Table 1.6 is a typical example of such a contingency table. In the table the welfare sample, mentioned in example 1.3, is cross–classfied according to income, in five income intervals, and wealth in five wealth intervals. The observed numbers are presented in table 1.6 as one often meet similar data, namely as percentages rowwise. The percentages immediately reveal the expected feature, that low incomes more often are connected with low wealth and high incomes more often with high wealth.

Many statistical methods for categorical data are concerned with describing the association between categorical variables. In case of the data in table 1.6, the problem is to express the obvious association between income and wealth in terms of parameters in a statistical model. △.

Table 1.6. Income and wealth cross–classified for a random sample in Denmark in 1974.

Income – 1000 Dkr –	Wealth – 1000 Dkr –					Total
	0	0–50	50–150	150–300	300–	
0–40	45	25	15	10	5	100
40–60	37	26	17	12	8	100
60–80	32	23	23	15	7	100
80–110	31	24	23	14	9	101
110–	23	18	21	18	21	101

Source: Hansen (1978). Table 6.H.32.

Example 1.5.

In an investigation of consumer behaviour a sample of 600 persons were confronted with 6 situations, where a purchased item did not live up to their expectations. For each situation, the interviewed persons were asked if they would complain to the shop or not. The answers to four of these questions are shown in table 1.7, where 1 stands for yes and 0 for no.

The purpose of collecting the data in table 1.7 was to evaluate the extent to which consumers can be graded on a consumer complain "scale" with persons who seldom complain on the lower end and persons who almost always complain on the upper end. If the persons can in fact be graded on a complain scale, one consequence would be that a positive answer to a question would indicate that the person is more likely to be a complainer than a non–complainer. Hence such a person would have a higher probability of complaining on another question than a person, who answered no to the first question. The answers to the questions are accordingly not independent. The existence of a complain scale for the persons can, therefore, partly be confirmed by showing that the answers to the four questions fail to be independent. To illustrate such an analysis consider the first response pattern 1111 in table 1.7. The probability of this response assuming independence is

$$P(1111) = p_1p_2p_3p_4,$$

where p_j is the probability that a randomly selected person answers yes to question number j. The estimates for p_1, p_2, p_3 and p_4 are the marginal frequencies

Table 1.7. The number of persons for each of the 16 possible response patterns on four questions concerning consumer complain behaviour.

Response pattern	Number of observations	Expected frequency given independence	Expected numbers given independence
1111	207	0.258	154.9
1110	72	0.185	110.8
1101	75	0.145	87.2
1100	76	0.104	62.3
1011	24	0.056	33.6
1010	24	0.040	24.0
1001	7	0.032	18.9
1000	20	0.022	13.5
0111	19	0.048	29.1
0110	22	0.035	20.8
0101	8	0.027	16.4
0100	14	0.019	11.7
0011	5	0.011	6.3
0010	11	0.008	4.5
0001	5	0.006	3.5
0000	11	0.004	2.5
Total	600	1.000	600

Source: Poulsen (1981).

$$\hat{p}_1 = 0.842$$
$$\hat{p}_2 = 0.822$$
$$\hat{p}_3 = 0.640$$
$$\hat{p}_4 = 0.583.$$

The expected frequency under independence with response pattern (1111) is thus $0.842 \cdot 0.822 \cdot 0.640 \cdot 0.583 = 0.258$. This number and the corresponding expected frequencies for the remaining 15 response patterns are shown in table 1.7. The corresponding expected numbers, obtained by multiplying with 600, are also shown in table 1.7. The main step of a statistical analysis of the data in table 1.7 is to evaluate the likelihood that the observed and the expected numbers for the various reponse patterns are equal apart from random errors. If this is the case the answers to the questions are independent and no complain scale exists. With the numbers in table 1.7 this likelihood is obviously very low. $\triangle$.

2. Preliminaries

2.1. Statistical models

In this chapter a short review is given of some basic elements of statistical theory which are necessary requisites for the theory and methods developed in subsequent chapters.

A **statistical model** is a specification of the probability distribution of the data. Let the **data set** consist of the observed numbers $x_1,...,x_n$. It is then assumed that there exist n **random variables** $X_1,...,X_n$ of which $x_1,...,x_n$ are the observed values. The joint probability

$$(2.1) \qquad f(x_1,...,x_n) = P\{(X_1=x_1)\cap...\cap(X_n=x_n)\}$$

then specifies the statistical model. In most cases we assume that the model belong to a family of models, which is indexed by one or more unknown **parameters**. The model is then written as

$$(2.2) \qquad f(x_1,...,x_n \,|\, \theta), \quad \theta \epsilon \Theta,$$

where the range space Θ is called the **parameter space**. The parameter space is usually a subset of a k–dimensional Euclidean space, in which case θ can be written $\theta=(\theta_1,...,\theta_k)$.

The probability (2.2) as a function of θ is called the **likelihood function** and is denoted by L, i.e.

$$L(\theta\,|\,x_1,...,x_n) = f(x_1,...,x_n\,|\,\theta).$$

Most statistical methods are based on properties of the likelihood function.

In many situations the random variables $X_1,...,X_n$ are independent and identically distributed, such that the likelihood function becomes

$$L(\theta\,|\,x_1,...,x_n) = f(x_1\,|\,\theta)...f(x_n\,|\,\theta)$$

with $f(x_i|\theta)=P(X_i=x)$.

A data set $(x_1,...,x_n)$ is often called **a sample**. This expression is derived from situations where the data are observed values from units in a sample drawn from a population. In accordance with this a function

$$t = t(x_1,...,x_n)$$

of the observations is called a **sample function**. Sample functions are important tools for drawing statistical conclusions from a sample. To stress their role as statistical tools, sample functions are often called **statistics**.

2.2. Estimation

A method for summarizing the information in a data set about a parameter is called an **estimation method**. The most commonly used method is the **maximum likelihood method**. A sample function which is used to estimate a parameter is called **an estimate**. The maximum likelihood (ML) estimate $\hat{\theta}$ is defined as the value of θ, which maximizes L, i.e.

$$L(\hat{\theta}|x_1,...,x_n) = \max_{\theta \in \Theta} L(\theta|x_1,...,x_n).$$

Obviously $\hat{\theta}$ is a sample function and, therefore, an estimate.

If Θ is a subset of the real line, the ML–estimator is in regular cases found as the solution to the **likelihood equation**

$$\frac{d\ln L(\theta|x_1,...,x_n)}{d\theta} = 0$$

The most important regular case is, when the statistical model forms an **exponential family**. Also for the so–called **parametric multinomial distribution** the regular cases can be identified. Both cases are described in chapter 3.

As a function $\theta(X_1,...,X_n)$ of random variables an estimate is called an **estimator**.

If Θ is a subset of a k–dimensional Euclidean space, the ML–estimator $\hat{\theta}=(\hat{\theta}_1,...,\hat{\theta}_k)$ is in regular cases found as a solution to the k **likelihood equations**

$$\frac{\partial \ln L(\theta_1,.., \theta_k | x_1,...,x_n)}{\partial \theta_j} = 0, \qquad j=1,...,k.$$

Two important properties of estimates are unbiasedness and consistency. An estimate $\hat{\theta}=\hat{\theta}(x_1,...,x_n)$ is said to be **unbiased** if it satisfies

$$E_\theta\left[\hat{\theta}(X_1,...,X_n)\right] = \theta,$$

where the mean value is taken with respect to the probability $f(x_1,...,x_n | \theta)$, i.e.

$$\sum_{x_1} .. \sum_{x_n} \hat{\theta}(x_1,...,x_n)f(x_1,...,x_n | \theta) = \theta.$$

An estimator is said to be **consistent** if it satisfies

$$\hat{\theta}(X_1,...,X_n) \overset{P}{\to} \theta \text{ as } n\to\infty \ .$$

From the law of large numbers it follows that $\hat{\theta}$ is consistent, if it is unbiased and

$$\text{var}\left[\hat{\theta}\right] \overset{P}{\to} 0 \text{ as } n\to\infty \ .$$

But an estimator can also be consistent under other conditions. In particular it does not need to be unbiased.

The basis for a description of the properties of an estimator is its probability distribution, i.e. the probability

$$g(t \mid \theta) = P(\hat{\theta}(X_1,...,X_n) = t \mid \theta),$$

for all values of θ.

The distribution of $\hat{\theta}$ is, however, in most cases so complicated, that it is of little practical use. In most cases statistical methods for practical use rely, therefore, on asymptotic properties of the estimator, i.e. approximations to the distribution of $\hat{\theta}$, which are valid for large sample sizes. In many important cases, to be considered later, it can be proved that $\hat{\theta}$ is asymptotically normally distributed, i.e. for a certain constant σ_θ,

$$P(\sqrt{n}\, \frac{\hat{\theta}-\theta}{\sigma_\theta} \leq u \mid \theta) \rightarrow \Phi(u) \text{ as } n \rightarrow \infty.$$

where $\Phi(u)$ is the cumulative distribution function for the standard normal distribution.

With an abbreviated notation, this is often written as

$$\hat{\theta} \overset{a}{\sim} N(\theta, \sigma_\theta^2/n).$$

This means that the distribution of $\hat{\theta}$ in large samples can be approximated by a normal distribution with mean value θ and variance σ_θ^2/n.

In the k–dimensional case it can be proved that the ML–estimator in regular cases has the asymptotic distribution.

$$(\hat{\theta}_1,...,\hat{\theta}_k) \overset{a}{\sim} N_k((\theta_1,...,\theta_k), \frac{1}{n}\Sigma_\theta),$$

where Σ_θ is non–negative definite matrix and N_k, the k–dimensional normal distribution.

Sometimes an **interval estimate** rather than a point estimate is preferred. The interval (θ_1, θ_2) is an interval estimate for θ, if there exist sample functions $\theta_1(x_1,...,x_n)$ and $\theta_2(x_1,...,x_n)$ such that

$$P(\theta_1(X_1,...,X_n) > \theta) = \alpha/2$$

and

$$P(\theta_2(X_1,...,X_n) < \theta) = \alpha/2$$

The interval $(\theta_1 \leq \theta \leq \theta_2)$ is then interpreted as an interval estimate or a **confidence interval** in the sense that the unknown parameter θ is contained in the interval with confidence level $1-\alpha$. Confidence intervals are usually justified by a so–called **frequency interpretation**. Suppose the confidence interval is computed for each sample in a long sequence of independent samples. The probability of both $\theta_1(X_1,...,X_n)$ being less than or equal to θ and $\theta_2(X_1,...,X_n)$ being larger than or equal to θ is then $1-\alpha$ in each of the samples. The relative frequency of the event $\{\theta_1 \leq \theta \leq \theta_2\}$ will therefore approach $1-\alpha$ in the long run due to the law of large numbers. With many independent calculations of confidence intervals the true parameter value θ will thus be in the interval with frequency $1-\alpha$. It can accordingly be claimed that the confidence level of θ being in the interval $[\theta_1,\theta_2]$ is $1-\alpha$. It is obvious that another word than "probability" must be used to describe our belief in the statement $\theta_1 \leq \theta \leq \theta_2$, as the event $\{\theta_1 \leq \theta \leq \theta_2\}$ for given values of θ_1 and θ_2 has probability either 1, if θ in fact is in the interval, or 0, if θ is in fact outside the interval.

Confidence intervals are very often obtained as approximate intervals based on an application of the central limit theorem. Suppose that $\hat{\theta}$ is an estimate for θ, e.g. the ML–estimate, and that

(2.3)
$$\hat{\theta} \overset{a}{\sim} N(\theta, \sigma_\theta^2/n),$$

It then follows that

$$P(-u_{1-\alpha/2} \leq \sqrt{n}(\hat{\theta}-\theta)/\sigma_\theta \leq u_{1-\alpha/2}) \simeq 1-\alpha$$

or

$$P(\hat{\theta}-u_{1-\alpha/2}\frac{\sigma_\theta}{\sqrt{n}} \leq \theta \leq \hat{\theta} + u_{1-\alpha/2}\frac{\sigma_\theta}{\sqrt{n}}) \simeq 1-\alpha$$

such that

$$\left[\hat{\theta}-u_{1-\alpha/2}\frac{\hat{\sigma}_\theta}{\sqrt{n}} \ , \ \hat{\theta} + u_{1-\alpha/2}\frac{\hat{\sigma}_\theta}{\sqrt{n}}\right]$$

where $\hat{\sigma}_{\theta}^2$ is an estimate of σ_{θ}^2, is a confidence interval with approximative confidence level $1-\alpha$.

2.3. Testing statistical hypotheses

A **statistical hypothesis** is a specification of the unknown value of a parameter. A typical statistical hypothesis is

$$H_0: \theta = \theta_0.$$

A **test** of a statistical hypothesis is a confrontation of H_0 with one or more alternatives. Typical alternatives are

$$H_1: \theta \neq \theta_0$$
$$H_1: \theta < \theta_0$$

or

$$H_1. \theta > \theta_0.$$

The purpose of the test is evaluate if the data supports H_0 or one of the alternatives. Usually the data is for use in test situations summarized in a **test statistic**.

Whether the hypothesis is supported by the data can be determined in two ways. One way is to divide the range space **T** of the test statistic in two regions. One region A is called the **acceptance region** and the remaining region $\overline{A}$ is called the **critical region**. The acceptance and critical regions must satisfy

$$T = A \cup \overline{A}$$

and

$$A \cap \overline{A} = \emptyset.$$

Since H_0 is rejected whenever $t \in \overline{A}$, the critical region should consist of all values of t, for which it is unlikely that H_0 is true. How $\overline{A}$ and A are chosen depend on the alternative. If thus $H_1: \theta \neq \theta_0$ and the test statistic $\hat{\theta} = t(x_1, ..., x_n)$ is an estimate of θ, the critical re–

gion can be chosen as those t–values for which $\hat{\theta}$ is not close to θ_0, i.e.

$$\overline{A} = \{t \,|\, |t-\theta_0| \geq c\}$$

for a certain number c. If the alternative is one–sided, for example $\theta > \theta_0$, the critical region is also chosen one–sided as

$$\overline{A} = \{t \,|\, t-\theta_0 \geq c\}$$

for a certain c. Note that the acceptance region and the critical region A and $\overline{A}$ are defined in the range space of the test statistic. The two subsets A^* and $\overline{A}^*$ in the range space $\mathbf{X}$ of the original observations defined through

$$A^* = \{x_1,...,x_n \,|\, t(x_1,...,x_n) \in A\}$$

and $\overline{A}^* = \mathbf{X} \backslash A^*$ are also referred to as the acceptance region and the critical region.

The extent to which the data supports the hypothesis can alternatively be measured by the **level of significance** defined as

$$p = P(T \geq t \,|\, \theta = \theta_0),$$

if the alternative is $H_1 : \theta > \theta_0$ and large values of t are indications of H_1 rather than H_0 being true. If the alternative is $\theta < \theta_0$, the level of significance is given by

$$p = P(T \leq t \,|\, \theta = \theta_0).$$

Under a two sided alternative $H_1 : \theta \neq \theta_0$ both large and small values of t should lead a rejection of H_0. Hence the level of significance is chosen as

$$p = 2\min\{P(T \geq t \,|\, \theta = \theta_0), P(T \leq t \,|\, \theta = \theta_0)).$$

A general method for selecting a test statistic is based on the **likelihood ratio**, defined for the hypothesis H_0: $\theta = \theta_0$ as

$$r(X_1,...,X_n) = \frac{f(X_1,...,X_n \mid \theta_0)}{f(X_1,...,X_n \mid \hat{\theta})} \quad ,$$

where $\hat{\theta}$ is the maximum likelihood estimator for θ. The observed value $r(x_1,...,x_n)$ of the likelihood ratio is a measure of the extent to which the given data set $(x_1,...,x_n)$ supports the null hypothesis, such that H_0 is rejected if r is small and accepted if r is close to its maximum value $r=1$. For a given observed value $r=r(x_1,...,x_n)$ of the likelihood ratio the significance level p of the data is accordingly

$$p = P(r(X_1,...,X_n) \leq r).$$

The critical region consist of values of $x_1,...,x_n$ for which

$$r(x_1,...,x_n) \leq c.$$

The critical level c is determined as

$$P(r(X_1,...,X_n) \leq c \mid \theta = \theta_0) = \alpha \quad ,$$

where α is the level of the test. In many cases there exist a **sufficient statistic** $T=t(X_1,...,X_n)$ for the parameter θ, for which the likelihood function can be factorized as

(2.4)
$$f(x_1,...,x_n \mid \theta) = h(x_1,...,x_n)g(t \mid \theta),$$

where h does not depend on θ and g only depend on the x's through t. The factorization (2.4) implies that the conditional probability

$$f(x_1,...,x_n \,|\, t) = P\{(X_1,...,x_1) \cap ... \cap (X_n = x_n) \,|\, T = t\}$$

has the form

$$f(x_1,...,x_n \,|\, t) = h(x_1,...,x_n) / \sum_{t(x_1,...,x_n) = t} ... \sum h(x_1,...,x_n),$$

which does not depend on θ. Hence T is a sufficient statistic for θ if the observed value of t contains all the information available in the data set about θ. It is easy to see that if T is sufficient for θ, then the likelihood ratio can be written

(2.5)
$$r(X_1,...,X_n) = \frac{g(T \,|\, \theta_0)}{g(T \,|\, \hat{\theta})} \quad,$$

such that the likelihood ratio test statistic only depends on T.

In case the model depends on k real valued parameters $\theta_1,...,\theta_k$ a test can in most cases not be based on simple calculations of levels of significance based on a sufficient statistic. The principles behind the likelihood ratio test are, however, still applicable. If all k parameter are specified under H_0, i.e.

$$H_0: \theta_1 = \theta_{10},...,\theta_k = \theta_{k0},$$

the level of significance for an observed value $r = r(x_1,...,x_n)$ of the likelihood ratio is given as

$$p = P(r(X_1,...,X_n) \leq r \,|\, \theta_1 = \theta_{10},...,\theta_k = \theta_{k0}),$$

with the likelihood ratio defined as

$$r(X_1,...,X_n) = \frac{f(X_1,...,X_n \,|\, \theta_{10},...,\theta_{k0})}{f(X_1,...,X_n \,|\, \hat{\theta}_1,...,\hat{\theta}_k)},$$

where $\hat{\theta}_1,...,\hat{\theta}_k$ are the ML–estimators for $\theta_1,...,\theta_k$. At level α, H_0 is rejected for

$$r(x_1,...,x_n) \leq r_\alpha,$$

where r_α is the α–percentile of the distribution of $r(X_1,...,X_n)$ under H_0.

If all the parameters of a model are specified under H_0, the hypothesis is called a **simple hypothesis**. In many situations, however, only some of the components of $\theta=(\theta_1,...,\theta_k)$ are specified under H_0. Suppose for example that the null hypothesis is

$$H_0:\ \theta_1=\theta_{10},...,\theta_r=\theta_{r0},$$

with $\theta_{r+1},...,\theta_k$ being unspecified. The likelihood ratio is then defined as

$$r(X_1,...,X_n) = \frac{f(X_1,...,X_n \mid \theta_{10},...,\theta_{r0},\tilde{\theta}_{r+1},...,\tilde{\theta}_k)}{f(X_1,...,X_n \mid \hat{\theta}_1,...,\hat{\theta}_k)}$$

where $\tilde{\theta}_{r+1},...,\tilde{\theta}_k$ are the ML–estimators for $\theta_{r+1},...,\theta_k$ in a model where $\theta_1,...,\theta_r$ have values $\theta_{10},...,\theta_{r0}$, and $\hat{\theta}_1,...,\hat{\theta}_k$ are the ML–estimators in a model, where all k parameters are unconstrained. A hypothesis, where only some of the parameters of the model are specified, is called a **composite hypothesis**.

For the composite hypothesis

$$H_0:\ \theta_1=\theta_{10},...,\theta_r=\theta_{r0},\ \text{with } \theta_{r+1},...,\theta_k \text{ unconstrained},$$

the level of significance for $r=r(x_1,...,x_n)$ is given by

$$p = P(r(X_1,...,X_n) \leq r \mid \theta_1=\theta_{10},...,\theta_r=\theta_{r0})$$

The level of significance depends for a composite hypothesis on the unknown values of $\theta_{r+1},...,\theta_k$.

In other situations the parameters $\theta_1,...,\theta_k$ are constrained through a common dependency on k–r new parameters $\tau_1,...,\tau_{k-r}$. The hypothesis, to be tested, can then be

(2.6)
$$
\begin{cases}
\theta_1 = h_1(\tau_1,...,\tau_{k-r}) \\
... \\
\theta_k = h_k(\tau_1,...,\tau_{k-r}) \quad ,
\end{cases}
$$

where $h_1,...,h_k$ are real valued functions of the τ's.

In this case the likelihood ratio is defined as

$$
r(X_1,...,X_n) = \frac{f(X_1,...,X_n \mid \tilde{\theta}_1,...,\tilde{\theta}_k)}{f(X_1,...,X_n \mid \hat{\theta}_1,...,\hat{\theta}_k)} ,
$$

where $\tilde{\theta}_j = h_j(\hat{\tau}_1,...,\hat{\tau}_{k-r})$ for j=1,...,k and $\hat{\tau}_1,...,\hat{\tau}_{k-r}$ are the ML–estimates for the τ's.

2.4. Checking the model

A **model check** is a procedure for evaluating to what extent the data supports the model.

The most direct way to check the model is by means of **residuals**. The residuals for the model

$$
f(x_1,...,x_n \mid \theta), \ \theta \epsilon \Theta
$$

are defined as

$$
\hat{e}_i = x_i - E[X_i \mid \hat{\theta}],
$$

where $\hat{\theta}$ is the ML–estimate for θ. Residuals are usually scaled such that they are measured in the same units relative to their standard error. The scaled residuals are de–

fined as

$$r_i = (x_i - E[X_i | \hat{\theta}]) / \sqrt{\text{var}[\hat{e}_i]} \ .$$

They are called **standardized residuals**. A model check based on residuals consists of making a **residual plot** of the standardized residuals. What the standardized residuals should be plotted against depend on the model in question.

For categorical data, there are usually many x–values, which have identical values. In this case the standardized residuals are derived as follows:

Suppose that $X_1,...,X_n$ are independent random variable, each of which can take one of m possible values $z_1,...,z_m$. Let further π_j be the probability that X_i takes the value z_j and define the random variables $Y_1,...,Y_m$ as

$$Y_j = \text{number of } X_i\text{'s equal to } z_j.$$

Then $(Y_1,...,Y_m)$ follows a multinomial distribution with parameters $(n,\pi_1,...,\pi_m)$, i.e.

(2.7)
$$f(y_1,...,y_m | \pi_1,...,\pi_m) = P(Y_1 = y_1,...,Y_m = y_m)$$

$$= \begin{bmatrix} n \\ y_1 \cdots y_m \end{bmatrix} \pi_1^{y_1} ... \pi_m^{y_m} ,$$

where the π_j's depend on θ. The residuals based on the y_j's are defined as

$$\hat{e}_j = y_j - E[Y_j | \pi_j(\hat{\theta})] = y_j - n \ \pi_j(\hat{\theta}).$$

The standardized residuals are

$$r_j = (y_j - n\pi_j(\hat{\theta})) / \sqrt{\text{var}[\hat{e}_j]} ,$$

where $\text{var}[\hat{e}_j]$ depends on the estimated π_j's.

The model is accepted as a description of the data based on a residual plot if the residuals are small and does not exhibit a systematic patterns. A model check based on residuals can be supplemented by a goodness of fit test.

If $(Y_1,...,Y_m)$ indeed follows a multinomial distribution with parameters $(n,\pi_1(\theta),...,\pi_m(\theta))$, it can be shown that the random variable

$$Q = \sum_{j=1}^{m} (Y_j - n\pi_j(\hat{\theta}))^2/(n\pi_j(\hat{\theta}))$$

is asymptotically χ^2-distributed with m−2 degrees of freedom.

The data obviously supports the model if the observed value q of Q is close to zero, while large values of q indicate model departures. Hence the model is rejected as a description of the data if the level of significance computed approximately as

$$p = P(Q \geq q) \simeq P(\chi^2(m-2) \geq q).$$

is large.

A test based on Q is called a **goodness of fit test** and Q is called the **Pearson test statistic**. If the given model depends on several real valued parameters $\theta_1,...,\theta_k$, the goodness of fit statistics becomes

$$Q = \sum_{j=1}^{m} (Y_j - n\pi_j(\hat{\theta}_1,...,\hat{\theta}_k))^2/(n\pi_j(\hat{\theta}_1,...,\hat{\theta}_k)),$$

and the asymptotic χ^2-distribution has m−k−1 degrees of freedom. The precise assumptions for the asymptotic χ^2-distribution of Q and proofs for some important special cases are given in chapter 3.4. The model (2.7), where the π_j's depend on a parameter θ is called the **parametric multinomial distribution**.

An alternative to the test statistic Q, is the **transformed likelihood ratio test statistic**. It can be derived as a test statistic for a parametric hypothesis. Assume first that the distribution of $(Y_1,...,Y_m)$ is extended to all multinomial distributions of dimension m with no constraints on the π_j's. Within this class of models consider the hypothesis

(2.8) $$H_0: \pi_j = \pi_j(\theta_1,...,\theta_k), \; j=1,...,m.$$

Since Y_j/n is the ML–estimate for π_j in an unconstrained multinomial distribution, the likelihood ratio for H_0 is

(2.9) $$r(Y_1,...,Y_m) = \frac{\left[\begin{smallmatrix} n \\ Y_1 \cdots Y_m \end{smallmatrix}\right](\pi_1(\hat{\theta}_1,\ldots,\hat{\theta}_k))^{Y_1} \cdots (\pi_m(\hat{\theta}_1,\ldots,\hat{\theta}_k))^{Y_m}}{\left[\begin{smallmatrix} n \\ Y_1 \cdots Y_m \end{smallmatrix}\right](\frac{Y_1}{n})^{Y_1} \cdots (\frac{Y_m}{n})^{Y_m}},$$

where $\hat{\theta}_1,...,\hat{\theta}_k$ are the ML–estimates for the θ_j's under the hypothesis.

The transformed likelihood ratio test statistic $-2\ln r(Y_1,...,Y_m)$ is due to (2.9) equal to

$$-2\ln r(Y_1,...,Y_m) = 2 \sum_{j=1}^{m} Y_j \ln(Y_j/(n\hat{\pi}_j))$$

with $\hat{\pi}_j = \pi_j(\hat{\theta}_1,...,\hat{\theta}_k)$. The test statistics Q and $-2\ln r(Y_1,...,Y_m)$ are asymptotically equivalent in the sense that

$$|Q-(-2\ln r(Y_1,...,Y_m))| \overset{P}{\to} 0,$$

and Q and $-2\ln r(Y_1,...,Y_m)$ have the same asymptotic χ^2–distribution. It is a matter of taste whether Q or $-2\ln r$ are preferred as a gooodness of fit test statistic. A general class of power divergence statistics, which include Q and $-2\ln r$ was introduced by Cressie and Read (1984).

The level of significance is defined as

$$p = P(r(Y_1,...,Y_m) \le r(y_1,...,y_m)),$$

where $r(y_1,..,y_m)$ is the observed value of $r(Y_1,...,Y_m)$. Since $\ln x$ is a monotone function, the level of significance can also be computed as

$$p = P(-2\ln r(Y_1,...,Y_m) \geq -2\ln r(y_1,...,y_m)) .$$

To test a statistical hypothesis can just as well be described as checking a new model against an already accepted model. Hence the words hypothesis and model are often synonomous. As an illustration consider the multinomial model (2.7). If no constraints are imposed on the τ_j's the model is called the **saturated model**.

Under the hypothesis (2.8), the likelihood is given by

(2.10)
$$L(\theta) = \left[y_1,...,y_m \right] \prod_{j=1}^{m} (\pi_1(\theta))^{y_1}...(\pi_m(\theta))^{y_m},$$

The likelihood function for the saturated model is

(2.11)
$$L(\pi_1,...,\pi_m) = \left[y_1,...,y_m \right] \pi_1^{y_1}...\pi_m^{y_m} .$$

A comparison of (2.10) and (2.11), shows that to test the parametric hypothesis

$$H_0: \pi_j = \pi_j(\theta) , j=1,...,m$$

is equivalent to compare the saturated model (2.11) and the parametric multinomial model (2.10).

In general to test the hypothesis

$$H_0: \theta = \theta_0$$

against the alternative

$$H_1: \theta \neq \theta_0$$

is equivalent to comparing the model with likelihood $L(\theta_0)$, containing just one distribution, with the model with unconstrained likelihood $L(\theta)$.

3. Statistical Inference

3.1. Log–linear models

The majority of interesting models for categorical data are **log–linear models**. A family of log–linear models is often referred to as an **exponential family**.

Consider n independent, identically distributed discrete random variables $X_1,...,X_n$ with common point probability

(3.1)
$$f(x|\theta_1,...,\theta_k) = P(X_i=x|\theta_1,...,\theta_k),$$

which depend on k real valued parameters $\theta_1,...,\theta_k$. The model

(3.2)
$$f(x_1,...,x_n|\theta_1,...,\theta_k) = \prod_{i=1}^{n} f(x_i|\theta_1,...,\theta_k)$$

is then called a log–linear model or is said to form an exponential family, if the logarithm of (3.1) has the functional form

(3.3)
$$\ln f(x|\theta_1,...,\theta_k) = \sum_{j=1}^{m} g_j(x)\varphi_j(\theta_1,...,\theta_k)+h(x)-K(\theta_1,...,\theta_k)$$

where g_j, φ_j, and h are all real valued functions of their arguments. The function K satisfies

(3.4)
$$K(\theta_1,...,\theta_k) = \ln\{\sum_x \exp(\sum_j g_j(x)\varphi_j(\theta_1,...,\theta_k)+h(x))\},$$

since $\sum_x f(x|\theta_1,...,\theta_k)=1$.

The dimension of the exponential family is the smallest integer m for which the representation (3.3) is possible. The dimension of an exponential family is less than the apparent dimension of the logarithmic form (3.3) if there are linear dependencies between either the g's or the φ's. Suppose for example that

$$\ln f(x \mid \theta_1,\ldots,\theta_k) = \sum_{j=1}^{5} g_j(x)\varphi_j(\theta_1,\ldots,\theta_k) + h(x) - K(\theta_1,\ldots,\theta_k).$$

but that for all values of the θ's

$$\varphi_1(\theta_1,\ldots,\theta_k) + \ldots + \varphi_5(\theta_1,\ldots,\theta_k) = 0.$$

Then $\varphi_5 = -\varphi_1 - \ldots - \varphi_4$, and we have

$$\ln f(x \mid \theta_1,\ldots,\theta_k) = \sum_{j=1}^{4} g_j^*(x)\varphi_j(\theta_1,\ldots,\theta_k) + h(x) - K(\theta_1,\ldots,\theta_k)$$

with $g_j^*(x) = g_j(x) - g_5(x)$ and the dimension m=4 rather than the apparent dimension m=5.

Under the log–linear model (3.3), the logarithm of the point probability of $x_1,\ldots,x_n$ can be written as

$$(3.5) \qquad \ln f(x_1,\ldots,x_n \mid \theta_1,\ldots,\theta_k) = \sum_{j=1}^{m} t_j\tau_j + \sum_{i=1}^{n} h(x_i) - nK(\theta_1,\ldots,\theta_k)$$

where

$$t_j = \sum_{i=1}^{n} g_j(x_i), \; j=1,\ldots,m$$

and

$$\tau_j = \varphi_j(\theta_1,\ldots,\theta_k), \;\; j=1,\ldots,m.$$

The parameters $\tau_1,\ldots,\tau_m$ are called the **canonical parameters**. The strategy for making statistical inference based on log–linear models is to formulate the statistical problem under consideration in terms of the canonical parameters, if at all possible. In case infer–

ence is needed about the original parameters, the relevant results are derived from those obtained for the canonical parameters.

Since the joint probability of the x's according to (3.4) and (3.5) has the multiplicative form

$$f(x_1,...,x_n \mid \theta_1,...,\theta_k) = G(t_1,...,t_m;\tau_1,...,\tau_m)H(x_1,...,x_n),$$

the t_j's form a set of sufficient statistics for the canonical parameters. This means that as regards statistical inference concerning the τ's, we can restrict attention to the joint distribution of the **sufficient statistics**

$$T_j = \sum_{i=1}^{n} g_j(X_i), \; j=1,...,m.$$

The joint distribution of $T_1,...,T_m$

$$f(t_1,...,t_m \mid \theta_1,...,\theta_k) = P(T_1=t_1,...,T_m=t_m)$$

can according to (3.5) be written as

$$f(t_1,...,t_m \mid \theta_1,...,\theta_k) = e^{\sum t_j \cdot \tau_j} \sum_{S(t_1,\cdots,t_m)} e^{\sum h(x_i)} e^{-nK(\theta_1,...,\theta_k)},$$

where $S(t_1,...,t_m) = \{x_1,...,x_n \mid \sum_i g_j(x_i)=t_j, \; j=1,...,m\}$.

Since K defined in (3.4) only depend on the θ's through the τ's, K can also be defined as

$$K(\tau_1,...,\tau_m) = \ln\{\sum_x \exp(\sum_j g_j(x)\tau_j + h(x))\}.$$

Hence with

$$h_1(t_1,...,t_m) = \ln_{S(t_1,...,t_m)} \Sigma \; e^{\Sigma h(x_i)},$$

the logarithm of $f(t_1,...,t_m|\tau_1,...,\tau_m)$ has the log–linear form

$$(3.6) \qquad \ln f(t_1,...,t_m|\tau_1,...,\tau_m) = \sum_{j=1}^{m} t_j\tau_j + h_1(t_1,...,t_m) - nK(\tau_1,...,\tau_m),$$

Here and in the following K is used both when the arguments are $\theta_1,...,\theta_k$ and when the arguments are $\tau_1,...,\tau_m$.

The concept of an exponential family goes far back. At one point in time its discovery was attributed to four people Fisher, Darmois, Pitman and Koopman. As a tool for making statistical inference, it was brought to prominence by Lehmann (1959) in his book on testing statistical hypothesis. The exact conditions for the validity of the commonly used statistical methods based on the likelihood function and a precise definition of the concept of an exponential distribution is due to Barndorff–Nielsen (1978). In a parallel development Haberman (1974b) gave rigorous proofs of the validity of a wide range of statistical results for log–linear models applied to contingency table data.

In the following a number of results concerning estimators and statistical tests in log–linear models are stated. These results are extensively used in contingency table theory and other theories for categorical data. Not all results in this chapter are supported by rigorous mathematical proofs, but in order to gain insight in the mathematical structure of the likelihood equations, on which the ML–estimates are based, and the structure of the asymptotic limits for the distributions of estimators and test statistics, the results are proved for the case of one real valued canonical parameter τ.

The results are stated for the case, where the x's are independent and identically distributed. Under certain extra conditions the results are also true for situations where $x_1,...,x_n$ are not identically distributed, as long as the model reduces to the form (3.6) for a set of m sufficient statistics. Assume thus that all the X_i's follow log–linear models (3.3), but that the probability

$$f_i(x \mid \theta_1, \ldots, \theta_k) = P(X_i = x \mid \theta_1, \ldots, \theta_k)$$

depends on i. Then

$$\ln f_i(x \mid \theta_1, \ldots, \theta_k) = \sum_{j=1}^{m} g_{ij}(x) \varphi_j(\theta_1, \ldots, \theta_k) + h_i(x) - K_i(\theta_1, \ldots, \theta_k).$$

It is assumed that the canonical parameters are the same for all i. Equation (3.5) then takes the form

$$\ln f(x_1, \ldots, x_n \mid \theta_1, \ldots, \theta_k) = \sum_{j=1}^{m} t_j \tau_j + \sum_{i=1}^{n} h_i(x_i) - \sum_{i=1}^{n} K_i(\theta_1, \ldots, \theta_k),$$

where

$$t_j = \sum_{i=1}^{n} g_{ij}(x_i)$$

Equation (3.6) is, however, unchanged if $h_1(t_1, \ldots, t_m)$ is defined as

$$h_1(t_1, \ldots, t_m) = \ln \sum_{S(t_1, \ldots, t_m)} e^{\sum_i h_i(x_i)}.$$

with $S(t_1, \ldots, t_m) = \{x_1, \ldots, x_n \mid \Sigma g_{ij}(x_i) = t_j, \ j=1, \ldots, m\}$, and $nK(\tau_1, \ldots, \tau_m)$ is defined as $\Sigma K_i(\theta_1, \ldots, \theta_k)$.

A discussion of estimation problems in exponential families with non–identically distributed random variables was given by Nordberg (1980).

3.2. The one–dimensional case

For m=1 the log–linear model (3.5) has the form

$$\ln f(x_1, \ldots, x_n \mid \theta) = t(x_1, \ldots, x_n) \varphi(\theta) + h(x_1, \ldots, x_n) - nK(\theta)$$

while (3.6) has the form

(3.7) $$\ln f(t \mid \tau) = t\tau + h_1(t) - nK(\tau).$$

Here $\tau = \varphi(\theta)$ is the canonical parameter and $t = t(x_1, \ldots, x_n)$ the sufficient statistic for τ.

Theorem 3.1
The likelihood equation

(3.8) $$\frac{d\ln f(t \mid \tau)}{d\tau} = 0$$

for a log–linear model (3.7) is equivalent to

$$E[T \mid \tau] = t,$$

and to

$$nK'(\tau) = t.$$

Proof

For $m=1$, $K(\tau) = \frac{1}{n}\ln\{\Sigma_t \exp(t\tau + h_1(t))\}$. Hence

(3.9) $$e^{nK(\tau)} = \Sigma_t e^{t\tau + h_1(t)},$$

and it follows by differentiation that

(3.10) $$nK'(\tau)e^{nK(\tau)} = \Sigma_t t e^{t\tau + h_1(t)}.$$

Dividing by $e^{nK(\tau)}$ on both sides in (3.10) and using that

$$f(t \mid \tau) = e^{t\tau + h_1(t)} e^{-nK(\tau)}$$

then yields

$$nK'(\tau) = E[T \mid \tau].$$

Since, however, according to (3.7)

$$\frac{d\ln f(t\,|\,\tau)}{d\tau} = t - nK'(\tau),$$

the theorem follows. $\square$.

In case the ML–estimator is a solution to the likelihood equation, it can thus be found by simply equating the observed value of the sufficient statistic and its mean value. Theorem 3.1 can be sharpend, if two important concepts for log–linear models are introduced, the domain and the support. The **domain** or the natural parameter space, D, is defined as the subset of the range space of τ for which

(3.11)
$$\sum_t e^{t\tau + h_1(t)} < \infty.$$

Since the domain is defined through condition (3.11), it can be a smaller set or a larger set than the parameter space for τ defined as

$$\{\tau\,|\,\tau = \tau(\theta),\ \theta \in \Theta\}.$$

It can be shown that the domain D is always an interval and that the function $K(\tau)$ is infinitely often differentiable with finite derivatives for τ in the interior of D.

The **support** T_0 of the log–linear model (3.7) is the set of all t–values with positive probability i.e.

$$T_0 = \{t\,|\,f(t\,|\,\tau) > 0\}.$$

According to theorem 3.1 the ML–estimate is found from the equation

(3.12)
$$nK'(\tau) = t$$

in regular cases. From (3.10) follows

$$nK''(\tau)e^{nK(\tau)} + n^2(K'(\tau))^2 e^{nK(\tau)} = \Sigma_t t^2 e^{t\tau + h_1(t)}.$$

Hence the second derivative $K''(\tau)$ of K satisfies

(3.13) $$nK''(\tau) = var[T \mid \tau],$$

since $nK'(\tau)=E[T \mid \tau]$. From (3.13) follows that $K'(\tau)$ is an increasing function and (3.12) has at most one solution $\hat{\tau}$, if $var[T \mid \tau]>0$ for all τ. The next theorem shows an even stronger result, namely that the likelihood equation has a unique solution if t is in the interior of the smallest interval covering the support T_0.

Theorem 3.2

If the domain D is an open interval, there exist a unique solution $\hat{\tau}$ to the likelihood equation

tion

$$t = E[T \mid \tau],$$

for each value of t in the interior of the smallest interval covering the support T_0. The solution $\hat{\tau}$ is the ML–estimate.

Proof

Let $\overline{K}$ and $\underline{K}$ be the upper and lower limits of $nK'(\tau)$ as τ ranges from the upper to the lower limit of the domain. Let in addition $\overline{t}$ and $\underline{t}$ be the upper and lower limits of the support. From the inequality

$$nK'(\tau) = e^{-nK(\tau)} \Sigma_t t e^{t\tau + h_1(t)} \leq \overline{t} e^{-nK(\tau)} \Sigma_t e^{t\tau + h_1(t)} = \overline{t}$$

when $\overline{t}<\infty$, then follows that $\overline{K}\leq\overline{t}$. A similar argument shows that $\underline{K} \geq \underline{t}$.

Suppose now that $\overline{K}<\overline{t}$, such that $\overline{K}<+\infty$ and that there exists a K with $\overline{K}<K<\overline{t}$. If $\overline{K}<\infty$, the domain D must have $+\infty$ as its upper limit, since $\tau\leq\overline{\tau}$ for all $\tau\in D$ would entail that $K(\tau)\to\infty$ as $\overline{\tau}\to\infty$ as can be seen from the equality

$$e^{nK(\tau)} = \underset{t}{\Sigma}e^{t\tau+h_1(t)}$$

and the definition of the domain. But if $K(\tau)\to\infty$ for $\tau\to\overline{\tau}\langle\infty$ then also $K'(\tau)\to\infty$. Thus a finite upper limit for $nK'(t)$ can only happen for $\overline{\tau}=+\infty$.

Consider now the inequality

$$K-nK'(\tau) = e^{-nK(\tau)}\underset{t}{\Sigma}(K-t)e^{t\tau+h_1(t)} = e^{\tau K-nK(\tau)}\underset{t}{\Sigma}(K-t)e^{(t-K)\tau+h_1(t)}$$

$$\leq e^{\tau K-nK(\tau)}\underset{t<K}{\Sigma}(K-t)e^{h_1(t)}$$

The last term is positive and independent of τ. Hence $K=\overline{K}$ if it can be proved that

$$e^{\tau K-nK(\tau)} \to 0 \text{ as } \tau\to\infty.$$

This convergence follows from the inequality

$$e^{nK(\tau)}e^{-\tau K} = \underset{t}{\Sigma}e^{\tau(t-K)+h_1(t)} \geq \underset{t\geq K-\epsilon}{\Sigma} e^{\tau(t-K)+h_1(t)} \geq e^{\epsilon\tau}\underset{t\geq K-\epsilon}{\Sigma} e^{h_1(t)},$$

the right hand side of which tends to infinity as $\tau\to\infty$. Since $K=\overline{K}$ for all K satisfying $\overline{K}\langle K\langle\overline{t}$, it follows that $\overline{K}=\overline{t}$ and the theorem is proved. $\square$.

Example 3.1

In order to illustrate the concept of a log–linear model and the use of theorems 3.1 and 3.2 consider n binary variables $X_1, \ldots, X_n$ with

$$P(X_i = x) = \begin{cases} \theta & \text{for } x = 1 \\ 1 - \theta & \text{for } x = 0 \end{cases}$$

The probability $f(x | \theta) = P(X_i = x)$ can then be written

$$f(x | \theta) = \theta^x (1 - \theta)^{1-x},$$

such that

$$\ln f(x | \theta) = x \ln \frac{\theta}{1 - \theta} + \ln(1 - \theta).$$

The model is thus log–linear of the form (3.3) with $m = 1$, $g_1(x) = x$,

$$\varphi_1(\theta) = \ln \frac{\theta}{1 - \theta},$$

$h(x) = 0$ and $K(\theta) = -\ln(1 - \theta)$. The canonical parameter is accordingly

$$\tau = \ln \frac{\theta}{1 - \theta} \ .$$

For n independent observations $x_1, \ldots, x_n$, $t = \Sigma x_i$ is the sufficient statistic for τ. Since the conditions for the binomial distribution are satisfied, T is binomially distributed, i.e.

$$P(T = t) = \binom{n}{t} \theta^t (1 - \theta)^{n-t}.$$

The log–linear expression (3.7) for the probability distribution of T is thus given by

$$\ln f(t | \tau) = t\tau + \ln \binom{n}{t} - nK(\tau)$$

with

$$K(\tau) = \ln(1+e^\tau),$$

since $(1-\theta) = 1/(1+e^\tau)$.

It then follows from theorem 3.1, that the likelihood equation is equivalent with

(3.14)
$$ne^\tau/(1+e^\tau) = t.$$

The solution to (3.14) is

$$\hat{\tau} = \ln\frac{t/n}{1-t/n}.$$

or since $\theta = e^\tau/(1+e^\tau)$

$$\hat{\theta} = \frac{t}{n}.$$

This last result can, however, be derived directly from the likelihood equation

$$t = E[T \mid \theta] = n\theta.$$

In this case the likelihood equation can be solved directly yielding an explicit expression for $\hat{\tau}$. It is instructive, however, to derive the domain and the support for the binomial model. The domain consist of all values τ for which

$$\sum_{t=0}^{n} \binom{n}{t} e^{t\tau} < \infty,$$

which is all τ–values, since the sum is finite. Being the complete real line, the domain is an open interval and it follows from theorem 3.2 that the ML–estimate is a unique solution to (3.12) for all values of t in the interior of the smallest interval containing the support. The support is the set

$$\{0,1,...,n\},$$

such that (3.12) has a solution for all values of t (integers or non–integers) in the open interval (0,n). Surprisingly t=0 and t=n are not included in spite of the fact that θ can be estimated also for these extreme values. These values correspond, however, to τ equal to

$+\infty$ or $-\infty$ in (3.12), which are not proper values of the canonical parameter. It is thus important that theorem 3.2 is formulated in terms of the canonical parameters. $\triangle$.

It may seem a complicated way to derive ML–estimates to take the detour of introducing log–linear models, domains, supports, etc. But the fact that the ML–estimator can be derived from a simple mean value equation and that precise rules for the solvability of this equation can be formulated, simplifies the treatment of more complex models with many parameters. In addition it is possible to derive strong results concerning the asymptotic distribution of ML–estimators and of goodness of fit test statistics if attention is restricted to log–linear models. These results are important in the many cases, where the likelihood equations do not yield explicit solutions.

Theorem 3.3.

If τ is in the domain, then the ML–estimator $\hat{\tau}$ converge in probability to τ as $n\to\infty$ and $\hat{\tau}$ is asymptotically normally distributed with asymptotic mean value τ and asymptotic variance $\frac{1}{n}(K''(\tau))^{-1}$, i.e.

$$P(\sqrt{n}\sqrt{K''(\tau)}(\hat{\tau}-\tau) \leq u) \to \phi(u) \text{ for } n\to\infty.$$

for all $u\epsilon R$, where ϕ is the cumulative distribution function for the normal distribution.

Proof

Since

$$T/n = \frac{1}{n}\sum_{i=1}^{n} g(X_i)$$

is an averaged of n independent random variables, the law of large numbers yields

(i) $$T/n \overset{P}{\to} E[g(X_i)] = K'(\tau)$$

and the central limit theorem yields

(ii) $$T/n \overset{a}{\approx} N(K'(\tau), K''(\tau)/n)$$

according to (3.13).

From (i) and (3.12) follows that

$$K'(\hat{\tau}) \overset{P}{\to} K'(\tau), \text{ as } n \to \infty.$$

But since $K'(\tau)$ is continuous and monotonely increasing it can then be concluded that

$$\hat{\tau} \overset{P}{\to} \tau, \text{ as } n \to \infty.$$

If $K'(\tau)$ is expanded in a Taylor series, one further gets

$$t/n - K'(\tau) = t/n - K'(\hat{\tau}) - (\tau - \hat{\tau})K''(\tau^*)$$

with $|\tau^* - \hat{\tau}| \leq |\tau - \hat{\tau}|$. Since $t = nK'(\hat{\tau})$ this means that $\tau - \hat{\tau}$ has the same distribution as $(t/n - K'(\tau))/K''(\tau^*)$. The result (ii) then implies

$$\sqrt{n}(\hat{\tau} - \tau)K''(\tau^*)/\sqrt{K''(\tau)} \overset{a}{\approx} N(0,1).$$

But $\hat{\tau} \overset{P}{\to} \tau$ implies that $\tau^* \overset{P}{\to} \tau$, and since K'' is continuous, it follows that

$$\sqrt{n}(\hat{\tau} - \tau)\sqrt{K''(\tau)} \overset{a}{\approx} N(0,1).$$

and the theorem is proved. $\square$.

Results similar to theorem 3.3 can be proved for models, which are not log–linear, but then more complicated conditions on the properties of the log–likelihood function are needed.

There are two immediate applications of theorem 3.3. Firstly a confidence interval with approximate confidence level $1-\alpha$ can be derived from the equation

$$P(-u_{1-\alpha/2} \leq \sqrt{n}\sqrt{K''(\tau)}(\hat{\tau}-\tau) \leq u_{1-\alpha/2} | \tau) = 1-\alpha$$

which is approximately valid for large values of n. The interval

$$\left[\hat{\tau}-u_{1-\alpha/2}/\left[\sqrt{n}\sqrt{K''(\hat{\tau})}\right], \quad \hat{\tau}+u_{1-\alpha/2}/\left[\sqrt{n}\sqrt{K''(\hat{\tau})}\right]\right]$$

is accordingly an approximate $1-\alpha$ confidence interval for τ.

Consider, secondly, the hypothesis

$$H_0: \tau=\tau_0$$

against the alternative

$$H_1: \tau \neq \tau_0 .$$

If the critical region is chosen as a set, where $|\hat{\tau}-\tau_0|$ is large, the level of significance is according to theorem 3.3 approximately

(3.15) $$p = P(|U| \geq \sqrt{n}\sqrt{K''(\tau_0)}|\hat{\tau}-\tau_0|)$$

where $U \sim N(0,1)$ and $\hat{\tau}$ is the observed value of the ML–estimator. The critical region for a test with approximate level α is

$$\{t \mid |\hat{\tau}-\tau_0| \geq u_{1-\alpha/2}/\sqrt{nK''(\tau_0)}\}.$$

For a log–linear model the likelihood ratio only depends on the sufficient statistic T for the canonical parameter. The logarithm to the likelihood ratio, or the **log–likelihood ratio** is according to (3.6) and (2.5) given by

$$\ln r(X_1,...,X_n) = \ln f(T\,|\,\tau_0) - \ln f(T\,|\,\hat{\tau}) = T\tau_0 - nK(\tau_0) - T\hat{\tau} + nK(\hat{\tau})$$

or

(3.16)
$$\ln r(T) = T(\tau_0 - \hat{\tau}) - n(K(\tau_0) - K(\hat{\tau}))$$

The following theorem shows that the level of significance for a likelihood ratio test can be approximated by a percentile in a χ^2–distribution.

Theorem 3.4.

If $X_1,...,X_n$ are independent, identically distributed random variables with a log–linear model (3.3), then the transformed likelihood ratio test statistic $-2\ln r(T)$ and $nK''(\tau_0)(\hat{\tau}-\tau_0)^2$ has under H_0 the same asymptotic distribution namely

$$Z = -2\ln r(T) \overset{a}{\sim} \chi^2(1).$$

Proof.

From (3.16) and a Taylor–expansion of $K(\tau_0)$ around $\hat{\tau}$ follows that

$$-2\ln r(T) = 2T(\hat{\tau}-\tau_0) + 2nK'(\hat{\tau})(\tau_0-\hat{\tau}) + nK''(\tau^*)(\tau_0-\hat{\tau})^2 = nK''(\tau^*)(\hat{\tau}-\tau_0)^2$$

where $|\tau^*-\hat{\tau}| \le |\tau_0-\hat{\tau}|$, since $T=nK'(\hat{\tau})$ when $\hat{\tau}$ is the ML–estimator. According to theorem 3.3 $\tau^* \overset{P}{\to} \tau_0$ and $\sqrt{n}\sqrt{K''(\tau_0)}(\hat{\tau}-\tau_0) \overset{a}{\sim} N(0,1)$ under H_0, such that

$$-2\ln r(T) \overset{a}{\sim} \chi^2(1). \quad \square .$$

If the X's are non–identically distributed, theorem 3.4 is true when the model for each X_i is log–linear and the sufficient statistics for τ is $T=\Sigma_i g_i(X_i)$.

Since the hypothesis $H_0:\tau=\tau_0$ is rejected for small values of the likelihood ratio, the level of significance for H_0 is according to theorem 3.4 computed approximately as

$$p \simeq P(Q \ge -2\ln r(t)),$$

where $Q \sim \chi^2(1)$.

It further follows from theorem 3.4 that the test with approximate level of significance (3.15), based on the asymptotic distribution of the ML–estimate is equivalent with the transformed log–likelihood ratio test.

Example 3.2:

Consider n independent Poisson distributed random variables $X_1,...,X_n$ with common parameter λ, i.e.

$$P(X_i=x) = \frac{\lambda^x}{x!} e^{-\lambda}.$$

This model is log–linear with canonical parameter $\tau=\ln\lambda$ and sufficient statistic $T=\Sigma X_i$, since

$$\ln f(x\,|\,\lambda) = x\ln\lambda-\ln x!-\lambda.$$

Since further $K(\tau)=\lambda=e^\tau$, the ML–estimate is given as the unique solution to

$$\Sigma x_i = E[T] = ne^\tau,$$

i.e. $\hat{\tau}=\ln\bar{x}$.

Note that $\lambda=0$ with $\tau=-\infty$ is a boundary point of the domain and that $\bar{x}=0$, or $x_1=...=x_n=0$, correspondingly a boundary point for the support.

For the hypothesis

$$H_0: \lambda=\lambda_0$$

or

$$H_0: \tau=\tau_0=\ln\lambda_0,$$

the transformed log–likelihood ratio is according to (3.16) given by

$$-2\ln r(t) = 2n\bar{x}(-\ln\lambda_0+\ln\bar{x})+2n(\lambda_0-\bar{x}).$$

According to theorem 3.4, $-2\ln r(t) \overset{a}{\approx} \chi^2(1)$, such that the level of significance is given approximately as

$$p = P(Q \geq 2n\bar{x}(\ln\bar{x} - \ln\lambda_0) - 2n(\bar{x}-\lambda_0))$$

where $Q \sim \chi^2(1)$. △.

If the model is parameterized by the original parameter θ rather than the canonical parameter τ, the results contained in theorems 3.1 to 3.4 are true in almost identical versions provided the function

$$\tau = \varphi(\theta)$$

is strictly monotone in the domain D. Without proof, we state briefly the equivalents of theorems 3.1 to 3.3.

Theorem 3.1A

The likelihood equation

$$\frac{d\ln f(x_1,\ldots,x_n \mid \theta)}{d\theta} = 0$$

is equivalent to

$$E[T \mid \theta] = t$$

or to

$$nK'(\theta)/\varphi'(\theta) = t,$$

where $K(\theta)=K(\varphi(\theta))$.

Let $D_\theta = \{\theta \mid \tau = \varphi(\theta) \in D\}$ be the domain of θ.

Theorem 3.2A

If D_θ is an open interval, then there exist a unique solution $\hat{\theta}$ to the likelihood equation

$$t = E[T \mid \theta]$$

for each value of t in the interior of the smallest interval covering the support T_0. The solution $\hat{\theta}$ is the ML–estimate.

Theorem 3.3A

If θ is in the domain D_θ, then the ML–estimator $\hat\theta$ converge in probability to θ as n→∞ and

$$\hat\theta \sim N(\theta,\sigma^2(\theta)/n),$$

where

$$\sigma^2(\theta) = [K''(\theta) - K'(\theta)\varphi''(\theta)/\varphi'(\theta)]^{-1}.$$

The variance of $\hat\theta$ in theorem 3.3A is derived by using the rules of differentiation to $K''(\tau)$ in theorem 3.3 as follows:

A Taylor expansion of $\hat\tau=\varphi(\hat\theta)$ yields

$$\hat\tau = \tau+(\hat\theta-\theta)\varphi'(\theta)+... \quad ,$$

such that $\mathrm{var}\left[\hat\tau\right]$ can be approximated by

$$\mathrm{var}[\hat\tau] \simeq (\varphi'(\theta))^2\mathrm{var}[\hat\theta].$$

Hence

$$\mathrm{var}[\hat\theta] \simeq \mathrm{var}[\hat\tau]/(\varphi'(\theta))^2.$$

Accordingly to theorem 3.3, however, $\mathrm{var}[\hat\tau]\simeq(n(K''(\tau))^{-1}$. Implicit differentiation of $K(\theta)$ now yields

$$K'(\theta) = K'(\tau)\varphi'(\theta)$$

and

$$K''(\theta) = K''(\tau)(\varphi'(\theta))^2+K'(\tau)\varphi''(\theta).$$

Hence

$$\mathrm{var}[\hat\theta] \simeq n^{-1}[K''(\theta) - \frac{K'(\theta)}{\varphi'(\theta)} \varphi''(\theta)]^{-1},$$

which is the result stated in theorem 3.3A.

Theorem 3.4 is valid also if the parameterization is in terms of a strictly monotone transformation $\theta=\varphi^{-1}(\tau)$ of the canonical parameter. In this case the likelihood ratio is defined as

$$r(T) = \frac{f(X_1,...,X_n | \theta_0)}{f(X_1,...,X_n | \hat{\theta})} \ ,$$

where $\hat{\theta}$ is the maximum likelihood estimate of θ.

3.3. The multi–dimensional case

Consider the log–linear model (3.6) with canonical parameters $\tau_1,...,\tau_m$ and sufficient statistics

$$T_j = \sum_{i=1}^{n} g_j(X_i), \quad j=1,...,m \ ,$$

in the i.i.d (independent identically distributed) case and

$$T_j = \sum_{i=1}^{n} g_{ij}(X_i) \ , \quad j=1,...m$$

in the non–i.i.d. case,

The analogue to theorem 3.1 is then

Theorem 3.5.

The likelihood equations

$$\partial \ln f(t_1,...,t_m | \tau_1,...,\tau_m)/\partial \tau_j = 0, \quad j=1,...,m$$

are equivalent to

$$E[T_j | \tau_1,...,\tau_m] = t_j, \quad j=1,...,m$$

or

$$n \partial K(\tau_1,...,\tau_m)/\partial \tau_j = t_j, \quad j=1,...,m \ .$$

In the multi–dimensional case, the domain is defined by

(3.17)
$$D = \{\tau_1,...,\tau_m \mid \sum_{t_1 \cdots t_m} ... \sum \exp(\sum_j t_j \tau_j + h(t_1,...,t_m)) \langle +\infty \}.$$

It can be proved that D is a **convex set**, i.e. if two points in R^m are both in D, then the complete line connecting the two points is also in D. A typical convex domain is shown in fig. 3.1.

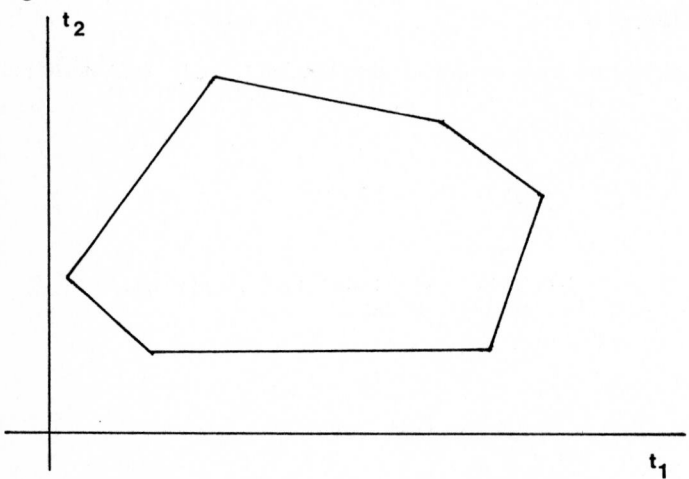

Figur 3.1. The domain for the canonical parameters in a two–dimensional log–linear model.

The **support** T_0 of the log–linear model is the subset of vectors $(t_1,...,t_m) \in R^m$ for which

$$f(t_1,...,t_m \mid \tau_1,...,\tau_m) > 0.$$

It can be shown that the support does not depend on the values of the τ's.

The function $K(\tau_1,...,\tau_m)$ is the key to a study of the properties of the ML–estimators for $\tau_1,...,\tau_m$. From theorem 3.5 follows that the mean values of the T_j's are equal to the partial derivatives of K, apart from the factor n, and that the likelihood equations are obtained by equating these derivatives with the observed values of the sufficient statistics. In addition (3.13) generalizes to

$$n\partial^2 K(\tau_1,...,\tau_m)/\partial \tau_j^2 = \mathrm{var}[T_j]$$

and

$$n\partial^2 K(\tau_1,...,\tau_m)/(\partial \tau_j \partial \tau_\ell) = \text{cov}(T_j, T_\ell).$$

The equivalent of theorem 3.2 is

Theorem 3.6:

If the domain D is an open set, then there exist a unique set of solutions $\hat{\tau}_1,...,\hat{\tau}_m$ to the likelihood equations

(3.18) $$t_j = E[T_j | \tau_1,...,\tau_m], \quad j=1,...,m,$$

whenever $(t,...,t_m)$ is an interior point in the smallest convex set enclosing the support. The solutions are the ML–estimators for $\tau_1,...,\tau_m$.

Theorem 3.3 generalizes as follows

Theorem 3.7:

If $\tau=(\tau_1,...,\tau_m)$ is in the domain, then the ML–estimator converge in probability to τ as $n\to\infty$ and

$$\sqrt{n}(\hat{\tau}-\tau) \overset{a}{\approx} N_m(0,M^{-1}),$$

where M is an m–dimensional square matrix with elements

$$m_{j\ell}(\tau_1,...,\tau_m) = \partial^2 K(\tau_1,...,\tau_m)/(\partial \tau_j \partial \tau_\ell).$$

This implies that

$$n(\hat{\tau}-\tau)'M(\hat{\tau}-\tau) \overset{a}{\approx} \chi^2(m).$$

In the multi–dimensional case the log–likelihood ratio for testing

$$H_0: \tau_1=\tau_{10},...,\tau_m=\tau_{m0}$$

against an alternative where the τ's are unconstrained is given by

$$\ln r(T_1,...,T_m) = \sum_{j=1}^{m} T_j(\tau_{j0}-\hat{\tau}_j)-n(K(\tau_{10},...,\tau_{m0})-K(\hat{\tau}_1,...,\hat{\tau}_m)).$$

Theorem 3.8:

The transformed likelihood ratio test statistic

$$Z = -2\ln r(T_1,...,T_m)$$

for the hypothesis

$$H_0: \tau_j=\tau_{j0}, \quad j=1,...,m$$

has asymptotic distribution

$$Z \overset{a}{\sim} \chi^2(m)$$

if $(\tau_{10},...,\tau_{m0})$ is a point in the interior of the domain.

The result in theorem 3.8 is extremely useful because the likelihood ratio test statistic or simple transformations thereof only in rare situations has a known distribution.

Example 3.3.

Let $X_1,...,X_k$ be multinomially distributed with parameters n and $p_1,...,p_k$, i.e.

$$f(x_1,...,x_k|p_1,...,p_k) = (x_1.\overset{n}{..}x_k)p_1^{x_1}...p_k^{x_k}.$$

Then

$$\ln f(x_1,...,x_k|p_1,...,p_k) = \sum_{j=1}^{k-1} x_j\ln(p_j/p_k)+n\ln p_k+\ln(x_1.\overset{n}{..}x_k)$$

and the model is log–linear with canonical parameters

$$\tau_j = \ln p_j - \ln p_k, \quad j=1,\ldots,k-1$$

and sufficient statistics

$$T_j = X_j, \quad j=1,\ldots,k-1.$$

The log–linear model is thus of dimension $m=k-1$, and the ML–estimates are given by the equations

$$x_j = E[X_j] = np_j, \quad j=1,\ldots,k-1.$$

The ML–estimates $\hat{p}_j = x_j/n$ for the original parameters are thus found directly. For the canonical parameters the ML–estimates are

$$\hat{\tau}_j = \ln(x_j/n) - \ln(x_k/n) = \ln(x_j/x_k).$$

Consider now the hypothesis

$$H_0: p_1 = p_{10}, \ldots, p_k = p_{k0},$$

In terms of the canonical parameters, H_0 has the form

$$H_0: \tau_j = \ln p_{j0} - \ln p_{k0}, \quad j=1,\ldots,k-1.$$

Since $\hat{p}_j = x_j/n_j$, the log–likelihood ratio

$$\ln r(x_1,\ldots,x_k) = \sum_{j=1}^{k} x_j(\ln p_{j0} - \ln(x_j/n))$$

such that the transformed likelihood ratio is given by

$$-2\ln r(x_1,\ldots,x_k) = 2\sum_{j=1}^{k} x_j(\ln x_j - \ln(np_{j0})).$$

According to theorem 3.8 this test statistic follows approximately a χ^2–distribution with k–1 of freedom, provided $\tau_{j0}=\ln p_{j0}-\ln p_{k0}$, $j=1,...k-1$ is in the interior of the domain for the canonical parameters.

The sum in the definition (3.17) of the domain for the present model is finite. Hence the domain consist of all values of $\tau_1,...,\tau_{k-1}$ in R^{k-1}. In terms of the p_j's only vectors $(p_1,...,p_k)$ for which at least one $p_j=0$ are thus excluded. This means that the approximation to a limiting χ^2–distribution is valid if none of the values $p_{10},...,p_{k0}$ are zero. It is essential that the true dimensionality m=k–1 of the log–linear model is established. For the present example the only linear ties between observations or parameters are

$$n = x_1+...+x_k$$

and

$$1 = p_1+...+p_k.$$

Both these are accounted for when the canonical parameters are chosen as

$$\tau_j = \ln p_j-\ln p_k$$

for all j.

Notice also in this example that the canonical parameters are only unique up to a constant. Thus also $\tau_j=\ln p_j-\frac{1}{k}\sum_{j=1}^{k}\ln p_j$ could have been used. △.

Example 3.4.

As an example, where the support of the log–linear model is non– trivial and the conditions for existence of finite solutions to the likelihood equations accordingly also non–trivial, consider a simple example of the logistic regression model, which we return to in chapter 9. Let $X_1,...,X_n$ be binary variables with point probabilities

$$P(X_i=x) = \begin{cases} p_i & \text{for x=1} \\ 1-p_i & \text{for x=0} \end{cases}$$

and

$$p_i = e^{\beta_0+\beta_1 z_i}/(1+e^{\beta_0+\beta_1 z_i})$$

or

$$\ln \frac{p_i}{1-p_i} = \beta_0 + \beta_1 z_i,$$

where $z_1,...,z_n$ have known values.

This is called a logistic regression model for binary variables, because the linear expression is in terms of a logistic transformation of the probabilities. The model is log–linear, since for $x=1$ and 0, $\ln f_i(x|\beta_0,\beta_1)$ can be written as

$$\ln f_i(x|\beta_0,\beta_1) = x\ln p_i+(1-x)\ln(1-p_i) = x\beta_0+z_i x\beta_1-\ln(1+e^{\beta_0+\beta_1 z_i}).$$

It is thus a case of non–identically distributed random variables. The canonical parameters are $\tau_1=\beta_0$ and $\tau_2=\beta_1$, and the corresponding sufficient statistics

$$T_1 = \sum_i X_i, \quad T_2 = \sum_i z_i X_i$$

Since

$$P(X_i=1) = e^{\beta_0+\beta_1 z_i}/(1+e^{\beta_0+\beta_1 z_i}),$$

the likelihood equations are according to theorem 3.5

$$t_1 = E[\sum_i X_i] = \sum_i e^{\beta_0+\beta_1 z_i}/(1+e^{\beta_0+\beta_1 z_i})$$

and

$$t_2 = E[\sum_i z_i X_i] = \sum_i z_i e^{\beta_0+\beta_1 z_i}/(1+e^{\beta_0+\beta_1 z_i}).$$

These equations do not have explicit solutions and must be solved by numerical methods. Hence it is of interest to determine for which observed values of t_1 and t_2 there are solutions. According to theorem 3.6 the likelihood equations have a unique set of solu–

tions, if (t_1, t_2) is an interior point of the convex extension of the support. In this case it is easy to derive the support. If, namely, $t_1 = i$, it follows that exactly i of the binary variables have the value 1. Hence if $z_{(1)} \leq z_{(2)} \leq ... \leq z_{(n)}$ are the z's in order of magnitude, the minimum and maximum values of t_2 are

$$z_{(1)} + ... + z_{(i)} \leq t_2 \leq z_{(n-i+1)} + ... + z_{(n)}.$$

It is further easy to see that the set obtained by connecting the 2n points $\{0,0\}$,

$$\{i, z_{(1)} + ... + z_{(i)}\}, \ i = 1, ..., n$$

and

$$\{i, z_{(n-i+1)} + ... + z_{(n)}\}, \ i = 1, ..., n-1$$

is a convex set. Hence the likelihood equations have solutions if (t_1, t_2) does not coincide with any of the 2n boundary points. For the case n=10 and $z_i = i$, $i = 1, ..., 10$ the convex extension of the support is shown in fig. 3.2.

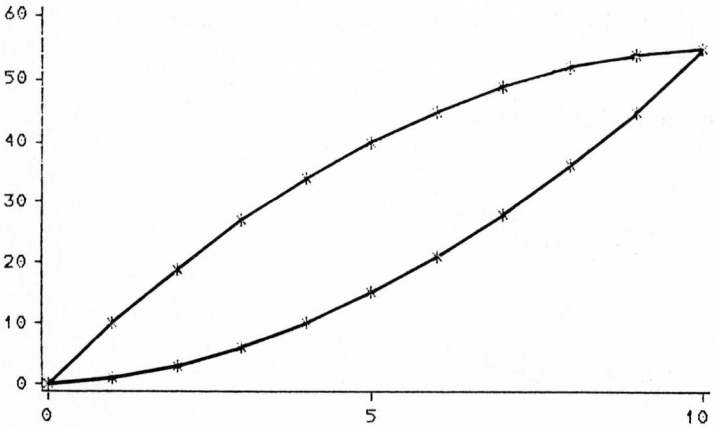

Fig. 3.2. The convex extension of the support for a logistic regression model with n=10 and $z_i = i$.

Thus if $t_1 = 4$, there are only solutions to the likelihood equations if t_2 is larger than 10 and less than 34. If $t_2 = 10$, we are in a situation where $x_1 = x_2 = x_3 = x_4 = 1$ and

$x_5=x_6=x_7=x_8=x_9=x_{10}=0$, and if one tries to solve the likelihood equations in this case, the estimation procedure would converge to an infinite value of either β_0 or β_1. On the other hand for $x_1=x_2=x_3=x_5=1$ and $x_4=x_6=x_7=x_8=x_9=x_{10}=0$, such that $t_2=11$, there are no problems with solving the likelihood equations. $\triangle$.

Example 3.5.

Consider a two–way contingency table with I rows and J columns, and observed count x_{ij} in row i and column j. Assume further that the counts are independent and Poisson distributed with parameters $\lambda_{11},...,\lambda_{IJ}$. The distribution of the observed counts is then

$$f(x_{11},...,x_{IJ}\,|\,\lambda_{11},...,\lambda_{IJ}) = \underset{i\ j}{\Pi\Pi}[(x_{ij}!)^{-1}\lambda_{ij}^{x_{ij}}e^{-\lambda_{ij}}].$$

The model is log–linear, since

$$\ln f(x_{11},...,x_{IJ}\,|\,\lambda_{11},...,\lambda_{IJ}) = \underset{i\ j}{\Sigma\Sigma}x_{ij}\ln\lambda_{ij}-\lambda_{..}-\underset{i\ j}{\Sigma\Sigma}\ln(x_{ij}!).$$

with sufficient statistics $x_{11},...,x_{IJ}$ and canonical parameters $\tau_{ij}=\ln\lambda_{ij}$, i=1,...,I, j=1,...,J. The likelihood equations are accordingly

$$x_{ij} = E[X_{ij}] = \lambda_{ij}, \quad i=1,...,I, \;\; j=1,...,J$$

and the ML–estimates become

$$\hat{\tau}_{ij} = \ln x_{ij}.$$

The likelihood ratio for the hypothesis

$$H_0: \lambda_{11} = \lambda_{11}^0,...,\lambda_{IJ} = \lambda_{IJ}^0$$

become

$$r(x_{11},...,x_{IJ}) = \underset{i\ j}{\Pi\Pi}[(\lambda_{ij}^0)^{x_{ij}}/x_{ij}^{x_{ij}}]\exp(\underset{i\ j}{\Sigma\Sigma}(x_{ij}-\lambda_{ij}^0)).$$

since $\hat{\lambda}_{ij} = x_{ij}$. Hence the transformed log–likelihood ratio is

$$-2\ln r(x_{11},...,x_{IJ}) = 2\sum_i\sum_j x_{ij}\ln(\frac{x_{ij}}{\lambda^0_{ij}}) + 2\lambda^0_{..} - 2x_{..} ,$$

which under H_0 is approximately χ^2–distributed with IJ degrees of freedom. If it is assumed that, $\lambda^0_{..} = x_{..}$, then

$$-2\ln r(x_{11},...,x_{IJ}) = 2\sum_i\sum_j x_{ij}(\ln x_{ij} - \ln\lambda^0_{ij}). \qquad \triangle$$

Theorems 3.5 to 3.7 can be formulated in terms of the original parameters of the model, if the concept of **identifiability** for a log–linear model is introduced. Model (3.3) is identifiable in terms of the parameters $\theta_1,...,\theta_m$ if $\theta_{m+1},...,\theta_k$ are functions of $\theta_1,...,\theta_m$ and the equations

$$\begin{cases} \tau_1 = \varphi_1(\theta_1,...,\theta_k) \\ \vdots \\ \tau_m = \varphi_m(\theta_1,...,\theta_k) \end{cases}$$

have a unique set of solutions

$$\begin{cases} \theta_1 = \varphi_1^{-1}(\tau_1,...,\tau_m) \\ \vdots \\ \theta_m = \varphi_m^{-1}(\tau_1,...,\tau_m) \end{cases},$$

and in addition, $\partial^2\varphi_j(\theta_1,...,\theta_k)/\partial\theta_q\partial\theta_p$ is uniformly bounded for all j, p and q less than or equal to m in any closed subset of D_θ, where

$$D_\theta = \{\theta_1,...,\theta_k \mid \tau_j = \varphi_j(\theta_1,...,\theta_k) \in D, \ j=1,...,m\},$$

and D is the domain. Since D_θ maps onto D under the transformation φ^{-1}, it follows that D_θ is open if and only if D is open. Theorems 3.5 to 3.7 now generalize as follows:

Theorem 3.5A.

If the log–linear model (3.3) is identifiable in terms of $\theta_1,...,\theta_m$ the likelihood equations

$$\partial \ln f(x_1,...,x_n \mid \theta_1,...,\theta_k)/\partial \theta_j = 0, \quad j=1,...,m$$

are equivalent to

(3.19)
$$t_j = E[T_j \mid \theta_1,...,\theta_k], \quad j=1,...,m$$

Note that there are only m equations in (3.19). In fact the k–m last θ's are functions of $\theta_1,...,\theta_m$ when the model is identifiable in $\theta_1,...,\theta_m$ and it is sufficient to find $\hat{\theta}_1,...,\hat{\theta}_m$.

Theorem 3.6A:

If the domain D_θ for the log–linear model (3.3) is open, then there exist a unique set of solutions $\hat{\theta}_1,...,\hat{\theta}_k$ to the likelihood equations whenever $(t_1,...,t_m)$ is in the interior of the smallest convex set enclosing the support. The solutions are the ML–estimates for $\theta_1,...,\theta_k$.

Note that D_θ is not necessarily the original parameter set Θ.

Theorem 3.7A:

If the model (3.3) is identifiable in terms of $\theta=(\theta_1,...,\theta_m)$ and $(\theta_1,...,\theta_k)$ is in the domain D_θ, then the ML–estimates $\hat{\theta}=(\hat{\theta}_1,...,\hat{\theta}_m)$ converge in probability to the true values for n→∞ and

$$\sqrt{n}(\hat{\theta}-\theta) \overset{a}{\approx} N_m(0,M^{-1}),$$

where the square matrix M has elements

$$m_{j\ell} = \partial^2 K(\theta_1,...,\theta_k)/\partial\theta_j\partial\theta_\ell - \Sigma\Sigma \frac{\partial K(\theta_1,...,\theta_k)}{\partial\theta_p} w_{pq} \frac{\partial^2 \varphi_q(\theta_1,...,\theta_k)}{\partial\theta_j\partial\theta_\ell}$$

and w_{pq} are the elements of the inverse to the matrix with elements $\partial\varphi_j(\theta_1,...,\theta_k)/\partial\theta_\ell$, $j,\ell=1,...,m$.

Theorem 3.8A:

The transformed likelihood ratio test statistic

$$Z = -2\ln r(T_1,...,T_m)$$

for the hypothesis

$$H_0: \theta_j = \theta_{j0}, \quad j=1,...,k$$

has asymptotic distribution

$$Z \overset{a}{\sim} \chi^2(m)$$

if the log–linear model is identifiable in terms of $\theta_1,...,\theta_m$ and $(\theta_{10},...,\theta_{k0})$ is an interior point of the domain D_θ.

Theorems 3.5A, 3.6A and 3.8A are formulated in terms of all k θ's even though the dimension of the log–linear model is m. As mentioned above the last k–m θ's are redundant, since indentifiability of the model in terms of $\theta_1,...,\theta_m$ means that $\theta_{m+1},...,\theta_k$ are functions of $\theta_1,...,\theta_m$. As an example of this consider the multinomial distribution, with $\theta_j=p_j$ and $\tau_j=\ln p_j-\ln p_k$, $j=1,...,k-1$. Then the likelihood equations are

(3.20) $$x_j = np_j, \quad j=1,...,k-1$$

with solutions $\hat{p}_j=x_j/n$, $j=1,...,k$ since

$$\hat{p}_k = 1 - \sum_{j=1}^{k-1} \hat{p}_j = 1 - \sum_{j=1}^{k-1} x_j/n = x_k/n.$$

In practice the equation system (3.20) is extended to k equations by adding $x_k = n\hat{p}_k$, which is automatically true if the equations (3.20) are satisfied, although the model is identifiable in terms of $p_1, \ldots p_{k-1}$.

3.4. Testing composite hypotheses

Consider first the composite hypothesis

(3.21) $$H_0: \tau_1 = \tau_{10}, \ldots, \tau_r = \tau_{r0},$$

where r of the m canonical parameters have given values and the remaining m−r canonical parameters are unspecified under the hypothesis.

Under H_0 the model is still log–linear with canonical parameters $\tau_{r+1}, \ldots, \tau_m$ and sufficient statistics $T_{r+1}, \ldots, T_m$ since from (3.6)

(3.22) $$\ln f(t_1, \ldots, t_m \mid \tau_{10}, \ldots, \tau_{r0}, \tau_{r+1}, \ldots, \tau_m) = \sum_{j=r+1}^{m} t_j \tau_j + h_1(t_1, \ldots, t_m) + \sum_{j=1}^{r} t_j \tau_{j0}$$

$$-nK(\tau_{10}, \ldots, \tau_{r0}, \tau_{r+1}, \ldots, \tau_m).$$

The likelihood equations for a ML–estimation of $\tau_{r+1}, \ldots, \tau_m$ are accordingly

(3.23) $$t_j = E[T_j \mid \tau_{10}, \ldots, \tau_{r0}, \tau_{r+1}, \ldots, \tau_m], \quad j=r+1, \ldots, m.$$

Let the solutions to (3.23) be $\tilde{\tau}_{r+1}, \ldots, \tilde{\tau}_m$. Conditions for a unique solution of (3.23) follow from theorem 3.6, with only $\tau_{r+1}, \ldots \tau_m$ being the canonical parameters of the log–linear model. In fact if $(\tau_{10}, \ldots, \tau_{r0}, \tau_{r+1}, \ldots, \tau_m)$ is in the domain D for the model (3.6), then $\tau_{r+1}, \ldots, \tau_m$ is in the domain D_r for the model (3.22). It is also easy to see that if

$(t_1,...,t_r,t_{r+1},...,t_m)$ is point in the support for the model (3.6), then $(t_{r+1},...,t_m)$ is also a point in the support for the model (3.22). Hence equations (3.23) have a set of unique solutions if equations (3.18) have a set of unique solutions.

For the composite hypothesis (3.21) the likelihood ratio test statistic is given by

$$(3.24) \qquad r(t_1,...,t_m) = \frac{f(t_1,...,t_m \mid \tau_{10}, \ldots, \tau_{r0}, \tilde{\tau}_{r+1},...,\tilde{\tau}_m)}{f(t_1, \ldots, t_m \mid \hat{\tau}_1,...,\hat{\tau}_m)}.$$

In many instances the parameters specified under the hypothesis is not a subset of the canonical parameters, but rather a subset of the original parameters $\theta_1,...,\theta_k$. In other cases the hypothesis specify certain dependencies between the original parameters or the canonical parameters. In order to derive the asymptotic distribution of the transformed likelihood ratio test statistic in such cases, it is necessary to establish an equivalence between the hypothesis under consideration and a hypothesis of the form (3.21).

Consider first a hypothesis where a set of r dependencies between the canonical parameters is specified, e.g.

$$(3.25) \qquad H_0: \begin{cases} \psi_1(\tau_1,...,\tau_m) = \psi_{10} \\ \cdots \\ \psi_r(\tau_1,...,\tau_m) = \psi_{r0} \end{cases}$$

where $\psi_1,...,\psi_r$ are continuous functions. The number of parameters specified under (3.25) is r, if there exists a parameterization $\psi_1,...,\psi_m$ with $\psi_1,...,\psi_r$ given by the left hand sides in (3.25) and m−r continuous functions

$$(3.26) \qquad H_0: \begin{cases} \psi_{r+1}(\tau_1,\ldots, \tau_m) = \psi_{r+1} \\ \cdots \\ \psi_m(\tau_1,...,\tau_m) = \psi_m \end{cases}$$

such that $\psi_1,...,\psi_m$ are identifiable in the log–linear model.

The log–likelihood ratio for (3.25) is given by

$$(3.27) \qquad r(t_1,...,t_m) = \frac{f(t_1,...,t_m | \tilde{\tau}_1,...,\tilde{\tau}_m)}{f(t_1,...,t_m | \hat{\tau}_1,...,\hat{\tau}_m)},$$

where $\tilde{\tau}_1,...,\tilde{\tau}_m$ are the values of $\tau_1,...,\tau_m$ that maximize the likelihood function under the constraints (3.25).

Assume secondly that the log–linear model is identifiable in terms of m of the original parameters $\theta_1,...,\theta_k$, for example $\theta_1,...,\theta_m$. Consider then the hypothesis

$$(3.28) \qquad H_0: \begin{cases} \theta_1 = \theta_{10} \\ \vdots \\ \theta_r = \theta_{r0} \end{cases}$$

Under (3.28) the likelihood ratio is given by

$$(3.29) \qquad r(x_1,...,x_n) = \frac{f(x_1,...,x_n | \theta_{10},...,\theta_{r0}, \tilde{\theta}_{r+1},...,\tilde{\theta}_k)}{f(x_1,...,x_n | \hat{\theta}_1,...,\hat{\theta}_k)}$$

where $\tilde{\theta}_{r+1},...,\tilde{\theta}_k$ are the ML–estimates under the constraints (3.28) and $\hat{\theta}_1,...,\hat{\theta}_k$ the unconstrained ML–estimates.

Assume finally that the model is identifiable in terms of $\theta_1,...,\theta_m$ and H_0 specifies r constraints between the identifiable parameters, i.e.

$$(3.30) \qquad H_0: \begin{cases} \psi_1 = (\theta_1,...,\theta_m) = \psi_{10} \\ \vdots \\ \psi_r = (\theta_1,...,\theta_m) = \psi_{r0} \end{cases}$$

The likelihood ratio is then given by

$$(3.31) \qquad r(x_1,...,x_n) = \frac{f(x_1,...,x_n | \tilde{\theta}_1,...,\tilde{\theta}_k)}{f(x_1,...,x_n | \hat{\theta}_1,...,\hat{\theta}_k)},$$

where $\tilde{\theta}_1,...,\tilde{\theta}_k$ are the ML–estimates under the constraints (3.30), and $\hat{\theta}_1,...,\hat{\theta}_k$ the unconstrained ML–estimates. Note that the constraints only concern the identifiable parameters $\theta_1,...,\theta_m$, but since the remaining parameters, $\theta_{m+1},...,\theta_k$ are functions of $\theta_1,...,\theta_m$ the constraints (3.30) also apply to $\theta_{m+1},...,\theta_k$.

We can collect all four cases, by saying that the parameters satisfy exactly r constraints if one of the hypotheses (3.21), (3.25), (3.28) or (3.30) holds under the given conditions. With this terminology, we have

Theorem 3.9

For a composite hypothesis, with exactly r constraints on the parameters, the transformed likelihood ratio test statistic has limiting distribution

$$(3.32) \qquad\qquad -2\ln r(T_1,...,T_m) \overset{a}{\approx} \chi^2(r).$$

The limiting distribution for $-2\ln r$ is thus a χ^2–distribution both when all canonical parameters are specified and under a composite hypothesis with r constraints on the parameters. The number of degrees of freedom are in both cases the number of parameters specified under the hypothesis.

In most cases composite hypotheses are formulated in terms of the original parameters as in (3.28) or (3.30). The role of the canonical parameters and the subset of identifiable parameters is then to determine the correct number of constraints between the parameters. If e.g. two of the equations in (3.30) are linearly dependent, then a set of m–r parameters $\psi_{r+1},...,\psi_k$ does not exist for which the model is identifiable in terms of $\psi_1,...,\psi_m$, and theorem 3.9 does not hold. Thus all linear dependencies between the equations in (3.30) must be accounted for in order to determine the degrees of freedom for the limiting χ^2–distribution.

Example 3.6.

Table 3.1 show the traffic accidents in Denmark in 1981 involving pedestrians distributed according to the week day on which the accident happend.

Table 3.1. Traffic accidents in Denmark in 1981 involving pedestrians distributed over weeks days.

Week day	Number of accidents	Expected numbers H_0	H_1
Monday	279	248.43	279.8
Tuesday	256	248.43	279.8
Wednesday	230	248.43	279.8
Thursday	304	248.43	279.8
Friday	330	248.43	279.8
Saturday	210	248.43	170.0
Sunday	130	248.43	170.0
Totals	1739	1739.01	1739.0

Source: Road traffic accidents. Publication 1982:8
Statistics Denmark. Table 5.3.

As a model for the data assume that all accidents happen independent of each other
and that, given a particular week day, the number of accidents follow a Poisson distribution. Let $X_1,...,X_7$ be the daily number of accidents. Then for day j

$$P(X_j=x_j) = \frac{\lambda_j^{x_j} e^{-\lambda_j}}{x_j!}$$

and $X_1,...,X_7$ are independent. Hence the log–likelihood of $x_1,...,x_7$ can be written

$$\ln f(x_1,...,x_7 \,|\, \lambda_1,...,\lambda_7) = \sum_{j=1}^{7} x_j \ln \lambda_j - \sum_{j=1}^{7} \ln x_j! - \sum_{j=1}^{7} \lambda_j.$$

It follows that the model is log–linear with canonical parameters $\tau_j=\ln\lambda_j$, $j=1,...,7$ and
sufficient statistics $x_1,...,x_7$. The ML–estimates for the λ_j's are given as solutions to

$$x_j = E[X_j] = \lambda_j$$

or simply as $\hat{\lambda}_j=x_j$.

Consider now the hypothesis

$$H_0: \lambda_1 = \ldots = \lambda_7 = \lambda,$$

to the effect that the expected number of accidents is independent of week day. Since the model is identifiable in terms of the λ_j's, H_0 is a hypothesis of the type (3.30) with

$$\begin{cases} \lambda_1 = \lambda_7 = 0 \\ \ldots \\ \lambda_6 - \lambda_7 = 0 \end{cases}$$

There are thus exactly 6 constraints. Theorem 3.9 then applies since the model is identifiable in terms of $\psi_1 = \lambda_1 - \lambda_7, \ldots, \psi_6 = \lambda_6 - \lambda_7$ and $\psi_7 = \lambda_7$. If all λ's are equal the log–likelihood function becomes

$$\ln f(x_1, \ldots, x_7 \mid \lambda, \ldots, \lambda) = \ln\lambda \sum_{j=1}^{7} x_j - \sum_{j=1}^{7} \ln x_j! - 7\lambda$$

and the likelihood equation for the estimation of λ is

$$\sum_{j=1}^{7} x_j = E[\sum_{j=1}^{7} X_j] = 7\lambda$$

with solution $\hat{\lambda} = \Sigma x_j / 7$. The likelihood ratio then with $x_\cdot = \Sigma x_j$ becomes

$$r(x_1, \ldots, x_7) = \frac{\Pi[(\frac{x_\cdot}{7})^{x_j} e^{-x_\cdot/7} / x_j!]}{\Pi[x_j^{x_j} e^{-x_j} / x_j!]}$$

such that

$$-2\ln r = 2\sum_j x_j [\ln x_j - \ln\frac{x_\cdot}{7}] .$$

According to theorem 3.9, $-2\ln r \sim \chi^2(6)$ and we reject H_0 if the observed value of $-2\ln r$ is large. The expected numbers x./7 under H_0 are shown as column 2 in table 3.1. The value of $-2\ln r$ computed from table 3.1 is $-2\ln r=115.56$, which is clearly significant. Consider next the hypothesis that the expected number of accidents is the same for Monday to Friday, and the same for Saturday and Sunday. Formally this is the hypothesis

$$H_1: \begin{cases} \lambda_1 = \ldots = \lambda_5 \\ \lambda_6 = \lambda_7 \end{cases}.$$

Written on the form (3.30) we have

$$\begin{cases} \lambda_1 - \lambda_5 = 0 \\ \vdots \\ \lambda_4 - \lambda_5 = 0 \\ \lambda_6 - \lambda_7 = 0 \end{cases}.$$

Since the model is identifiable in terms of $\psi_1=\lambda_1-\lambda_5$, $\psi_2=\lambda_2-\lambda_5$, $\psi_3=\lambda_3-\lambda_5$, $\psi_4=\lambda_4-\lambda_5$, $\psi_5=\lambda_6-\lambda_7$, $\psi_6=\lambda_5$ and $\psi_7=\lambda_7$, theorem 3.9 applies to H_1 with r=5. Under H_1, the log–likelihood ratio is given by

$$\ln f(x_1,\ldots,x_7|\lambda_1,\lambda_1,\lambda_1,\lambda_1,\lambda_1,\lambda_6,\lambda_6) = \ln\lambda_1 \sum_{j=1}^{5} x_j + \ln\lambda_6 \sum_{j=6}^{7} x_j - \sum_{j=1}^{7} \ln x_j! - 5\lambda_1 - 2\lambda_6.$$

Hence the ML–estimates for λ_1 and λ_6 are obtained from

$$\sum_{j=1}^{5} x_j = 5\lambda_1$$

and

$$x_6 + x_7 = 2\lambda_6$$

with solutions

$$\tilde{\lambda}_1 = \sum_{j=1}^{5} x_j/5$$

and

$$\tilde{\lambda}_6 = (x_6 + x_7)/2.$$

The transformed likelihood ratio becomes

$$z = -2\ln r(x_1,...,x_7) = 2 \sum_{j=1}^{7} x_j[\ln x_j - \ln\tilde{\lambda}_j].$$

According to theorem 3.9, z is approximately χ^2-distributed with 5 degrees of freedom. The values of $\tilde{\lambda}$ are shown in column 3 of table 3.1. With these values $-2\ln r$ is computed to 41.8. The corresponding level of significance $P(Q \geq 41.8)$ is less than 0.0005 when $Q \sim \chi^2(5)$. Neither H_0 nor H_1 can thus be accepted for the data in table 3.1 and it seems that a more complex model is needed to describe the data. $\triangle$.

3.5. The parametric multinomial distribution

Many discrete statistical models are based on the multinomial distribution. When, as in section 2.4, the cell probabilities depend on one or more parameters, the distribution is called a **parametric multinomial distribution.**

The multinomial distribution can be generated by n independent, discrete random variables $X_1,...,X_n$, which can attain the values 1,...,m with probabilities

$$\pi_j = P(X_i = j), \quad j=1,...,m.$$

If these probabilities do not depend on i and if

$$Y_j = \text{number of } X_i\text{'s with observed value } j, \quad j=1,...,m,$$

then the vector $(Y_1,...,Y_m)$ follows a multinomial distribution with count parameter n og probability parameters $\pi_1,...,\pi_m$. The joint point probability $P(Y_1=y_1,...,Y_m=y_m)$ is given by

$$(3.33) \qquad f(y_1,...,y_m) = \begin{bmatrix} n \\ y_1 \cdots y_m \end{bmatrix} \pi_1^{y_1} \cdots \pi_m^{y_m}.$$

As shown in example 3.3 the multinomial distribution is log–linear with canonical

parameters

$$\tau_j = \ln \pi_j - \ln \pi_m, \quad j=1,...,m-1$$

and sufficient statistics $y_1,...,y_{m-1}$.

For the parametric multinomial distribution, $(Y_1,...,Y_m)$ has joint point probability

(3.33) and the π's are functions

$$(3.34) \qquad \pi_j = \pi_j(\theta_1,...,\theta_k), \quad j=1,...,m$$

of k parameters $\theta_1,...,\theta_k$, where $k<m-1$.

The validity of the usual asymptotic results applied to the parametric multinomial

distribution depend on the model being identifiable in its parameters. Fortunately the

conditions for identifiability are well studied in this case, cf. for example Andersen

(1980a), p.95, Bishop, Feinberg and Holland (1975), p.510 or Rao (1973), p.359–360.

The key assumptions for the following theorems are:

Regularity: The functions $\pi_j(\theta_1,...,\theta_k)$, $j=1,...,m$ are positive and differentiable with

continuous derivatives and the square matrix **M** with elements

$$(3.35) \qquad m_{pq} = \sum_{j=1}^{m} \frac{1}{\pi_j(\theta_1,...,\theta_k)} \frac{\partial \pi_j(\theta_1,...,\theta_k)}{\partial \theta_p} \frac{\partial \pi_j(\theta_1,...,\theta_k)}{\partial \theta_q},$$

has rank k.

Identifiability: The model must be identifiable in terms of $\boldsymbol{\theta} = \theta_1,...,\theta_k$.

The exact definition of identifiability is given in the references above. If the matrix **M** has rank less than k, the model is not identifiable.

The matrix **M** with elements (3.35) is generally knowns as the **information matrix,** because it describes the precision with which the parameters can be estimated and hence the strength of information in the data concerning the values of the parameters.

Theorem 3.10:

Under regularity and identifiabily conditions, the asymptotic distribution of the ML–estimates in the model (3.33), (3.34) **is given by**

$$(3.36) \qquad \sqrt{n}(\hat{\boldsymbol{\theta}} - \boldsymbol{\theta}) \overset{a}{\sim} N_k(\mathbf{0}, \mathbf{M}^{-1}).$$

From theorem 3.12 follows that the asymptotic variance of $\hat{\theta}_p$ is

$$(3.37) \qquad \mathrm{var}[\hat{\theta}_p] \overset{a}{=} \frac{1}{n} m^{pp}, \quad p=1,...,k,$$

where m^{pp} is the p'th diagonal element of $\mathbf{M}^{-1}$. From (3.37) a confidence interval for θ_p with approximate confidence level $1-\alpha$ can be constructed as

$$(3.38) \qquad (\hat{\theta}_p - u_{1-\alpha/2}\sqrt{\widehat{m^{pp}}}/\sqrt{n},\ \hat{\theta}_p + u_{1-\alpha/2}\sqrt{\widehat{m^{pp}}}/\sqrt{n})$$

where $\widehat{m}^{pp}$ is the p'th diagonal element of $\mathbf{M}^{-1}$ with the θ's replaced by their ML–estimates.

Theorem 3.11:

Under regularity and identifiability conditions, the test statistic

$$(3.39) \qquad Z = 2 \sum_{j=1}^{m} Y_j(\ln Y_j - \ln(n\hat{\pi}_j)),$$

with

$$\hat{\pi}_j = \pi_j(\hat{\theta}_1,...,\hat{\theta}_k)$$

and $\hat{\theta}_1,...,\hat{\theta}_k$ being the ML–estimates, has the asymptotic distribution

$$Z \overset{a}{\sim} \chi^2(m-1-k) .$$

The result in theorem 3.11 is usually applied in connection with a **goodness of fit test**. Suppose the assumed model for the data is a multinomial distribution with cell probabilities (3.34), then (3.39) can be used to test whether the assumed model fits the data. If H_0 is the composite hypothesis that the multinomial probabilities $\pi_1,...,\pi_m$ are constrained by their common dependencies (3.34) on $\theta_1,...,\theta_k$, then the log–likelihood under H_0 is

$$\ln\tilde{L} = \ln\left[y_1\cdots y_m\right] + \sum_{j=1}^{m} y_j \ln(\pi_j(\hat{\theta}_1,...,\hat{\theta}_k)).$$

Without H_0 the ML–estimators for the π's are $\hat{\pi}_j = y_j/n$, $j=1,...,m$. Hence the log–likelihood without H_0 is

$$\ln\hat{L} = \ln\left[y_1\cdots y_m\right] + \sum_{j=1}^{m} y_j \ln(y_j/n).$$

It follows that the log–likelihood ratio for testing H_0 is

$$\ln r = \ln\tilde{L} - \ln\hat{L} = \sum_{j=1}^{m} Y_j[\ln(\pi_j(\hat{\theta}_1,...,\hat{\theta}_k)) - \ln(Y_j/n)],$$

and H_0 is rejected if $\ln r$ is small. This is, however, equivalent to rejecting H_0 if

$$-2\ln r = 2 \sum_{j=1}^{m} Y_j(\ln Y_j - \ln(n\hat{\pi}_j))$$

is large, where $\hat{\pi}_j = \pi_j(\hat{\theta}_1,...,\hat{\theta}_k)$. The data supports the model if the observed value of Z is small, while a large observed value of Z is an indication of a lack of fit between data and model. The level of significance is accordingly approximateoly equal to $P(Q > -2\ln r)$, where $Q \sim \chi^2(m-1-k)$.

Many statisticians prefer the **Pearson test statistic**, defined as

$$Q = \sum_{j=1}^{m} (Y_j - n\hat{\pi}_j)^2/(n\hat{\pi}_j),$$

to Z. The two quantities are, however, asymptotically equivalent under H_0. Also Q is thus asymptotically χ^2–distributed with $m-1-k$ degrees of freedom under the conditions of theorem 3.11.

With Z replaced by Q, theorem 3.11 was first rigorously proved by Birch (1964), and the theorem is widely refered to as **Birch's theorem.**

Under regularity conditions, where $n\pi_j > 0$ for all j, the expected numbers $n\hat{\pi}_j$ will with probability one be bounded away from zero. If, therefore, theorem 3.11 is used as basis for approximating the distribution of Z or Q, it is a critical condition for the validity of the approximation that the estimated expected numbers $n\hat{\pi}_j$ are not too close to zero. Cases where an observed multinomial distribution has small expected numbers are referred to as **sparse multinomials or sparse tables.** Sparse tables are discussed in Haber-man (1977a), Dale (1986) and Koehler (1986).

If a test based on Z or Q has revealed that the data does not support the model, it can be of interest to study which data points contribute to the lack of fit. For such a study **residual analysis** is often helpful. The residuals for the multinomial distribution are defined as the differences

$$Y_j - n\hat{\pi}_j$$

between observed and expected numbers. In order to judge if a given residual contribute significantly to a lack of fit, it is common practice to standardize the residuals by dividing them with their standard errors. Hence we need the variance of $Y_j - n\hat{\pi}_j$.

Theorem 3.12:

Under regularity and identifiability conditions the asymptotic variance of the residual $Y_j - n\hat{\pi}_j$ is given by

$$(3.40) \qquad \text{var}[Y_j - n\hat{\pi}_j] \overset{a}{=} n\pi_j(1-\pi_j - \Sigma\Sigma \frac{1}{\pi_j} \frac{\partial\pi_j}{\partial\theta_p} \frac{\partial\pi_j}{\partial\theta_q} m^{pq})$$

where $\hat{\pi}_j = \pi_j(\hat{\theta}_1,...,\hat{\theta}_k)$ and m^{pq} is the element in row p and column q of M^{-1}.

Theorem 3.12 can be derived from theorem 3.10 rather easily. Details are given in Rao (1973), p.393, which also seems to be the first time the formula appears in the litterature.

Often the squared roots

$$r_j^* = (y_j - n\hat{\pi}_j)/\sqrt{n\hat{\pi}_j}$$

of the individual terms in Pearsons test statistic Q are referred to as standardized residuals. Theorem 3.12 shows that these residuals do not have unit variance or even the same variance. Residuals with unit variance are given by

$$(3.41) \qquad r_j = (y_j - n\hat{\pi}_j)/\sqrt{n\hat{\pi}_j(1-\hat{\delta}_j)},$$

where

$$(3.42) \qquad \hat{\delta}_j = \hat{\pi}_j + \Sigma\Sigma \frac{1}{\pi_j} \frac{\partial\hat{\pi}_j}{\partial\theta_p} \frac{\partial\hat{\pi}_j}{\partial\theta_q} \hat{m}^{pq},$$

and $\hat{m}^{pq}$ is the (p,q)–element of the inverse of M evaluated at $(\hat{\theta}_1,...,\hat{\theta}_k)$. We shall call the quantities (3.41) **standardized residuals**.

Since $\hat{\delta}_j$ is positive, r_j is always larger than the **Pearson residuals**. Hence r_j^* will not often enough point to critical model departures.

How to choose residuals has attracted much attention in recent years. The main interest is in residuals of the type (3.41) or normalized versions of the individual terms in the test statistic (3.39). Key references are Cox and Snell (1968), (1971) and Pierce and Schafer (1986).

Consider next the case of L independent multinomial distributions, where for each $l=1,\ldots,L$, the vector $(Y_{11},\ldots,Y_{lm})$ has a multinomial distribution with parameters n_l and $\pi_{l1},\ldots,\pi_{lm}$. The likelihood funtion is then given by

$$(3.43) \qquad L(\pi_{11},\ldots,\pi_{Lm}) = \prod_{l=1}^{L} \begin{bmatrix} n_l \\ y_{11}\cdots y_{lm} \end{bmatrix} \prod_{j=1}^{m} \pi_{lj}^{y_{lj}} .$$

We assume that for each $l=1,\ldots,L$, $n_l \to \infty$ in such a way that the ratio n_l/n does not tend to zero, where $n = n_1 + \ldots + n_L$.

One such situation is survey sampling, where units are sampled by simple random sampling within L strata. The total sample of size n is then composed of the subsamples from the strata of a priori fixed sizes $n_1, \ldots, n_L$. Within a given stratum the n_l units are sampled at random. The distribution of a categorical variable over its categories can then be described by the multinomial distribution within each stratum and provided the L subsamples are independent, the total likelihood will be (3.43).

Consider now the parametric models

$$(3.44) \qquad \pi_{lj} = \pi_{lj}(\theta_1,\ldots,\theta_k), \; l=1,\ldots,L, \; j=1,\ldots,m$$

for the π's, and let **M** be the matrix with elements

$$(3.45) \qquad m_{pq} = \sum_{l=1}^{L} \sum_{j=1}^{m} \frac{n_l}{n} \frac{1}{\pi_{jl}} \frac{\partial \pi_{jl}}{\partial \theta_p} \frac{\partial \pi_{jl}}{\partial \theta_q}, \; p,q=1,\ldots,k.$$

Theorems 3.10 to 3.12 then take the following forms:

Theorem 3.13

Under regularity and identifiability conditions, the asymptotic distribution of the ML-estimates is given by

(3.46)
$$\sqrt{n}\,(\hat{\boldsymbol{\theta}} - \boldsymbol{\theta}) \overset{a}{\sim} N_k(0, M^{-1}).$$

The confidence limits (3.38) accordingly also applies to the model (3.43), (3.44) when the information matrix is defined with elements (3.45) rather than (3.35).

Theorem 3.14:

Under regularity and identifiability conditions the test statistic

(3.47)
$$z = 2 \sum_{l=1}^{L} \sum_{j=1}^{m} Y_{lj}(\ln Y_{lj} - \ln(n_l \hat{\pi}_{lj})),$$

where

$$\hat{\pi}_{lj} = \pi_{lj}(\hat{\theta}_1, \dots, \hat{\theta}_k),$$

and $\hat{\theta}_1, \dots, \hat{\theta}_k$ are the ML-estimates, has the asymptotic distribution

$$Z \sim \chi^2(L(m-1)-k).$$

Note that in (3.43) the number of free parameters without constraints on the π's is $L(m-1)$, since there are L independent multinomial distributions with $m-1$ degrees of freedom each.

Theorem 3.15

Under regularity and identifiability conditions the asymptotic variance of the residual $y_{lj} - n_l \hat{\pi}_{lj}$ is

$$(3.48) \qquad \text{var}[Y_{1j} - n_1\hat{\pi}_{1j}]^a = n_1\pi_{1j}(1-\pi_{1j} - \sum_p\sum_q \frac{n_1}{n}\frac{1}{\pi_{1j}}\frac{\partial\pi_{1j}}{\partial\theta_p}\frac{\partial\pi_{1j}}{\partial\theta_q}m^{pq}),$$

where $\hat{\pi}_{1j} = \pi_{1j}(\hat{\theta}_1,...,\hat{\theta}_k)$ and m^{pq} is the (p,q)–element of $\mathbf{M}^{-1}$.

For the model (3.44) the standardized residuals are thus

$$(3.49) \qquad r_{1j} = (y_{1j} - n_1\hat{\pi}_{1j})/\sqrt{n_1\hat{\pi}_{1j}(1-\hat{\delta}_{1j})},$$

where

$$(3.50) \qquad \hat{\delta}_{1j} = \hat{\pi}_{1j} + \sum_p\sum_q\frac{n_1}{n}\times\frac{1}{\pi_{1j}}\frac{\partial\hat{\pi}_{1j}}{\partial\theta_p}\frac{\partial\hat{\pi}_{1j}}{\partial\theta_q}\hat{m}^{pq}$$

and $\hat{m}^{pq}$ is the (p,q)–element of $\mathbf{M}^{-1}$ evaluated at $(\hat{\theta}_1,...,\hat{\theta}_k)$.

3.6. Generalized linear models

Related to the theory of the exponential family is the theory of **generalized linear models**. Generalized linear models in their modern formulation is due to Nelder and Wedderburn (1972). A comprehensive account can be found in Nelder and McCullogh (1983).

A generalized linear model is a generalized form of the exponential family introduced in section 3.1, for non–identically distributed observations and canonical parameters, which are linear functions of a set of **covariates**.

For each observation x_i, let $z_{i1},...,z_{ip}$, be a set of p covariates and assume that the probability

$$P(X_i = x_i) = f(x_i|\theta_i,\delta)$$

has the generalized exponential form

$$(3.51) \qquad \ln f(x_i|\theta_i,\delta) = g(x_i)\varphi(\theta_i)/a(\delta) - K(\theta_i)/a(\delta) + h(x_i,\delta),$$

where θ_i is a model parameter and δ is a parameter, which may or may not be known. This model is called a generalized linear model if the canonical parameters $\tau_1,...,\tau_n$ has the linear form

$$(3.52) \qquad \tau_i = \varphi(\theta_i) = \sum_{j=1}^{p} z_{ij}\beta_j \,,$$

where $\beta_1,...,\beta_p$ are unknown regression parameters. The log–likelihood function for a generalized linear model with n independent observations $x_1,...x_n$ is given by

$$(3.53) \qquad \ln L(\tau_1,...,\tau_n,\delta) = \sum_i y_i \tau_i / a(\delta) - \sum_i K(\tau_i)/a(\delta) + \sum_i h(x_i,\delta),$$

where $K(\tau_i) = K(\theta_i)$ and $y_i = g(x_i)$.

The factor $1/a(\delta)$ does not change the exponential form, if δ is known. If δ is unknown the model may or may not be an exponential family. If, however, only the τ's are of interest simple calculations show, that the likelihood equations derived from (3.53) are the same with arbitrary $a(\delta)$ as with $a(\delta)=1$. Hence $a(\delta)$ acts as an unimportant scaling factor as regards inferences concerning the τ's. The factor $a(\delta)$ is called the **dispersion factor** and δ the **dispersion parameter**. For discrete models $a(\delta)$ is almost always known. Hence in the following $a(\delta)=1$.

There is a canonical parameter τ_i for each observation y_i, but if the regression function (3.52) is inserted in the log–likelihood function and $a(\delta)$ is set to 1, the log–likelihood in terms of the β's become

$$(3.54) \qquad \ln L(\beta_1,...,\beta_p) = \sum_{j=1}^{p} \beta_j \sum_{i=1}^{n} y_i z_{ij} - \sum_i K_i(\beta_1,...,\beta_p) + \sum_i h(y_i),$$

where

$$K_i(\beta_1,...,\beta_p) = K(\sum_j z_{ij}\beta_j)$$

and

$$h(y_i) = h(x_i, \delta).$$

A generalized linear model for discrete data is thus an exponential family with cano-nical parameters $\beta_1, \ldots, \beta_p$ and sufficient statistics

$$T_j = \sum_i Y_i z_{ij} \, , \; j = 1, \ldots, p.$$

Accordingly the likelihood equations are

$$t_j = E[T_j] = \sum_i z_{ij} E[Y_i]$$

or

(3.55) $$\sum_i z_{ij}(y_i - E[Y_i]) = 0, \qquad j = 1, \ldots, p.$$

In section 3.7 below it is discussed how such equations are solved by numerical methods.

The support of the exponential family depends on the values of the covariates $z_{i1}, \ldots, z_{ip}$, $i = 1, \ldots, n$. This means that conditions for a unique solution to the likelihood e-quations depend on the actual values of the covariates, and one has to be very careful to check for singularities.

A general ized linear model can also be defined if

(3.56) $$\eta_i = \sum_{j=1}^{p} z_{ij} \beta_j \, ,$$

which is called the **linear predictor**, is not equal to the canonical parameter τ_i.

The linear predictor is connected with the mean value function

$$\mu_i = E[Y_i]$$

through the **link function**

(3.57)
$$\eta_i = G(\mu_i),$$

which is assumed to be strictly monotone with inverse

$$\mu_i = G^{-1}(\eta_i).$$

In order to formulate a generalized linear model with dispersion factor 1 it is necessary to define

(1)   a distribution (3.51) belonging to the exponential family,

(2)   a linear predictor (3.56) with corresponding covariates

and

(3)   a link function (3.57).

If $\eta_i = \tau_i$ the link funktion is called a **canonical link function**.

If the link function is not canonical the canonical parameter τ_i is derived from the link function by solving the equation

$$\mu_i = K'(\tau_i) = G^{-1}(\eta_i).$$

The linear predictor and the link function together with the distribution

$$\ln f(x \mid \tau_i) = g(x_i)\tau_i - K(\tau_i) + h(x_i$$

then specifies the model.

Example 3.7.

The logistic regression model in example 3.4 is a typical example of a generalized linear model. The model is binomial and hence an exponential family with canonical parameter

$$\tau_i = \ln \frac{p_i}{1-p_i},$$

where $p_i = \theta_i$ is the model parameter. Since

$$\tau_i = \beta_0 + \beta_1 z_i ,$$

the link function is canonical. The dimension is $p=2$ with $z_{i1}=1$ and $z_{i2}=z_i$. Hence the likelihood equations are

$$\sum_i x_i = \sum_i E[Y_i] = \sum_i p_i$$

and

$$\sum_i x_i z_i = \sum_i z_i E[Y_i] = \sum_i p_i z_i$$

with

$$p_i = e^{\beta_0 + \beta_1 z_i} / (1 + e^{\beta_0 + \beta_1 z_i}). \qquad \triangle .$$

3.7. Solution of likelihood equations

The likelihood equations for a log–linear model with k real–valued parameters $\theta_1,...,\theta_k$ has the form (3.19). If the mean value function is denoted

$$E[T_j \mid \theta_1,...,\theta_k] = \psi_j(\theta_1,...,\theta_k),$$

the ML–estimates are, therefore, found by solving the equations

(3.58) $$t_j = \psi_j(\theta_1,...,\theta_m), \quad j=1,...,m$$

with respect to the θ's. In (3.58) the θ's are either the canonical parameters or a set of m parameters, which are identifiable in the sense introduced in section 3.3.

If the equations (3.58) do not have explicit solutions, numerical procedures are called for, e.g. the so–called **Newton–Raphson procedure**. The Newton-Raphson procedure is based on a Taylor expansion of (3.58) up to first order terms, i.e.

$$t_j = \psi_j(\theta_1^o,\ldots,\theta_m^o) + \sum_{l=1}^{m} (\theta_l - \theta_l^o) \frac{\partial \psi_j(\overset{*}{\theta}_1,\ldots,\overset{*}{\theta}_m)}{\partial \theta_l}$$

where $|\overset{*}{\theta}_1 - \theta_1^o| \leq |\theta_1 - \theta_1^o|$, l=1,...,m. On matrix form this can be written

$$t = \psi(\theta_1^o,\ldots,\theta_m^o) + (\theta - \theta^o)\overset{*}{\Psi},$$

where $t=(t_1,\ldots,t_m)$, $\psi(\theta_1,\ldots,\theta_m)=(\psi_1(\theta_1,\ldots,\theta_m),\ldots,\psi_m(\theta_1,\ldots,\theta_m))$, $\theta=(\theta_1,\ldots,\theta_m)$ and $\overset{*}{\Psi}$ is the a square matrix with elements

$$\frac{\partial \psi_j(\overset{*}{\theta}_1,\ldots,\overset{*}{\theta}_m)}{\partial \theta_l}, \quad j=1,\ldots,m, \quad l=1,\ldots,m.$$

If θ_l is close to θ_l^o for all l, $\overset{*}{\theta}_1$ is also close to θ_1^o and we get for θ close to θ^o the approximation

(3.59)
$$t \simeq \psi(\theta_1^o,\ldots,\theta_m^o) + (\theta - \theta^o)\Psi,$$

where Ψ is $\overset{*}{\Psi}$ with $\overset{*}{\theta}$ replaced by θ^o. Solving (3.59) yields

(3.60)
$$\theta = \theta^o + (t - \psi(\theta^o))\Psi^{-1}.$$

Equation (3.60) is an algorithm for obtaining a new approximation to the solution of (3.58) with the elements of θ^o as initial values. Repeated applications of this algorithm can be expected to converge if the initial values $\theta_1^o,\ldots,\theta_m^o$ are reasonable close to the solutions. The algorithm based on (3.60) is called the Newton–Raphson procedure. It normally converge rapidly if the initial values are close to the solution, but if this is not the case it may not converge at all. A disadvantage is that it requires the computation and inversion of the matrix Ψ, which in some cases can be very time consuming. In such cases one may either look for simpler procedures or for approximations to Ψ.

A time saving device, which sometimes works, is only to compute Ψ in the first one or two steps of the iterative procedure and then use the inverse of this matrix in all remaining steps. Another possibility is only to use the diagonal elements of Ψ, in which case the inversion is trivial and the time necessary for computing and inverting Ψ is greatly reduced.

It is instructive to consider the Newton–Raphson procedure for m=1. The likelihood equation is then

$$t = \psi(\theta)$$

and the equation (3.60) becomes

$$\theta = \theta^{o} + (t-\psi(\theta^{o}))[\partial\psi(\theta^{o})/\partial\theta]^{-1}.$$

As illustrated in figure 3.3 this means that new approximations to θ according to the Newton–Raphson procedure are obtained by searching along a tangent to $\psi(\theta)$.

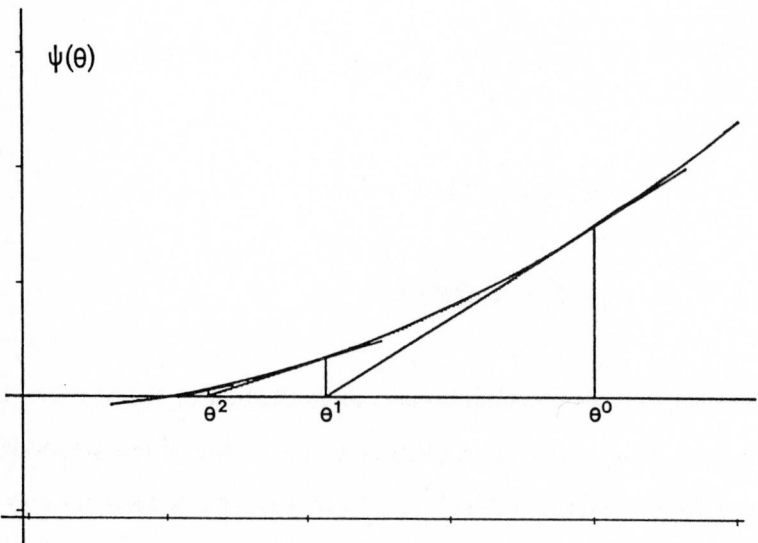

Fig.3.3. The Newton–Raphson procedure for the case k=m=1.

The Newton–Raphson procedure has the further advantage, that it as a by–product pro-vide estimates of the asymptotic variances and covariances of the ML–estimates. This follows from the following theorem

Theorem 3.16:

If the likelihood equations have the form (3.58), then the asymptotic covariance matrix of $\hat{\theta}_1,...,\hat{\theta}_m$ **is** M^{-1}, **where M has elements**

$$m_{jl} = \frac{\partial \psi_j(\theta_1,...,\theta_m)}{\partial \theta_l}, \quad j=1,...,m, \ l=1,...,m.$$

Proof:

Let for simplicity $\theta_1=\tau_1,...,\theta_m=\tau_m$ be the canonical parameters. Then the likelihood e-quations are

$$t_j = n\partial K(\tau_1,...,\tau_m)/\partial \tau_j.$$

The theorem then follows immediately from theorem 3.7. If $\theta_1,...,\theta_m$ are not the canonical parameters, the likelihood equations are

$$t_j = n\partial K(\theta_1,...,\theta_m)/\partial \theta_p \cdot w_{pj}, \quad j,p=1,...,m,$$

where w_{pj} is defined in theorem 3.7A and theorem 3.6 now follows from theorem 3.7A.

☐.

The practical use of theorem 3.16 is that the matrix Ψ^{-1} obtained in the last step of the iterative procedure is an estimate of the asymptotic covariance matrix of the ML–esti-mates.

The particular form (3.58) of the likelihood equations allow in many cases for very

78

simple iterative procedures with attractive properties. One method known as the **Deming–Stephan method** or the **iterative proportional fitting method** is widely used for log–linear models of the type treated in chapters 4,5 and 6. The use of this method require that the right hand side of (3.58) is a sum of basic parameters from which the θ's can be obtained by direct calculation. If these basic parameters are called $\delta_1,...,\delta_I$, it is thus assumed that (3.58) has the form

$$(3.61) \qquad t_j = \sum_{i=1}^{I} w_{ji}\delta_i, \quad j=1,...,m,$$

where w_{ji} are known constants. The θ's are functions of the δ's, i.e.

$$(3.62) \qquad \theta_j = \theta_j(\delta_1,...,\delta_I), \quad j=1,...,m.$$

Let now $\delta_1^o,...,\delta_I^o$, be a set of initial values. A new set of δ's is then derived by changing the initial set of δ's proportionally to satisfy (3.61) for j=1, i.e.

$$\delta_i^1 = \frac{t_1}{\sum_i w_{1i}\delta_i^o} \delta_i^o .$$

Obviously now

$$\sum_i w_{1i}\delta_i^1 = t_1.$$

Next, the δ_i^1's are adjusted proportionally to satisfy (3.61) for j=2, i.e.

$$\delta_i^2 = \frac{t_2}{\sum_i w_{2i}\delta_i^1} \delta_i^1 .$$

When all m t_j's have been adjusted the algorithm goes back to adjusting t_1, etc. The iterations are stopped, when further adjustments do not change the values of the t_j's within the required accuracy. Finally the θ s are obtained from (3.62). The procedure may require a large number of steps, but each step only involve simple calculations. The convergence of the procedure is often insensitive to the choice of initial values.

The Deming–Stephan procedure is widely used in connection with log–linear model for contingency tables, because the parameters of interest are the expected values in the cells, which are the δ's of the Deming–Stephan procedure. By iterative proportional marginal fitting the expected cell numbers under the hypothesis are thus derived directly.

Other situations, where the form of the likelihood equations calls for simple iterative procedures, are latent structure models, treated in chapter 12. For such models (3.58) typically has the form

$$t_j = \delta(\theta_j)D_j(\theta_1,...,\theta_m), \ j=1,...,m,$$

where $\delta_j = \delta(\theta_j)$ is a monotone function with an explicit solution

$$\theta_j = \delta^{-1}(\delta_j)$$

while $D_j(\theta_1,...,\theta_k)$ has a rather limited variation as a function of the θ's.

From initial values $\theta_1^o,...,\theta_m^o$ improved estimates can then be obtained as

$$\theta_j^1 = \delta^{-1}(t_j/D_j(\theta_1^o,...,\theta_m^o)), \ j=1,...,m.$$

This procedure, is known as **the partial solution method**.

In generalized linear models of the form (3.54), solutions to the likelihood equations can be obtained by a weighted least square method. Consider the Newton–Raphson method (3.60) where θ is the vector of β 's and t is the vector

$$t_j = \sum_i z_{ij} y_i, \ j=1,...,p$$

of sufficient statistics. The matrix $\mathbf{\Psi}$ in (3.60) then has elements

$$\Psi_{jl} = \frac{\partial E[T_j]}{\partial \beta_l}, \qquad j,l=1,...,p.$$

Since, however,

$$E[T_j] = \sum_i z_{ij} K'(\tau_i),$$

we get from (3.52)

$$\Psi_{jl} = \sum_i z_{ij} K''(\tau_i) \frac{\partial \tau_i}{\partial \beta_l} = \sum_i z_{ij} z_{il} K''(\tau_i).$$

Equation (3.59) for the generalized linear model thus takes the form

$$(3.63) \qquad t_j \simeq E[T_j | \beta_{10},...,\beta_{p0}] + \sum_{l=1}^{p} (\beta_l - \beta_{l0}) \sum_i z_{ij} z_{il} K''(\tau_{i0})$$

with

$$\tau_{i0} = \sum_j z_{ij} \beta_{j0}.$$

On the other hand,

$$E[T_j | \beta_{10},...,\beta_{p0}] = \sum_i z_{ij} K'(\tau_{i0})$$

and

$$t_j = \sum_i z_{ij} y_i,$$

such that (3.63) as an equality has the form

$$(3.64) \qquad \sum_i z_{ij} \left[\frac{y_i - K'(\tau_{i0})}{w_i} + \sum_{l=1}^{p} \beta_{l0} z_{il} \right] w_i = \sum_{l=1}^{p} \beta_l \sum_i z_{ij} z_{il} w_i, \qquad j=1,...p$$

where

$$w_i = K''(\tau_{i0}).$$

If we put

$$(3.65) \qquad y_{i0} = \frac{y_i - K'(\tau_{i0})}{w_i} + \sum_{l=1}^{p} \beta_{l0} z_{il} \,,$$

(3.64) are the estimation equations for the weighted least square estimates of $\beta_1,\dots,\beta_p$ in a linear regression model with y_{i0} as response variable, the covariates $z_{i1},\dots,z_{ip}$ as explanatory variables and w_i as weights, i.e.

$$(3.66) \qquad y_{i0} = \sum_{j=1}^{p} \beta_j z_{ij} + e_i \,,$$

where e_i is an error term.

It follows that the new set of approximations to the ML–estimates for $\beta_1,\dots,\beta_p$ in each step of the Newton–Raphson procedure are obtained as weighted least squares estimates in the model (3.66). It follows that the Newton Raphson procedure is equivalent to the following **iterative weighted least square** method suggested by Nelder and Wedderburn (1972).

(i)   Given initial values $\beta_{10},\dots,\beta_{p0}$, form the pseudo response variables $y_{10},\dots,y_{n0}$ as (3.65).

(ii)   Obtain the weighted least squares estimates for $\beta_1,\dots,\beta_p$ from the linear regression model (3.66) with $w_i = K''(\tau_{i0})$ as weights.

(iii) repeat (i) and (ii) with the new estimates as initial values until convergence is obtained.

This method is widely used for the regression models in chapter 8 and 9. It is the main numerical tool in the GLIM software package, cf. Baker and Nelder (1978). Jørgensen (1984) extended the method and coined the phrase a **delta algorithm**. The method of iterative weighted least squares has been generalized to distributions outside the exponential family by Green (1984), who also discussed robust alternatives. For further reading the reader is referred to Nelder and McCullogh (1983) or Jørgensen (1984).

3.8. Exercises

3.1. Two players A and B play each other in n plays. The winner of each play receive 1 dollar from the loser. The plays are assumed to be independent and the probability that A wins is θ.

Let X_i be A's gain/loss in play number i, i.e. 1 dollar if he wins and -1 dollar if he loses.

(a) Show that the log–likelihood function is

$$\ln L(\theta) = \frac{n+\Sigma x_i}{2} \ln\theta + \frac{n-\Sigma x_i}{2} \ln(1-\theta).$$

Why does it only depend on Σx_i?

(b) Find the canonical parameter τ and the sufficient statistic.

(c) Specify and solve the likelihood equation.

(d) What is the connection to the binomial distribution.

3.2. A random variable with point probability

$$f(x) = \theta^x \frac{1-\theta}{1-\theta^{k+1}} \,, \; x=0,...,k \,, \; 0 < \theta < 1,$$

is said to follow the truncated geometric distribution.

Let $x_1,...x_n$ be n independent observations from this distribution.

(a) Show that the distribution of X forms an exponential family.

(b) Identify the canonical parameter τ and the sufficient statistic T.

(c) Construct the support and identify those values t of T for which theorem 3.2 does not ensure a unique solution to the likelihood equation.

(d) Write down the likelihood equation for θ.

3.3. A random variable X with point probability

$$f(x) = \frac{1}{c(\theta)} x^{\theta}, \quad x=1,2,...$$

is said to follow the zeta–distribution. The function $c(\theta)$ is Riemann's zeta–function. It can be shown that the domain is $(-\infty,-1)$.

(a)   Show that the distribution of X forms an exponential distribution.

(b)   Identify the canonical parameter and the sufficient statistic for n independent observation $x_1,...,x_n$.

(c)   Derive the likelihood equation and use theorem 3.1 to verify the formula

$$\sum_{x=1}^{\infty} \ln x \cdot x^{\theta} = C'(\theta).$$

(d)   Identify those values of $x_1,...,x_n$ for which theorem 3.2 does not guarantee a unique solution to the likelihood equation.

3.4.   A random variable with point probability

$$f(x) = [-\ln(1-\theta)]^{-1} \theta^x/x, \quad x=1,2,...$$

is said to follow a logarithmic series distribution.

(a)   Find the domain of the distribution.

(b)   Derive the likelihood equations for n independent observations.

(c)   Determine the asymptotic variance of the ML–estimator as a function of θ.

(d)   Suppose $\bar{x}=3.5$. Determine the ML–estimate $\hat{\theta}$ by the Newton–Raphson method and give an estimate of its standard error.

3.5. A random variable X with point probability

$$f(x) = \begin{bmatrix} x+k-1 \\ k-1 \end{bmatrix} \theta^k (1-\theta)^x, \quad x=0,1,2,...,$$

where k is an integer, is said to follow a Pascal distribution. Let $x_1,...,x_n$ be n independent observations from this distribution.

(a) Derive and solve the likelihood equation for θ.

(b) Identify the function $K(\theta)$ and use theorem 3.3A to find the asymptotic variance of θ.

(c) Let k=4, n=5 and consider the following observations $x_1,...,x_5=3,2,5,3,2$. Test then the hypothesis

$$H_0: \theta = 0.5.$$

by a likelihood ratio test.

3.6. Let X_1, X_2, X_3 follow a trinomial distribution with

$$f(x_1,x_2,x_3) = \begin{bmatrix} n \\ x_1 x_2 x_3 \end{bmatrix} p_1^{x_1} p_2^{x_2} p_3^{x_3}.$$

(a) What is the dimension of the exponential family.

(b) Find the canonical parameters.

(c) Derive the likelihood ratio test for $H_0: p_1=p_2=p_3$.

(d) If $x_1=4$, $x_2=3$, $x_3=5$ compute the value of the transformed likelihood ratio test statistic and test H_0.

3.7. The table below show for three municipalities in Denmark the persons, which was interviewed in connection with the Danish Welfare Study, cross–classified according to the household income (rounded to 10.000 Dkr.) and whether the household has a swimming pool or a freezer. If analyzed by a logistic regression model, check for each data set, whether there are unique solutions to the likelihood equations.

Municipality	Income (10.000 Dkr.)	Sample size	Number of household with Pool	Number of household with Freezer
Fredensborg	1	1	0	1
	4	2	0	1
	6	3	0	2
	7	1	0	1
	8	2	0	2
	10	1	0	1
	15	1	0	1
Total		11	0	9
Karlebo	0	1	0	1
	1	2	0	2
	2	1	0	1
	3	1	0	0
	4	3	1	3
	5	1	0	1
	7	2	0	2
	8	4	1	3
Total		15	2	13
Stenløse	0	2	0	2
	3	3	1	3
	6	3	0	3
	7	3	0	3
	8	2	0	2
	10	1	0	1
Total		14	1	14

3.8. In exercise 3.2 suppose we have observed 10 observations with average value $\bar{x}=1.2$.

(a) Find $\hat{\theta}$ using the Newton–Raphson method.

(b) Test the hypothesis

$$\theta = 0.5$$

by a likelihood ratio test.

3.9. Between October 1961 and December 1964, there were born 98 twins at a hospital in Melbourne. The distribution of these according to sex is shown below

2 boys	2 girls	1 boy, 1 girl
29	36	33

(a) Suppose these data are described by a trinomial distribution with parameters $(98,p_1,p_2,p_3)$. Show that if the sex of twin number two is independent of the sex of twin number one, then

$$p_1 = p_2 = \frac{1}{4}, \, p_3 = \frac{1}{2}.$$

(b) Test the hypothesis in (a).

(c) Twins can be classified as monozygotes and dizygotes. For monozygotes the sex of the two twins is the same. If the probability of observing a monocygotic pair of twins is θ, then show that

$$p_1 = p_2 = \frac{1+\theta}{4}, \, p_3 = \frac{1-\theta}{2}.$$

(d) Derive the likelihood ratio test for the hypothesis in (c) and use the observations from Melbourne to test the hypothesis.

3.10. Let $X_1,...,X_n$ be independent of Poisson distributed with common mean value λ_1. Let further $Y_1,...,Y_n$ be independent Poisson distribution with common mean value λ_2. Suppose we are interested in the parameter

$$\theta = \lambda_1/\lambda_2.$$

(a) Show that the model is identifiable in $\theta_1=\theta$ and $\theta_2=\lambda_2$.

(b) Derive the canonical parameters and the likelihood equations.

(c) Derive the transformed likelihood ratio test statistic

$$Z = -2\ln r(T_1,T_2)$$

for the hypothesis

$$H_0: \theta = 1.$$

(d) Would you accept H_0 if $n=50$, $\bar{x}=20$ and $\bar{y}=30$?

3.11. In example 3.6 consider the hypothesis

$$H_0: \lambda_1 = \lambda_2 = \lambda_3, \ \lambda_4 = \lambda_5$$

(a)   Show that the model is identifiable in terms of $\lambda_1, \lambda_4, \lambda_6$ and λ_7.

(b)   Formulate H_0 in terms of the canonical parameters.

(c)   Test H_0 based on the data in table 3.1.

3.12. Pairwise comparison are observations of a set of variables jointly two by two. Consider for example four variables indexed 1,2,3,4. Data consist of the number of times x_{ij} variable i is prefered to variable j. It is assumed that any of n individuals prefers variable i to variable j with probability

$$p_{ij} = e^{\epsilon_i - \epsilon_j}/(1 + e^{\epsilon_i - \epsilon_j})$$

(a)   Show that if $x_{11}, \ldots, x_{I-1.I}$ are independent, then the log–likelihood become

$$\ln L = \sum_{i<j}\sum (\epsilon_i - \epsilon_j)x_{ij} - n \sum_{i<j}\sum \ln(1 + e^{\epsilon_i - \epsilon_j}).$$

(b)   How many parameters are identifiable.

(c)   Derive the likelihood equations.

(d)   Find conditions for the solvability of the likelihood equations.

3.13. In exercise 3.7 consider the number of households in Karlebo with and wihtout freezer as a function of income.

(a)   Show that formulàted as a generalized linear model, a logistic regression model has

$$\tau_i = \beta_0 + \beta_1 z_i$$

and

$$K(\tau_i) = \ln(1 + e^{\tau_i}).$$

(b)   Describe how the method of iterative weighted least squares work for the house holds with and without freezers in Karlebo.

(c) Choose initial values for β_0 and β_1, for example from a plot and work through the calculations for two iterations.

4. Two–way Contingency Tables

4.1. Three models

A two–way contingency is a number of observed counts set up in a matrix with I rows and J columns. Data are thus given as a matrix

$$X = \begin{bmatrix} x_{11}...x_{1J} \\ \vdots \\ x_{I1}...x_{IJ} \end{bmatrix}$$

The statistical model for such data depends on the way the data are collected. A great variety of such tables can, however, be treated by three closely connected statistical models. Let the random variables corresponding to the contingency table be $X_{11},...,X_{IJ}$. Then in the first model the X_{ij}'s are assumed to be independent with

$$X_{ij} \sim Ps(\lambda_{ij}),$$

i.e. X_{ij} is Poisson distributed with parameter λ_{ij}. The likelihood function for this model is

(4.1) $$f(x_{11},...,x_{IJ} \mid \lambda_{11},...,\lambda_{IJ}) = \prod_{i=1}^{I} \prod_{j=1}^{J} \frac{\lambda_{ij}^{x_{ij}}}{x_{ij}!} e^{-\lambda_{ij}}.$$

The log–likelihood is accordingly given by

(4.2) $$\ln L(\lambda_{11},...,\lambda_{IJ}) = \sum_i \sum_j x_{ij} \ln \lambda_{ij} - \sum_i \sum_j \ln x_{ij}! - \sum_i \sum_j \lambda_{ij}.$$

The model is thus a IJ–dimensional log–linear model with canonical parameters $\ln \lambda_{11},...,\ln \lambda_{IJ}$ and sufficient statistics $T_{ij}=X_{ij}$, $i=1,...,I$, $j=1,...,J$.

The Poisson model (4.1) covers many applications in traffic research, where the observed counts are traffic accidents. The contingency table can, for example, represent traffic accidents cross–classified according to type of accident and time of year.

The Poisson model is not the model most frequently met in practice. Consider a situation, where a sample of size n is drawn at random from a population of size N. For each sampled unit two categorical variables A and B (e.g. sex and social rank) are observed. Let then x_{ij} be the number of persons, which belong to category i according to variable A and category j according to variable B. If N is sufficiently large the distribution of $x_{11},...,x_{IJ}$ can be described by a multinomial distribution with parameters $(n,p_{11},...,p_{IJ})$, where p_{ij} is the probability that a randomly sampled unit falls jointly in categories i and j. The probability of observing the data is

$$(4.3) \qquad f(x_{11},...,x_{IJ}|p_{11},...,p_{IJ}) = \binom{n}{x_{11}...x_{IJ}} \Pi_i \Pi_j p_{ij}^{x_{ij}},$$

such that the log–likelihood function become

$$(4.4) \qquad \ln L(p_{11},...,p_{IJ}) = \ln \binom{n}{x_{11}...x_{IJ}} + \sum_i \sum_j x_{ij} \ln p_{ij}.$$

This is a log–linear model with canonical parameters $\ln p_{11},...,\ln p_{IJ}$ and sufficient statistics $T_{ij}=X_{ij}$. Since, however, the p_{ij}'s are constrained by

$$\sum_{i=1}^{I} \sum_{j=1}^{J} p_{ij} = 1,$$

the dimension of the exponential family is less than IJ.

In fact (4.4) can be rewritten as

$$(4.5) \qquad \ln L(p_{11},...,p_{IJ-1}) = \ln \binom{n}{x_{11}...x_{IJ}} + \sum_{(i,j)\neq(I,J)} x_{ij}(\ln p_{ij}-\ln p_{IJ}) + n\ln p_{IJ},$$

showing that the canonical parameters are $\ln p_{11} - \ln p_{IJ}, \ldots, \ln p_{IJ-1} - \ln p_{IJ}$, and the dimension of the model is $IJ-1$.

In most situations it is not necessary to explicitly write down the reduced form (4.5) as long as we bear in mind that only $IJ-1$ parameters are identifiable. Thus the p_{ij}'s can be estimated and the asymptotic distributions of the transformed log–likelihood ratio can be derived directly in terms of the p_{ij}'s.

The conditional probability distribution of $X_{11}, \ldots, X_{IJ}$ given $X.. = x..$ is under the Poisson model (4.1) given by

$$(4.6) \qquad f(x_{11}, \ldots, x_{IJ} \mid x..) = \left[\begin{matrix} x.. \\ x_{11} \cdots x_{IJ} \end{matrix} \right] \prod_{i=1}^{I} \prod_{j=1}^{J} (\lambda_{ij} / \lambda..)^{x_{ij}},$$

with $\lambda.. = \sum_i \sum_j \lambda_{ij}$. Model (4.3) can thus be derived from model (4.1) by conditioning on the total $X..$.

At least formally models (4.3) and (4.6) are identical .This means that by fixing the total at a given level $x.. = n$, the Poisson model can be treated as the multinomial model (4.3). The equivalents of the cell probabilities p_{ij} are the proportions $\lambda_{ij}/\lambda..$. This means that if one is only interested in the relative magnitudes of the λ_{ij}'s, inference can be drawn from the multinomial model (4.6). In a sense our lack of interest in the over–all level $\lambda..$ is reflected in a conditioning on $x..$. Statistical methods developed for the multinomial model (4.3) thus also apply to the Poisson model (4.1) as long as only estimates of ratios like $\lambda_{ij}/\lambda..$ and hypotheses concerning relative magnitudes of λ_{ij}'s are of interest.

It is not trivial that statistical properties derived from the Poisson model (4.1) are automatically also true for the derived conditional model (4.6). In this case conditions must be placed on the values of the conditioning statistic. In order for the asymptotic properties of goodness of fit test statistics to be valid, it is thus necessary that $x..$ approach infinity.

A third model is connected with **stratified sampling**, where the population is divided in I strata of sizes $N_1, \ldots, N_I$. From these strata independent random samples of sizes

$n_1,...,n_I$ are drawn. The observed counts $x_{i1},...,x_{iJ}$ in row i of the contingency tables are then the number of units among the n_i drawn from stratum i that belong to the J categories of a categorical variable B. If the stratum sizes are large, the data from such sampling can be described as I independent vectors each with a multinomial distribution

$$f(x_{i1},...,x_{iJ} \mid p_{1\mid i},...,p_{J\mid i}) = \left[x_{i1}...x_{iJ} \right] \prod_{j=1}^{J} p_{j\mid i}^{x_{ij}}.$$

The likelihood function is given by

(4.7)
$$L(p_{1\mid 1},...,p_{J\mid I}) = \prod_{i=1}^{I} \left[\left[x_{i1}...x_{iJ} \right] \prod_{j=1}^{J} p_{j\mid i}^{x_{ij}} \right].$$

The parameter $p_{j\mid i}$ is here the probability that a person in stratum i belongs to category j of variable B. Accordingly the $p_{j\mid i}$ satisfy

(4.8)
$$\sum_{j=1}^{J} p_{j\mid i} = 1 , \qquad i=1,...,I.$$

The model is thus a product of I independent multinomial distributions of sizes $n_1,...,n_I$.

The log–likelihood function has the log–linear form

(4.9)
$$\ln L(p_{1\mid 1},...,p_{J\mid I}) = \sum_i \ln \left[x_{i1}...x_{iJ} \right] + \sum_i \sum_j x_{ij} \ln p_{j\mid i}.$$

Due to the constraints (4.8) the dimension of the exponential family is $I \cdot (J-1)$ and from the reduced form

(4.10)
$$\ln L(p_{1\mid 1},...,p_{J\mid I}) = \sum_i \ln \left[x_{i1}...x_{iJ} \right] + \sum_{i=1}^{I} \sum_{j=1}^{J-1} x_{ij} (\ln p_{j\mid i} - \ln p_{J\mid i}) + \sum_{i=1}^{I} n_i \ln p_{J\mid i},$$

follows that the canonical parameters are $\ln p_{1|1} - \ln p_{J|1}, ..., \ln p_{J-1|I} - \ln p_{J|I}$.

Model (4.7) can be obtained from (4.1) by conditioning, since the conditional distribution of $X_{i1},...,X_{iJ}$ given $X_{i.} = x_{i.}$ under the Poisson model is equal to

$$(4.11) \qquad f(x_{i1},...,x_{iJ}|x_{i.}) = \left[\begin{array}{c} x_{i.} \\ x_{i1} \cdots x_{iJ} \end{array} \right] \prod_{j=1}^{J} \left(\frac{\lambda_{ij}}{\lambda_{i.}} \right)^{x_{ij}}.$$

Another analogue to (4.7) can be obtained from (4.3) by conditioning on the row marginals $x_{1.},...,x_{I.}$. In fact the distribution of $X_{i1},...,X_{iJ}$ given $X_{i.} = x_{i.}$, if (4.3) is the model, is

$$(4.12) \qquad f(x_{i1},...,x_{iJ}|x_{i.}) = \left[\begin{array}{c} x_{i.} \\ x_{i1} \cdots x_{iJ} \end{array} \right] \prod_{j=1}^{J} \left[\frac{p_{ij}}{p_{i.}} \right]^{x_{ij}}.$$

Both under the conditioning (4.11) and the conditioning (4.12), the log–likelihood has the form

$$(4.13) \qquad \ln L(q_{11},...,q_{IJ}) = \sum_i \ln \left[\begin{array}{c} n_i \\ x_{i1} \cdots x_{iJ} \end{array} \right] + \sum_i \sum_j x_{ij} \ln q_{ij},$$

where $n_i = x_{i.}$, $q_{ij} = \lambda_{ij}/\lambda_{i.}$ in (4.11) and $q_{ij} = p_{ij}/p_{i.}$ in (4.12). Thus at least formally (4.13) is identical with (4.9) and statistical methods developed for models (4.1) and (4.3) also apply to the case of stratified sampling if one is only interested in parameters derivable from $\lambda_{ij}/\lambda_{i.}$ or parameters derivable from $p_{ij}/p_{i.}$. The correspondances between models (4.1), (4.3) and (4.7) noted above are important, because they allow us to concentrate on one type of model, when we develope the necessary statistical methodologies. It is important to keep in mind, what parameters, it is possible to draw inference about, however. Thus if the Poisson model is reduced to model (4.13) by conditioning, one cannot draw inference about the over–all level $\lambda_{..}$ or about the row levels $\lambda_{i.}$, $i=1,...,I$.

In epidemiology the difference between models (4.3) and (4.7) are often described in terms of the experimental design. Model (4.3) is connected with what is called a **cross–**

sectional study, where a sample is taken from the population and the number of stages of a disease are recorded as they manifest themselves in various groups. In its most simple form the observations can be if the disease is found or not at individuals in various age groups. For a sample of size n, x_{i1} is then the number of individuals in age group i, who has the disease and x_{i2} the number in age group i, who do not have the disease. Model (4.7) on the other hand is an example of a **prospective study**. From each age group a sample is drawn, and among the n_i selected in age group i, it is recorded how many x_{i1}, who have the disease, and how many x_{i2}, who do not have the disease. For so–called **retrospective studies** the model is derived from (4.1) by conditioning. In such studies a number n_1, which have the disease and a number n_2 who do not have the disease are selected from a file of all individuals. Among the n_1 with the disease, it is then recorded how many individuals $x_{11},...x_{1J}$ are found in each of the J age groups. The distribution (4.11) then describes the distribution over age groups of those $x_{1.}=n_1$ selected with the disease.

4.2. The 2x2 table

A contingency table with I=J=2 is usually refered to as a **two by two table** or 2x2 table. This most simple of all contingency table is well suited for explaining some of the basic hypotheses to be considered in contingency tables.

It is most convenient to start by model (4.3). Let thus the model for the 2x2 table

$$\begin{bmatrix} x_{11} & x_{12} \\ x_{21} & x_{22} \end{bmatrix}$$

be

(4.14)
$$f(x_{11},x_{12},x_{21},x_{22}) = \begin{bmatrix} n \\ x_{11}...x_{22} \end{bmatrix} p_{11}^{x_{11}} p_{12}^{x_{12}} p_{21}^{x_{21}} p_{22}^{x_{22}} .$$

Due to the constraint $p_{11}+p_{12}+p_{21}+p_{22}=1$, there are three free parameters. If the table is formed by cross–classification of a random sample of size n according to two binary vari–

ables A and B, the interpretation of p_{ij} is

p_{ij} = P{a randomly sampled unit belongs to category i according to variable A and category j according to variable B.}

If the sampling is completely random and the population size N is large, p_{ij} can also be interpreted as the proportion of the population belonging jointly to categories i and j. Without further knowledge about the values of the p_{ij}'s, the ML–estimate of p_{ij} is

$$\hat{p}_{ij} = x_{ij}/n.$$

Hence inference about the distribution of the population over the four cells of the table can be drawn from the observed frequencies

In a 2x2 table the hypothesis of primary interest is

(4.15) $$H_0: p_{ij} = p_{i.} p_{.j}, \quad i,j=1,2,$$

also known as the **independence hypothesis**.

In order to interprete H_0, let A_1 and A_2 with

A_i = {a randomly sampled unit belongs to category i of variable A}, i=1,2

be the two events connected with variable A. Analogously let B_1 and B_2 with

B_j = {a randomly sampled unit belongs to category j of variable B}, j=1,2

be the events connected with variable B. The three terms in (4.15) can then be inter-preted as $p_{ij} = P(A_i \cap B_j)$, $p_{i.} = P(A_i)$ and $p_{.j} = P(B_j)$.

Hence H_0 is an independence hypothesis in the exact sense that the events A_i and B_j connected with the two binary variables are independent.

The formulation of H_0 in (4.15) is convenient for interpretation purposes, but it is useful also to have a formulation of H_0 in terms of the original four parameters p_{11}, p_{12}, p_{21} and p_{22} of the multinomial model. This is obtained by writing (4.15) for i=1 and j=1 as

$$p_{11} = (p_{11}+p_{12})(p_{11}+p_{21}),$$

multiplying the left hand side by $1 = p_{11}+p_{12}+p_{21}+p_{22}$ and reducing, which yields

$$(4.16) \qquad\qquad p_{11}p_{22} = p_{12}p_{21}.$$

It is easy to see that we get exactly the same condition for all other combinations of i and j in (4.15). Hence H_0 is equivalently expressed by (4.16). One consequence is that H_0 is satisfied if and only if the **odds ratio**

$$(4.17) \qquad\qquad \rho = \frac{p_{11}p_{22}}{p_{12}p_{21}}$$

has the value 1, i.e. (4.15) can be expressed as

$$(4.18) \qquad\qquad H_0 : \rho = 1.$$

The term odds ratio is derived from betting for example on horses, where the chance of winning is measured by the ratio of the chance of winning and the chance of loosing, called the odds in favour of ones bet. Independence between A and B thus means that the odds of variable being observed at level 1 is the same whether variable B is at level one or two. In a betting situation, we thus would bet on the events of A independently of any knowledge as regards the levels of variable B.

An even more popular formulation of H_0 is in terms of the **log–odds ratio**, i.e.

$$(4.19) \qquad\qquad H_0 : \ln\rho = 0,$$

where ρ is given by (4.17).

Since the ML–estimate of p_{ij} is x_{ij}/n, the empirical equivalent of ρ is

$$(4.20) \qquad r = \frac{x_{11}x_{22}}{x_{12}x_{21}} ,$$

which is extremely easy to compute in practice. In addition the statistical uncertainty of r is known under the multinomial model (4.14). In fact it can be shown (cf. for example Bishop, Holland and Fienberg (1977), p.497 or Andersen (1980a), p.167) that

$$\text{var}[\ln R] \simeq \frac{1}{n}\left(\frac{1}{p_{11}} + \frac{1}{p_{12}} + \frac{1}{p_{21}} + \frac{1}{p_{22}}\right) ,$$

where $R = X_{11}X_{22}/(X_{12}X_{21})$.

Hence confidence limits for $\ln\rho$ with approximate level of confidence $1-\alpha$ are given by

$$(4.21) \qquad \ln r \pm u_{1-\alpha/2}\sqrt{\frac{1}{x_{11}} + \frac{1}{x_{12}} + \frac{1}{x_{21}} + \frac{1}{x_{22}}} .$$

These confidence limits are relatively precise even for moderate values of n, because the asymptotic distribution of $\ln r$ is almost symmetric. Since r is a ratio, the distribution of r can be very skew for values of r close to 0. Confidence limits for ρ are obtained, therefore, by transforming the limits in (4.21) exponentially. Confidence limits for ρ with level of confidence $1-\alpha$ are thus given by

$$(4.22) \qquad \exp\left\{\ln r + u_{1-\alpha/2}\sqrt{\frac{1}{x_{11}} + \frac{1}{x_{12}} + \frac{1}{x_{21}} + \frac{1}{x_{22}}}\right\}.$$

and

$$(4.23) \qquad \exp\left\{\ln r - u_{1-\alpha/2}\sqrt{\frac{1}{x_{11}} + \frac{1}{x_{12}} + \frac{1}{x_{21}} + \frac{1}{x_{22}}}\right\}.$$

The confidence limits (4.22) and (4.23) for the odds ratio are very useful. In order to check the independence hypothesis, the appropriate first step is to compute r by (4.20) to get an impression of how strong the dependency is. If r is close to unity, the confidence limits (4.22) and (4.23) are computed to check if the hypothetical value 1 is between these limits and thus in agreement with the given data.

Formulae (4.21), (4.22) and (4.23) also apply under the models (4.1) and (4.7), but the interpretations of H_0 are of course different. Under (4.1), H_0 has the form

$$H_0: \lambda_{ij} = \lambda_{i.} \lambda_{.j} / \lambda_{..},$$

which states that the mean values λ_{ij} apart from a normalization factor $1/\lambda_{..}$ are products of a row effect $\lambda_{i.}$ and a column factor $\lambda_{.j}$. For model (4.7), H_0 takes the form

$$H_0: p_{j|1} = p_{j|2}, \quad j=1,2,$$

i.e. the probability of being at level j of variable B is the same, whether the sampled unit is from stratum 1 or stratum 2. But again the odds ratio r given by (4.20) is a measure of how close the data is to what should be expected under H_0. The theoretical odds ratio under model (4.7) is

$$\rho = \frac{p_{1|1} p_{2|2}}{p_{1|2} p_{2|1}}$$

and under model (4.1)

$$\rho = \frac{\lambda_{11} \lambda_{22}}{\lambda_{12} \lambda_{21}}.$$

It is typical for contingency tables that models may be different, depending on the sampling design, but that the statistical tools for checking the hypothesis H_0 are the same for all models.

Example 4.1:

A retrospective study of cancer of the ovary was carried out in Denmark in 1973. 300 women were selected for the study, 150 who had survived a cancer operation by 10 years and 150, who did not survive the operation by 10 years. One record was lost for a woman, who did not survive. For the remaining 299 women it was recorded whether the cancer at the time of operation was at an early or at an advanced stage. The resulting data are shown in table 4.1.

Table 4.1. 299 women in 1973 cross–classified according to survival and stage of cancer.

Stage of cancer	Survival by ten years		Total
	No	Yes	
Early	31	127	158
Advanced	118	23	141
Total	149	150	299

Source: Obel (1975).

The observed odds ratio for these data is

$$r = 0.0476,$$

showing that the odds of survival is very low when the cancer is operated at an advanced stage and that the odds of survival is high when the cancer is operated at an early stage. That the observed value of r does in fact point to the true odds ratio being different from 1 and hence that H_0 can not be true is seen by computing the confidence limits (4.22) and (4.23). For the data i table 4.1, the 95% confidence limits are

$$0.026 \leq \rho \leq 0.086,$$

showing that it is extremely unlikely that $\rho=1$.

Note that the appropriate model for these data is the model (4.7), since the women are sampled in two strata. Stratum one consists of those who have survived, and stratum two of those, who did not survive. From these two strata $n_1=149$ are selected at random from stratum 1 and $n_2=150$ are selected at random from stratum 2. In fact it was decided to sample 150 from each stratum, but 1 record was lost 1. The sampling design determines what quantities can be estimated. Thus in this case the probability, which we can assess is the chance of having the cancer at an early stage at the time of the operation, given that the patient has survived, and the corresponding three other conditional probabilities. We cannot, however, assess the chance of survival given that the cancer is at an early stage. We can check if it is different from the chance of survival given the cancer is at an advanced stage, but not the actually magnitude of the probability. This would require that the design is prospective, i.e. that a certain number of women with their cancer at an early stage and a certain number with the cancer at an advanced stage had been selected and it then was recorded how many survived in each group. $\triangle$.

Example 4.2.

As an example of model (4.3) consider the data in table 4.2 showing for a random sample of 4229 individuals in Denmark in 1965 whether they returned the postal questionnaire. The table also shows the distribution according to sex. A non—return is denoted in table 4.3 as a non—response

Table 4.2. A random survey in Denmark in 1965 cross classified according to sex and non—response.

	Male	Female	Total
Response	1893	1838	3731
Non—response	240	258	498
Total	2133	2096	4229

Source: Unpublished data from the Danish National Institute for Social Research.

The data in table 4.2 is an example of model (4.3) since the total n=4229 is fixed and the appropriate parameters are p_{ij}, with for example

$$p_{11} = P\{\text{a sampled person responds and is a male}\}.$$

The observed odds ratio for the data in table 4.2 is

$$r = 1.107,$$

which is rather close to 1, hence one would suspect that H_0 is true. The rate of non–response thus seems to be the same for men and women. In order to evaluate the degree to which the data supports H_0, one can compute the confidence limits (4.22) and (4.23). These limits show that the true odds ratio is between

$$0.918 \leq \rho \leq 1.335$$

with level of confidence 0.95, such that the data strongly supports $\rho=1$. Under $H_0:\rho=1$, we can in addition estimate the over–all rate of response, which is

$$\frac{3731}{4229} = 0.882$$

or 88.2%, and the percentage of men in the population, which is

$$\frac{2133}{4229} = 0.504$$

or 50.4%. These figures show that the response rate is satisfactory high and that the sample is well balanced sexually. △.

Example 4.3.

The data in table 4.3 are from an investigation in Sweden i 1961 and 1962 over the effects of speed limitations. The table shows for 18 weeks in 1961 and 1962 with a speed limit enforced and for 18 weeks in 1961 and 1962 without speed limits, the number of killed in the traffic.

Table 4.3. Number of persons killed in the traffic on main roads and secondary roads for periods of the same lenght without and with speed limitations in 1961 and 1962.

Speed limit	Main roads	Secondary roads	Total
90 km/hour	19	79	98
Free	102	175	277
Total	121	254	375

Source. Unpublished data from the Swedish Road Authorities.

For the data in table 4.3, model (4.1) is the appropriate one, since no totals are given beforehand. The parameters of interest are acordingly λ_{11}, λ_{12}, λ_{21} and λ_{22}, with for example

$$\lambda_{11} = \text{expected number of killed persons on main roads under a}$$
$$\text{speed limit of 90 km/hour in a period of 18 weeks.}$$

Hence the expected number of killed persons per week under a speed limitation on main roads can be estimated from table 4.3 as

$$\frac{19}{18} = 1.056,$$

while the expected number of killed persons on secondary roads is

$$\frac{79}{18} = 4.389.$$

These numbers does not mean that it is more dangerous to drive on secondary roads, since the total length of secondary roads i Sweden is many times larger than the total length of main roads. The Swedish authorities reported the accidents for both main roads and secondary roads because they wanted to check if a speed limitation is equally effective on main and secondary roads. This would be the case if the obvious drop in expected number of killed persons from free to limited speed is the same for main and for secondary roads. In terms of the parameters of the model this would be the case if

$$\frac{\lambda_{11}}{\lambda_{21}} = \frac{\lambda_{12}}{\lambda_{22}}$$

or

$$\rho = \frac{\lambda_{11}\lambda_{22}}{\lambda_{12}\lambda_{21}} = 1.$$

The problem is thus equivalent with testing H_0. The odds ratio, has observed value

$$r = 0.413.$$

The effect of a speed limit thus seems (as expected) to be much larger on main roads. That r=0.413 is not in statistical agreement with $\rho=1$ can be seen from the limits (4.22),

(4.23) which yield that

$$0.236 \le \rho \le 0.720.$$

with level of confidence 95%. △.

The hypothesis H_0 can also be evaluated by a formal test. Consider for example model (4.7). Here the 2x2 table reduces to a comparison of two binomial distributions, with likelihood function

(4.24)
$$L(p_{1|1}, p_{1|2}) = \prod_{i=1}^{2} \begin{bmatrix} n_i \\ x_{i1} \end{bmatrix} p_{1|i}^{x_{i1}} (1-p_{1|i})^{x_{i2}},$$

while H_0 has the form

(4.25)
$$H_0 = p_{1|1} = p_{1|2}.$$

The ML–estimates for the parameters are

$$\hat{p}_{1|1} = x_{11}/n_1$$

and

$$\hat{p}_{1|2} = x_{21}/n_2.$$

A test statistic for H_0 can thus be based on the difference

(4.26)
$$x_{11}/n_1 - x_{21}/n_2.$$

Since x_{11} and x_{21} are binomially distributed and independent, the variance of (4.26) is

$$\text{var}\begin{bmatrix} X_{11} \\ n_1 \end{bmatrix} - \frac{X_{21}}{n_2} = \frac{p_{1|1}(1-p_{1|1})}{n_1} + \frac{p_{1|2}(1-p_{1|2})}{n_2},$$

which under H_0 become

$$\text{var}\left[\frac{X_{11}}{n_1} - \frac{X_{21}}{n_2}\right] = p_1(1-p_1)\,(\frac{1}{n_1} + \frac{1}{n_2}),$$

where $p_1 = p_{1|1} = p_{1|2}$.

It follows that the test quantity

(4.27)
$$U = \frac{X_{11}/n_1 - X_{21}/n_2}{\sqrt{\hat{p}_1(1-\hat{p}_1)\,(\frac{1}{n_1} + \frac{1}{n_2})}}$$

where $\hat{p}_1$ is the ML–estimate of p_1, is approximately distributed as a standard normal deviate.

It is easily seen that

$$\hat{p}_1 = \frac{X_{11} + X_{21}}{n_1 + n_2}.$$

The test statistic (4.27) is very useful, but its use is limited to situations, where both n_1 and n_2 are so large that the approximation provided by the limiting distribution of the difference (4.26) is valid. In most cases it is required that both n_1 and n_2 are at least 10, but the validity of the approximation depends also on the value of p_1, such that neither $n_1 p_1$, $n_1(1-p_1)$, $n_2 p_1$ nor $n_2(1-p_1)$ must be too small.

In small samples one must either derive the exact distribution of (4.27) or rely on an important **conditional test** due to Fisher (1935).

Assume that (4.25) for the model with likelihood (4.24) is true. We can then derive the conditional distribution of X_{11} given $X_{11} + X_{21} = m_1$. Since under H_0, $X_{11} + X_{21}$ has a binomial distribution with parameters $n_1 + n_2$ and p_1, it follows from (4.24) that

$$(4.28) \qquad f(x_{11}|x_{11}+x_{21}=m_1) = \begin{bmatrix} n_1 \\ x_{11} \end{bmatrix} \begin{bmatrix} n_2 \\ x_{21} \end{bmatrix} / \begin{bmatrix} n_1+n_2 \\ m_1 \end{bmatrix},$$

which is a **hypergeometric distribution** with parameters n_1, m_1 and n_1+n_2.

In the original 2x2 table $n_1 = x_{1.}$, $m_1 = x_{.1}$ and $n_1+n_2 = x_{..}$. Hence the distribution (4.28) is simply the distribution of x_{11} given the marginals of the table. The same distribution is, therefore, valid under model (4.3) if we condition upon $x_{1.}$ and $x_{.1}$ and under model (4.1) if we condition is upon $x_{1.}, x_{.1}$ and $x_{..}$.

In order to test H_0, an observed value x_{11} is judged to be in disagreement with H_0, if the difference between x_{11} and the expected value $n_1 m_1/(n_1+n_2)$ is large. If the alternative to H_0 is

$$H_1: p_{1|1} > p_{1|2},$$

the level of significance become

$$p = P(X_{11} > x_{11}),$$

where the probability is computed in the hypergeometric distribution (4.28). This test is known as **Fisher's exact test**.

4.3. The log–linear parameterization

Since the data collected from different sample designs can be treated based on the same basic model, it is to expected that also parametric hypothesis in different models can be treated within a common parametric framework. This framework is provided by the **log–linear parameterization** introduced for contingency tables by Birch (1963), which is essentially a reparameterization in terms of the canonical parameters of the model (4.1). Consider thus the reparameterization

$$(4.29) \qquad \ln\lambda_{ij} = \tau_{ij}^{AB} + \tau_i^A + \tau_j^B + \tau_0 \;,$$

of the λ_{ij}'s, where the τ's satisfies the linear constraints

(4.30)
$$\sum_{i=1}^{I} \tau_{ij}^{AB} = \sum_{j=1}^{J} \tau_{ij}^{AB} = \sum_{i=1}^{I} \tau_{i}^{A} = \sum_{j=1}^{J} \tau_{j}^{B} = 0.$$

This means that only $(J-1)(I-1)$ of the τ_{ij}^{AB}'s, $I-1$ of the τ_{i}^{A}'s and $J-1$ of the τ_{j}^{B}'s have a free variation, and that the model is identifiable in terms of $\tau_{11}^{AB}, \ldots, \tau_{I-1,J-1}^{AB}, \tau_{1}^{A}, \ldots, \tau_{I-1}^{A}, \tau_{1}^{B}, \ldots, \tau_{J-1}^{B}, \tau_{0}$. That (4.29) is in fact a reparameterization follows from the expressions

(4.31)
$$\begin{cases} \tau_{ij}^{AB} = \mu_{ij}^{*} - \bar{\mu}_{i.}^{*} - \bar{\mu}_{.j}^{*} + \bar{\mu}_{..}^{*} \\ \tau_{i}^{A} = \bar{\mu}_{i.}^{*} - \bar{\mu}_{..}^{*} \\ \tau_{j}^{B} = \bar{\mu}_{.j}^{*} - \bar{\mu}_{..}^{*} \\ \tau_{0} = \bar{\mu}_{..}^{*}, \end{cases}$$

where $\mu_{ij}^{*} = \ln\lambda_{ij}$, $\bar{\mu}_{i.}^{*} = \sum_{j}\mu_{ij}^{*}/J$, $\bar{\mu}_{.j}^{*} = \sum_{i}\mu_{ij}^{*}/I$ and $\bar{\mu}_{..}^{*} = \sum_{i}\sum_{j}\mu_{ij}^{*}/(IJ)$. In (4.29) τ_{0} is called the **overall level**, τ_{i}^{A} is called a **row effect** and τ_{j}^{B} a **column effect**. Hence τ_{ij}^{AB} is a residual which measures that part of the logarithmic mean values which cannot be attributed to the over-all level, an isolated row effect or an isolated column effect. The τ_{ij}^{AB}'s thus measure the degree of interaction between the rows and the columns in the expected counts of the table. Accordingly they are called **interaction parameters** or just **interactions**.

In section 4.2 we introduced the odds ratio. It is easy to verify that the interaction parameters for a 2x2 table are connected with the odds ratio through

$$\ln\rho = 4\tau_{11}^{AB}.$$

In fact

$$\ln\rho = \ln\lambda_{11} - \ln\lambda_{12} - \ln\lambda_{21} + \ln\lambda_{22}$$

and

$$\tau_{11}^{AB} = \ln\lambda_{11} - \frac{1}{2}\left(\ln\lambda_{11} + \ln\lambda_{12}\right) - \frac{1}{2}\left(\ln\lambda_{11} + \ln\lambda_{21}\right)$$

$$+\frac{1}{4}\left(\ln\lambda_{11} + \ln\lambda_{12} + \ln\lambda_{21} + \ln\lambda_{22}\right) = \frac{1}{4}\left(\ln\lambda_{11} - \ln\lambda_{12} - \ln\lambda_{21} + \ln\lambda_{22}\right).$$

Hence an analysis of dependency in a 2x2 table can be based on an analysis of the interaction parameters. Note that for a 2x2 table the constraints $\tau^{AB}_{i.}=\tau^{AB}_{.j}=0$ for i,j=1,2 implies that

$$\tau^{AB}_{22} = \tau^{AB}_{11}$$

and

$$\tau^{AB}_{12} = \tau^{AB}_{21} = -\tau^{AB}_{11}$$

such that only τ^{AB}_{11} needs to be specified by the hypothesis.

There are various reasons for prefering the parameterization (4.29). Since the canonical parameters in (4.1) are $\ln\lambda_{ij}$, i=1,...,I, j=1,...,J, the log–linear parameters are essentially equal to the canonical parameters. In subsequent sections it will in addition become obvious that many important hypotheses have more convenient formulations in terms of the τ's than in terms of the λ's. It is a further advantage in relation to models (4.3) and (4.7) that the log–linear parameterization for these models are obtained from (4.1) by simply omitting some of the parameters. Thus (4.3) is parametrized by the interactions, the row effects and the column effects without the main effect, since

(4.32)
$$\lambda_{ij} = \exp\{\tau^{AB}_{ij}+\tau^{A}_{i}+\tau^{B}_{j}+\tau_{0}\}$$

in (4.1), entails that

$$p_{ij}=\lambda_{ij}/\lambda_{..} =\exp\{\tau^{AB}_{ij}+\tau^{A}_{i}+\tau^{B}_{j}\}/ \sum_{i=1}^{I} \sum_{j=1}^{J} \exp\{\tau^{AB}_{ij}+\tau^{A}_{i}+\tau^{B}_{j}\}.$$

The model (4.7) is parameterized by the interactions and the column effects only since the parameters according to (4.11) become

$$\lambda_{ij}/\lambda_{i.} =\exp\{\tau^{AB}_{ij}+\tau^{B}_{j}\}/ \sum_{j=1}^{J} \exp\{\tau^{AB}_{ij}+\tau^{B}_{j}\}.$$

4.4. The hypothesis of no interaction

The hypothesis of no interaction between the rows and the columns in a two–way contingency table is

$$(4.33) \qquad H_0: \tau_{i\,j}^{AB} = 0, \ i=1,...,I, \ j=1,...,J.$$

Actually (4.33) needs only to be specified for i=1,...,I–1 and j=1,...,J–1 due to the constraints (4.30). The Poisson model (4.1) is under H_0 known as the **multiplicative Poisson model**, since λ_{ij} under (4.33) can be written as

$$(4.34) \qquad \lambda_{ij} = \gamma \epsilon_i \delta_j,$$

with $\epsilon_i = \exp\{\tau_i^A\}$, $\delta_j = \exp\{\tau_j^B\}$ and $\gamma = \exp\{\tau_0\}$. Under model (4.3), H_0 is an **independence hypothesis**. The marginal probability that a randomly chosen person belongs to row i is

$$p_{i.} = \sum_{j=1}^{J} p_{ij}$$

and the corresponding marginal probability for column j is

$$p_{.j} = \sum_{i=1}^{I} p_{ij}.$$

Under (4.33) $p_{i.}$ and $p_{.j}$ become

$$p_{i.} = \lambda_{i.}/\lambda_{..} = \exp\{\tau_i^A\}/\sum_{i=1}^{I} \exp\{\tau_i^A\}$$

and

$$p_{.j} = \lambda_{.j}/\lambda_{..} = \exp\{\tau_j^B\}/\sum_{j=1}^{J} \exp\{\tau_j^B\}.$$

Hence

$$P_{ij} = \lambda_{ij}/\lambda_{..} = \exp\{\tau_i^A + \tau_j^B\}/\sum_{i=1}^{I}\sum_{j=1}^{J}\exp\{\tau_i^A + \tau_j^B\}$$

such that

(4.35)
$$P_{ij} = P_{i.}P_{.j}.$$

Under H_0, the probability of observing a person in cell (i,j) is thus the product of the marginal probabilities of observing a person in row i and in column j.

H_0 given by (4.33) is identical with H_0 given by (4.16) for a 2x2 table. In fact since $4\tau_{11}^{AB} = \ln\rho$, the odds ratio for a 2x2 table is 1 if and only if all four interaction parameters τ_{ij}^{AB} are zero.

Under the model (4.7), the hypothesis of no interaction corresponds to

(4.36)
$$P_{j|i} = P_{.j}, \quad j=1,...,J,$$

since $p_{j|i} = \lambda_{ij}/\lambda_{i.}$ in the multinomial distribution (4.11) and if $\tau_{ij}^{AB} = 0$ then

$$P_{j|i} = \exp\{\tau_i^A + \tau_j^B + \tau_0\} / \sum_{j=1}^{J}\exp\{\tau_i^A + \tau_j^B + \tau_0\} = \exp\{\tau_j^B\}/\sum_{j=1}^{J}\exp\{\tau_j^B\}.$$

The hypothesis (4.36) is also known as the **homogeneity hypothesis**, referring to the fact that under (4.36) the distribution over column categories is the same for each row.

The test statistic for H_0 is most easily obtained for model (4.3), which under H_0 is a parametric multinomial distribution with parameters n and

$$P_{ij} = \frac{1}{n}\exp(\tau_i^A + \tau_j^B + \tau_0), \quad i=1,...,I \quad j=1,...,J.$$

The log–likelihood is then

$$\ln L(p_{11},...,p_{IJ}) = \ln\left[x_{11}\cdots x_{IJ}\right]^n + \sum_i\sum_j x_{ij}(\tau_i^A + \tau_j^B + \tau_0) - n\ln(n)$$

$$= \ln\left[x_{11}\cdots x_{IJ}\right]^n + \sum_i x_{i.}\tau_i^A + \sum_j x_{.j}\tau_j^B + x_{..}\tau_0 - n\ln(n).$$

From this expression follows that the likelihood equations are

$$x_{i.} = np_{i.} , \quad i=1,...,I$$

and

$$x_{.j} = np_{.j} , \quad j=1,....,J.$$

Since the expected values under H_0, are

$$np_{ij} = np_{i.}p_{.j},$$

is not necessary to derive the ML–estimates for the τ's. The likelihood equations yields $\hat{p}_{i.} = x_{i.}/n$ and $\hat{p}_{.j} = x_{.j}/n$, such that the estimated expected values become

$$\hat{np}_{ij} = \frac{x_{i.}x_{.j}}{n}.$$

Hence the transformed likelihood ratio test statistic is according to (3.39) given by

$$(4.37) \qquad -2\ln r = 2\sum_i\sum_j X_{ij}\left[\ln X_{ij} - \ln(X_{i.}X_{.j}/n)\right].$$

Since there are $IJ-1$ parameters in model (4.3) and $I-1+J-1$ free parameters under (4.33), $Z=-2\ln r$ is, according to theorem 3.13, approximately χ^2–distributed with $IJ-1 -I-J+2=(I-1)(J-1)$ degrees of freedom. The hypothesis of no interaction is rejected for large values of

$$(4.38) \qquad z = 2\sum_i\sum_j x_{ij}\left(\ln x_{ij} - \ln\frac{x_{i.}x_{.j}}{n}\right).$$

When in (4.38) n is replaced by $x_{..}$, the statistic also covers model (4.1) and when $x_{i.}$ is replaced by n_i also model (4.7).

The level of significance can be computed approximately as

$$p = P(Q \geq z),$$

where $Q \sim \chi^2((I-1)(J-1))$.

Alternatively to (4.37), one can use the Pearson test statistic

$$Q = \sum_i \sum_j (X_{ij} - \frac{X_{i.} X_{.j}}{n})^2 / (\frac{X_{i.} X_{.j}}{n})$$

the distribution of which can also be approximated by a χ^2–distribution with $(I-1)(J-1)$ degrees of freedom.

In general the limiting χ^2–distribution is only valid when the X_{ij}'s are independent Poisson distributed random variable, or if they are jointly multinomially distributed. These assumptions are not satisfied under complex sampling schemes or when there are non–trivial dependencies between the cells of the table. The behaviour of Z and Q in such non–regular cases has received much attention in the literature, cf. e.g. Gleser and Moore, (1985) Tavare and Altham (1983) and Rao and Scott (1981).

In order to derive the ML –estimates for the log–linear parameters in the saturated model consider the Poisson model. The likelihood function is

$$\ln L = \sum_i \sum_j x_{ij} \tau_{ij}^{AB} + \sum_i x_{i.} \tau_i^A + \sum_j x_{.j} \tau_j^B + x_{..} \tau_0$$

$$- \sum_i \sum_j \ln x_{ij}! - \sum_i \sum_j \exp\{\tau_{ij}^{AB} + \tau_i^A + \tau_j^B + \tau_0\}.$$

Hence the likelihood equations are

(4.39) $x_{ij} = E[X_{ij}]$, i=1,...,I–1, j=1,...,J–1

(4.40)
$$x_{i.} = E[X_{i.}] , \quad i=1,\dots,I-1,$$

(4.41)
$$x_{.j} = E[X_{.j}] , \quad j=1,\dots,J-1,$$

and

(4.42)
$$x_{..} = E[X_{..}].$$

These equations have the same solutions as

(4.43)
$$x_{ij} = E[X_{ij}], \quad i=1,\dots,I, \quad j=1,\dots,J,$$

where the indices i and j run over all values including I and J. From

$$E[X_{ij}]=\exp(\tau_{ij}^{AB}+\tau_i^{A}+\tau_j^{B}+\tau_0)$$

and the constraints (4.30) then follow

$$\hat{\tau}_{ij}^{AB} = \ln x_{ij} - \frac{1}{J}\Sigma_j \ln x_{ij} - \frac{1}{I}\Sigma_i \ln x_{ij} + \frac{1}{IJ}\Sigma_i\Sigma_j \ln x_{ij}$$

$$\hat{\tau}_i^{A} = \frac{1}{J}\Sigma_j \ln x_{ij} - \frac{1}{IJ}\Sigma_i\Sigma_j \ln x_{ij}$$

$$\hat{\tau}_j^{B} = \frac{1}{I}\Sigma_i \ln x_{ij} - \frac{1}{IJ}\Sigma_i\Sigma_j \ln x_{ij}$$

$$\hat{\tau}_0 = \frac{1}{IJ}\Sigma_i\Sigma_j \ln x_{ij}.$$

Under H_0, $\tau_{ij}^{AB}=0$ and the likelihood equations become (4.40), (4.41) and (4.42) with solutions

$$\hat{\tau}_i^A = \ln x_{i.} - \frac{1}{I}\Sigma \ln x_{i.}$$

$$\hat{\tau}_j^B = \ln x_{.j} - \frac{1}{J}\Sigma \ln x_{.j}$$

$$\hat{\tau}_0 = \frac{1}{I}\Sigma_i \ln x_{i.} + \frac{1}{J}\Sigma_j \ln x_{.j} - \ln x_{..}$$

since $E[X_{ij}] = \exp(\tau_i^A + \tau_j^B + \tau_0)$ under H_0.

Example 4.4:

In 1968 there was a lively debate in Denmark over the effects of air pollution in the city of Fredericia, which is dominated by a large fertilizer plant, cf. Andersen (1974). In an attempt to study the effect of the suspected air pollution, the number of lung cancer cases was observed for each of the years 1968 to 1971 for Fredericia and three other cities close to Fredericia and of about the same size. These data are shown in table 4.4. In the table the lung cancer cases are also distributed over 6 age groups. Finally the table shows the marginal number of inhabitants for each age group and for each city.

Table 4.4. Observed number of lung cancer cases for four Danish cities, 1968 to 1971, distributed according to age.

Age	Fredericia	Horsens	City Kolding	Vejle	Total	Number of inhabitants
40–54	11	13	4	5	33	11.600
55–59	11	6	8	7	32	3.811
60–64	11	15	7	10	43	3.367
65–69	10	10	11	14	45	2.748
70–74	11	12	9	8	40	2.217
over 75	10	2	12	7	31	2.665
Total	64	58	51	51	224	
Number of inhabitants	6.294	7.135	6.983	6.026		26.408

Source: Clemmensen et al. (1974).

The ML–estimates of the log–linear parameters for the data in table 4.4 are shown in table 4.5.

Table 4.5. Log-linear parameters for the data in table 4.4.

$\hat{\tau}^{AB}_{ij}$	j=1	2	3	4		
i= 1	+0.199	+0.641	−0.529	−0.311		
2	+0.135	−0.196	+0.100	−0.038		
3	−0.150	+0.435	−0.318	+0.033		
4	−0.318	−0.042	+0.062	+0.298		
5	−0.102	+0.261	−0.018	−0.141		
6	+0.236	−1.098	+0.703	+0.159		

$\hat{\tau}^{A}_{i}$	i=1	2	3	4	5	6
	−0.167	−0.103	+0.182	+0.254	+0.133	−0.300

$\hat{\tau}^{B}_{j}$	j=1	2	3	4		
	+0.210	−0.066	−0.075	−0.069		

$\hat{\tau}_{0}$	+2.156					

The hypothesis of interest for the data in table 4.4 is that the risk of getting lung cancer is the same in all four cities. The alternative hypothesis is that the risk is higher in Fredericia. Marginal hypotheses concerning the cities can only be tested in a meaningful way if there is no interactions in the table. Such an interaction would namely imply that the risk of getting lung cancer for a cititizen of, say, Horsens would depend on an individuals age and no general statements of differences between the cities as regards cancer risks are possible. Consider the following model for the data

$$\begin{cases} X_{ij} \sim P_s(\lambda_{ij}), \\ X_{11}, \dots, X_{64} \qquad \text{are independent} \\ \lambda_{ij}/\lambda_{..} = \theta_{.j} N_{ij}/N_{..} \end{cases}$$

where X_{ij} is the number of lung cancer cases in city j and age group i, and N_{ij} is the number of inhabitants in city j and age group i. Assume in addition that the age distribution is the same in all four cities, i.e.

$$N_{ij}/N_{.j} = N_{i.}/N_{..} .$$

Since $E[X_{ij}]=\lambda_{ij}$, θ_{ij} is the individual risk of getting lung cancer in city j and age group i. The hypothesis of no interaction between city and age as regards lung cancer risk can, therefore, be expressed as

$$\theta_{ij} = \varphi_i \psi_j ,$$

which corresponds to

(4.44)
$$\lambda_{ij} = \lambda_{..} \varphi_i \psi_j N_{i.} N_{.j}/N_{..}^2$$

or (4.34) with $\gamma=\lambda_{..}$, $\epsilon_i=\varphi_i N_{i.}/N_{..}$ and $\delta_j=\psi_j N_{.j}/N_{..}$. The hypothesis of no interaction can accordingly be tested by the test statistic (4.37), which for the data in table 4.4 has observed value

$$z = 20.67.$$

The level of significance is approximately $P(Q{\geq}z)=0.148$, where $Q{\sim}\chi^2(15)$. The hypothesis of no interaction (4.44) is thus accepted. $\triangle$.

Consider the hypotheses

(4.45)
$$H_1: \tau_i^A = \tau_{i0}^A , \quad i=1,...,I$$

and

(4.46)
$$H_2: \tau_j^B = \tau_{j0}^B , \quad j=1,...,J,$$

where τ_{10}^A and τ_{j0}^B are known constants. These hypothesis are only relevant in case the hypothesis (4.33) of no–interaction has been accepted. The test statistic for H_1 can be derived from the distribution of $X_{1.},...,X_{I.}$, since (4.45) under H_0 is equivalent to

(4.47)
$$H_1: p_{i.} = e^{\tau_{i0}^A} / \Sigma_i e^{\tau_{i0}^A}$$

under model (4.3). In fact, under (4.33) and H_1

$$p_{i.} = e^{\tau_{i0}^A} \Sigma_j e^{\tau_j^B + \tau_0}.$$

But since

$$1 = p_{..} = \Sigma_i e^{\tau_{i0}^A} \Sigma_j e^{\tau_j^B + \tau_0},$$

(4.47) follows. In the same way (4.46) is under H_0 equivalent to

(4.48)
$$H_2: p_{.j} = e^{\tau_{j0}^B} / \Sigma_j e^{\tau_{j0}^B}.$$

From (4.47) and (4.48) follow that H_1 can be tested in the marginal distribution of $X_{1.},...,X_{I.}$ and that H_2 can be tested in the marginal distribution of $X_{.1},...,X_{.J}$. The same result is true for model (4.1). For model (4.7) only H_2 makes sense since the values of the row marginals are fixed.

Since H_1 is a hypothesis of fully specified probabilities in the multinomial distribution of $X_{1.},...,X_{I.}$, it follows that the appropriate test statistic is

(4.49)
$$Z_1 = 2 \sum_{i=1}^{I} X_{i.} (\ln X_{i.} - \ln(np_{i0})),$$

where p_{i0} is the right hand side of (4.47). H_1 can be specified in terms of the τ_i^A's or directly in terms of the $p_{i.}$'s. For the special case

$$H_1: \tau_i^A = 0,$$

$p_{i0} = 1/I$ and the test statistic becomes

$$(4.50) \qquad Z_1 = 2 \sum_{i=1}^{I} X_{i.} (\ln X_{i.} - \ln \tfrac{n}{I}).$$

H_1 is rejected if the observed value z_1 of Z_1, i.e. if $P(Z_1 \geq z_1)$ computed in a χ^2–distribution with $I-1$ degrees of freedom is small.

For H_2, the test statistic is

$$(4.51) \qquad Z_2 = 2 \sum_{j=1}^{J} X_{.j} (\ln X_{.j} - \ln(n p_{0j})),$$

where p_{0j} is the right hand side of (4.48) or a directly specified value of $p_{.j}$. The special case

$$H_2: \tau_j^B = 0,$$

yields $p_{0j} = 1/J$ and the test statistic becomes

$$(4.52) \qquad Z_2 = 2 \sum_{j=1}^{J} X_{.j} (\ln X_{.j} - \ln \tfrac{n}{J}).$$

H_2 is rejected if the observed value of Z_2 is large compared with a χ^2–distribution with $J-1$ degrees of freedom.

Under the Poisson model, the hypotheses H_1 and H_2 become

$$(4.53) \qquad H_1: \lambda_{i.}/\lambda_{..} = \epsilon_{i0},$$

where ϵ_{i0} is the right hand side of (4.47), and

$$(4.54) \qquad H_2: \lambda_{.j}/\lambda_{..} = \delta_{j0},$$

where δ_{j0} is the right hand side of (4.48).

As for the multinomial model H_1 and H_2 can be specified in terms of the τ_i^A's and the τ_j^B's or directly in terms of the ratios $\lambda_{i.}/\lambda_{..}$ and $\lambda_{.j}/\lambda_{..}$.

The fact that all hypotheses under H_0 can be studied in the distributions of the marginals of the table is called **collapsability**. In chapter 7 we return to conditions under which a table can be collapsed onto its lower dimensional marginals.

Example 4.4. (Continued)

The hypothesis of main interest for the lung cancer data in table 4.4 is that the risk of getting lung cancer is the same for all four cities. On parametric form this hypothesis is

$$H_2: \psi_j = \psi, \qquad j=1,...,4.$$

From (4.44) follows then that H_2 is equivalent to

$$\lambda_{.j}/\lambda_{..} = N_{.j}/N_{..} .$$

Hence H_2 can be tested by means of (4.48) with $p_{0j}=N_{.j}/N_{..}$. The observed value of Z_2 is

$$z_2 = 3.5$$

with level of significance

$$p \simeq P(Q \geq 3.5) = 0.32$$

where $Q \sim \chi^2(3)$.

It follows that a hypothesis of equal lung cancer risk in the four cities can not be rejected based on the available data. Note that H_2 is formulated directly in terms of the λ's and not in terms of the τ_j^B. △

Example 4.4 shows that it need not be necessary to derive the log–linear parameters in order to compute the test statistic. This is due to the fact that the transformed likelihood ratio test statistic always takes the form

(4.55)
$$Z = 2 \sum_{i=1}^{I} \sum_{j=1}^{J} X_{ij}[\ln X_{ij} - \ln \hat{\mu}_{ij}],$$

where $\hat{\mu}_{ij}$ are the expected values under the hypothesis. The test statistic (4.37) directly has this form. That the test statistics (4.49) and (4.51) also have this form can be seen as follows. Under the hypotesis (4.47), $\hat{\mu}_{ij}$ becomes

$$\hat{\mu}_{ij} = p_{i0} X_{.j},$$

since $p_{i.}$ is specified under H_1 and $\hat{p}_{.j} = X_{.j}/n$. Hence (4.55) become

(4.56)
$$Z = 2 \sum_{i=1}^{I} \sum_{j=1}^{J} X_{ij}[\ln X_{ij} - \ln p_{i0} - \ln X_{.j}]$$

$$= 2 \sum_{i=1}^{I} \sum_{j=1}^{J} X_{ij}[\ln X_{i.} - \ln(\frac{X_{i.} X_{.j}}{n})] + 2 \sum_{i} X_{i.}[\ln X_{i.} - \ln(np_{i0})].$$

The first term is the test statistic (4.37) for the independence hypothesis (4.35). This term is approximately χ^2–distributed with $(I-1)(J-1)$ degrees of freedom. The second term is the test statistic (4.49) for H_1, which is approximately χ^2–distributed with $(I-1)$ degrees of freedom. The sum (4.56) is thus approximately χ^2–distributed with $(I-1)(J-1)$ $+(I-1) = (I-1)J$ degrees of freedom. This was to be expected since (4.56) is the test statistic for the hypothesis H^* that both (4.35) and (4.47) hold. The number of degrees of freedom is IJ minus the number $J-1+1$ of parameters estimated under H^*, or

$$df = IJ-J = J(I-1).$$

That the direct test statistic (4.56) for both (4.35) and (4.47) split up in two terms corresponding to the two hypotheses under consideration is referred to as a **decomposition** of the test statistic.

Note that (4.47) only is meaningful if (4.35) is true. This means that hypotheses concerning the main effects only make sense if the interaction parameters τ_{ij}^{AB} are zero. The hypotheses H_0 and H_1 are thus **nested** with H_1 being dependent on H_0 being true. The term to be used later for nested hypotheses is **hierarchical hypotheses**. Thus H_0 and H_1 form a set of two hierarchical ordered hypotheses.

4.5. Residual analysis

If the model fails to describe the data, a study of possible directions for model departures can be based on the differences

$$X_{ij} - \hat{\mu}_{ij},$$

where $\hat{\mu}_{ij}$ are the expected numbers estimated under the given model. Haberman (1978), cf. also Haberman (1983), has shown that the **standardized residuals**

$$(4.57) \qquad r_{ij} = (x_{ij} - \hat{\mu}_{ij})/\sqrt{\hat{\mu}_{ij}(1 - x_{i.}/n)(1 - x_{.j}/n)}$$

are approximately normally distributed with mean 0 and variance 1. Hence an inspection of the standardized residuals can reveal, which cells contribute significantly to model departures.

Many other methods have been proposed for the identification of deviations from independence in a two–way table. Several of these are based on an inspection of the 2x2 subtables of the contingency table. One proposal is to plot selected values of the form

$$r'_{ijts} = \ln x_{ij} - \ln x_{is} - \ln x_{tj} + \ln x_{ts}$$

for $i \neq t$, $j \neq s$ in a suitable diagram. Under the independence hypothesis the expected value of r'_{ijts} is 0. Hence if $|r'_{ijts}|$ is large for many combinations of t and s, cell (i,j) represents a model departure. As regards such methods see for example Kotze and Hawkins (1984).

Example 4.5

Table 4.6 shows the opinions of a random sample of 838 persons in 1982 concerning "early retirement", a system which has been in effect since 1979 and a new system "partial pension" proposed in 1982 to the Danish Parliament.

Table 4.6. 838 persons cross–classified according to their views on two pensions systems.

Partial pension	Good system	Early retirement Relatively good	Bad system	Do not know
Good proposal	377	75	38	19
Maybe good proposal	92	25	15	8
Bad proposal	84	17	16	4
Do not know	34	17	6	11

Source: Olsen (1984). Table 3.12.

As a model for the data assume the multinomial model (4.3). The test statistic (4.37) for the independence hypothesis (4.35) has observed value

$$z = 27.22, \quad df = 9.$$

with level of significance p=0.001. Hence independence must be rejected. The dependencies of the table can be illustrated by the standardized residuals (4.57) shown in table 4.7.

Table 4.7. Standardized residuals for the data in table 4.6 under independence.

Partial pension	Good system	Early retirement Relatively good	Bad system	Do not know
Good system	+3.16	−1.23	−1.87	−2.11
Maybe good system	−1.23	+0.66	−0.80	+0.42
Bad system	−0.16	−0.63	+1.78	−0.93
Do not know	−3.77	+2.11	−0.04	+4.40

Table 4.7 shows that most of the model departures are connected with persons, which do not have an opinion of the new system. In addition there are significantly more persons than expected with the opinion that both systems are good. △.

4.6. Exercises

4.1. In the Danish Welfare Study the interviewed were classified as renters (if they rented their dwelling) or owners (if they owned their house or apartment). The table shows how many among renters and owners, who have a freezer.

	Has a freezer	
	Yes	No
Renter	2584	300
Owner	1096	795

(a) Formulate a model for the data.

(b) Estimate the odds ratio and construct 95% confidence limits for it.

(c) Does the data support that renters has a freezer as often as owners?

4.2. In order to test a hypothesis to the effect that alcohol abuse is hereditary data were collected on a large number of twins in Sweden and Norway. The table shows for monocygotic as wells as dicygotic twins, where at least one has a monthly consumption of more than 500 g. alcohol, in how many of these cases both have this high consumption.

		Both abuse	Only one abuse	Number of twins
Finland:	Monocygotes	159	1102	1216
	Dicygotes	220	2696	2916
Sweden:	Monocygotes	132	1171	1303
	Dicygotes	165	1756	1921

(a) Formulate a model for each of the two data sets.

(b) Test both in the Finish data set and in the Swedish data set, whether coinciding abuse is more frequent among monocygotes than among dicygotes.

(c) Compare the results from Sweden and Finland and try to draw a more general conclusion.

4.3. The tabel below from the Danish Welfare Study shows the number of broken marriages or permanent relationships cross–classified with sex. The sample only include those persons who are socio–economic active.

	Broken marriage or permanent relationship	
	Yes	No
Men	240	1099
Women	232	1133

(a) Discuss whether one of the model suggested in section 4.1 can be used to analyze this data set.

(b) Whatever conclusion reached in (a), perform a statistical analysis based on the odds ratio

4.4. The table shows for the total sample in the Danish Welfare Study how many among men and women, which often suffer from headaches.

(a) Estimate the odds ratio and construct 95% confidence limits for it. What does these limits tell us.

(b) Does the table supports a claim to the effect that the sample in the Welfare Study is representative of the Danish population as regards sex?

Sex	Suffer often from headaches	
	Yes	No
Men	379	2177
Women	620	1975

4.5. The table shows the answer to the broken marriage question in exercise 4.3 cross–classified according to social rank.

(a) Assume that a multinomial model describes the data. Express the independence hypothesis in terms of marginal or conditional probabilities of a given marriage or relationship ending up being broken.

(b) Test the independence hypothesis.

(c) Describe the departures from independence (if any) by suitable diagnostics, e.g. standardized log–linear parameters or standardized residuals.

Social rank group	Broken marriage or relationship	
	Yes	No
I	28	127
II	62	230
III	79	443
IV	181	850
V	124	582

4.6. The Danish survey company OBSERVA conducts political polls monthly. From 8 such polls in late 1983 and early 1984, the table shows the unweighted returns.

(a) Formulate a model for the data.

(b) Has the expected frequency of non–voters changed.

(c) Describe the changes in expected votes for the different parties between June 83 and January 84, if any.

(d) The following two blocks are sometimes identified in Denmark:

Socialist or leaning socialist: A,F,K,Y.

Conservative/liberal or leaning conservative/liberal: B,C,E,M,Q,V,Z

Has the balance between these blocks changed between August 83 and January 84?

Party Code

Poll	A	B	C	E	F	K	M	Q	V	Y	Z	Other	Intend to vote	Does not intend to vote
June 1983	416	45	338	13	131	8	47	20	129	22	76	2	1247	51
August 1983	396	43	341	12	121	9	40	20	128	20	76	5	1211	56
Sept. 1983	430	46	341	10	126	12	48	16	123	19	67	3	1241	51
Oct. 1983	420	53	353	12	124	8	42	20	129	21	70	6	1258	58
Nov. 1983	435	51	387	18	127	9	49	20	127	25	70	3	1321	59
Dec. 1983(5)	426	59	369	10	140	9	48	19	128	17	76	6	1307	59
Dec. 1983(30)	370	51	347	21	106	6	49	18	126	22	51	1	1168	34
Januar 1984	345	52	328	19	112	4	56	16	114	17	54	0	1117	22

4.7. In 1985 Radio Denmark conducted a survey regarding the interest among TV viewers with the Saturday afternoon broadcast called "Sportslørdag". In the sample of size n=958, 635 said that they had seen "Sportslørdag" at least one. These 635 were asked about their preferences regarding the lenght of the broadcast. The answers were grouped in four time intervals. The sample was collected in four age/sex strata in such a way that the relative sample sizes from the strata corresponded to actual sizes of the strata in the adult Danish population.

Age/sex stratum	Wishes to the length of "Sportslørdag"				Have seen "Sports- lørdag" at least once	Sample size
	Less than 2 hours	$2\frac{1}{2}$ to $3\frac{1}{2}$ hour	4 hour or more	Do not know		
Men, above 40	65	63	59	5	192	234
Women, above 40	77	39	32	4	152	225
Men, under 40	81	50	30	2	163	235
Women, under 40	80	38	6	4	128	264
Total	303	190	127	15	635	958

(a) Formulate a model for the data.

(b) Does the wishes to the length of "Sportslørdag" depend on age and sex among those who watch the broadcast?

(c) Describe the way the wishes depend on age and sex. Are there any strata for which the wishes are similar.

4.8. Consider again the data in exercise 4.7.

In Denmark in 1985 the distribution of the adult population on the four strata was as follows (in 1000 persons).

	Under 40	Above 40
Men	1033	1035
Women	985	1178

(a)   Is the claim correct that the samples from the strata were constructed to match the actual stratum sizes?

(b)   Are those who watch "Sportslørdag" representative of the population as regards sex and age?

(c)   In what direction goes the lack of representativeness?

4.9. In a 1982 study of attitude towards early retirement and partial pension a special index for working environment was also reported for each interviewed person. The index takes values from 0 to 10 with 0 representing an excellent working environment and 10 a very bad working environment. The table shows the sample cross–classfified according to this index and according to attitude towards early retirement.

Attitude towards Early retirement	Working environment index		
	0–2	3–6	7–10
Good system	267	152	52
Maybe good system	68	30	11
Bad system	26	24	10
Do not know	15	9	6
Total	376	215	79

(a)   Does the attitude towards early retirement have a connection with the working environment.

The attitude towards early retirement was also cross–classified with an index for health. The health index could take values from 0 to 22 with 0 being very bad health. The observed numbers are shown below

Attitude towards Early retirement	Health index		
	0–3	4–9	10–22
Good system	276	232	73
Maybe good	69	52	15
Bad system	41	26	13
Do not know	20	13	8
Total	406	323	109

(b)   Does the attitude towards early retirement have a connection with a persons health.

(c)   Compute the estimates of the log–linear parameters under independence for both the data sets above.

4.10. Two Danish Survey Companies OBSERVA and AIM both conducted political polls in June 1983. The table shows the actual number of persons in each sample who claimed that they intended to vote for the different political parties. Also shown is the actual percentage, who voted for the parties in the latest Danish parlamentary election in December 1981.

Party	OBSERVA	AIM	Election December 1981 %
A	416	268	32.9
B	45	22	5.1
C	338	160	14.5
E	13	6	1.4
F	131	66	11.3
K	8	10	1.1
M	47	16	8.3
Q	20	8	2.3
V	129	92	11.3
Y	22	9	2.7
Z	76	32	8.9
Other	–	–	0.2
Total	1245	689	100.0

Copyright: OBSERVA and AIM.

(a)   Suppose the distribution over parties is compared for the two survey companies. What hypothesis is tested in this way?

(b)   Compare the results from the polls with the latest election results. What hypothesis is tested here?

(c)   Are there any connection between the analyses in (a) and (b).

4.11. The table shows for the Danish survey company AIM, the planned sample, how many were at home at the time of the interview and how many refused to participate. The dates in the upper part of the table are all from 1983.

Personal interview	2–8/5	16–22/5	6–16/6	15–22/8	Poll 12–19/9	20–27/10	7–20/11
Planned sample	1307	1404	1436	1433	1470	1413	2690
Not at home	276	307	269	272	285	302	554
Refused to participate	186	210	187	192	187	196	385
Actual sample	845	887	980	969	998	915	1751

Telephone interview	Poll 13–14/12–1983	7/1–1984
Planned sample	1085	1018
Not at home	264	264
Refused to participate	101	41
Actual sample	720	713

(a) Has the number not at home or the number who refused to participate changed in the period covered.

(b) Does the data indicate a change in the ratio between actual and planned sample after the introduction of telephone interviewing.

4.12. The table show for 1271 Danish school children between 16 and 19 years of age in 1983–1985 the attitude among boys and girls towards having sport at school jointly with the other sex.

(a) Do boys and girls have different views on sport jointly with the other sex.

(b) If different views, characterize the differences.

Joint sport at school	Boys	Girls
Very good idea	168	346
Good idea	240	342
Neither good nor bad idea	175	147
Bad idea	47	51
Very bad idea	18	12

5. Three–way Contingency Tables

5.1. The log–linear parameterization

Consider a three–way contingency table $\{x_{ijk}, i=1,...I, j=1,...,J, k=1,...,K\}$. As model for such data, it may be assumed that the x's are observed values of random variables X_{ijk}, $i=1,...,I, j=1,...,J, k=1,...,K$ with a multinomial distribution

$$(5.1) \qquad\qquad X_{111},...,X_{IJK} \sim M(n,p_{111},...,p_{IJK}).$$

This would be the model, if x_{ijk} is the observed number in cell (i,j,k) after a simple random sample of size n has been drawn from a population and p_{ijk} is the proportion of individuals in the population belonging to cell (i,j,k). The p_{ijk}'s can also be interpreted as the probability that a randomly drawn individual will belong to cell (i,j,k). Alternatively it may be assumed that all X_{ijk}'s are independent Poisson distributed random variables

$$(5.2) \qquad\qquad X_{ijk} \sim Ps(\lambda_{ijk}),$$

with the λ_{ijk}'s being positive parameters. Model (5.1) can be obtained from model (5.2) by conditioning on $X_{...}$, since the conditional distribution of $X_{111},...,X_{IJK}$ given that $X_{...}=n$ is multinomial with parameters n and $\lambda_{111}/\lambda_{...},...,\lambda_{IJK}/\lambda_{...}$. The parameters of (5.1) and (5.2) are thus connected through

$$p_{ijk} = \lambda_{ijk}/\lambda_{...}.$$

A three–way contingency table is often pictured as a block with I rows, J columns and K layers as show in fig. 5.1.

We shall in general regard the observed numbers in a three–way contingency table as the observed responses on three categorical variables. Thus x_{ijk} is the observed number of individuals who have responded in category i on variable A, in category j on variable B and in category k on variable C.

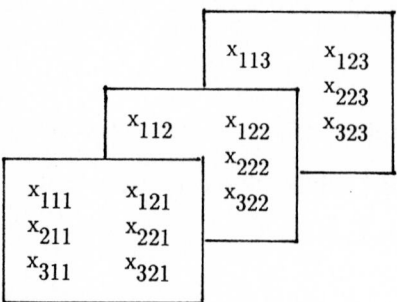

Fig. 5.1. Picture of a three–way contingency table.

The log–linear parameterization of (5.2) is as follows

(5.3)
$$\ln E[X_{ijk}] = \ln\lambda_{ijk} = \tau_0 + \tau_i^A + \tau_j^B + \tau_k^C + \tau_{ij}^{AB} + \tau_{ik}^{AC} + \tau_{jk}^{BC} + \tau_{ijk}^{ABC},$$

with the linear constraints

(5.4)
$$\tau_{.}^A = \tau_{.}^B = \tau_{.}^C = 0.$$

(5.5)
$$\tau_{i.}^{AB} = \tau_{i.}^{AC} = \tau_{.j}^{AB} = \tau_{j.}^{BC} = \tau_{.k}^{AC} = \tau_{.k}^{BC} = 0$$

and

(5.6)
$$\tau_{ij.}^{ABC} = \tau_{i.k}^{ABC} = \tau_{.jk}^{ABC} = 0.$$

Thus any summation over an index of a log–linear parameter is zero. The parameter τ_{ijk}^{ABC} is called a **three–factor interaction**, while $\tau_{ij}^{AB}, \tau_{ik}^{AC}$ and τ_{jk}^{BC} are called **two–factor interactions** and τ_i^A, τ_j^B and τ_k^C are called **main effects**. The log–linear parameterization and the name "interaction" are due to Birch (1963). That (5.3) is a reparameterization of (5.2) is straight forward to verify by expressing the τ's in terms of $\mu_{ijk}^* = \ln\lambda_{ijk}$, i=1,...,I, j=1,...,J, k=1,...,K. For example the three–factor interaction τ_{ijk}^{ABC} is given as

(5.7)
$$\tau_{ijk}^{ABC} = \bar{\mu}_{ijk}^* - \bar{\mu}_{ij.}^* - \bar{\mu}_{i.k}^* - \bar{\mu}_{.jk}^* + \bar{\mu}_{i..}^* + \bar{\mu}_{.j.}^* + \bar{\mu}_{..j}^* - \bar{\mu}_{...}^*,$$

the two factor interaction τ^{AB}_{ij} is given as

$$\tau^{AB}_{ij} = \bar{\mu}^*_{ij.} - \bar{\mu}^*_{i..} - \bar{\mu}^*_{.j.} + \bar{\mu}^*_{...},$$

etc. In these expressions a bar indicates an average and a dot a summation, e.g.

$$\bar{\mu}^*_{ij.} = \frac{1}{K} \sum_{k=1}^{K} \mu^*_{ijk}.$$

The constraints mean that the model can be parametrized in terms of the τ's for $i=1,...,I-1$, $j=1,...,J-1$, $k=1,...,K-1$ with the remaining τ's given implicitly through (5.4), (5.5) and (5.6), e.g.

$$\tau^{AC}_{iK} = - \sum_{k=1}^{K-1} \tau^{AC}_{ik}.$$

That the number of τ's match the number of λ's in the Poisson model (5.2) is demonstrated in table 5.1. For the multinomial model, there are IJK-1 p's, but here τ_0 is redundant.

Table 5.1. Number of free parameters in the parameterization (5.3).

Parameter	Number of free parameters
τ_0	1
τ^A_i	I-1
τ^B_j	J-1
τ^C_k	K-1
τ^{AB}_{ij}	(I-1)(J-1)
τ^{AC}_{ik}	(I-1)(K-1)
τ^{BC}_{jk}	(J-1)(K-1)
τ^{ABC}_{ijk}	(I-1)(J-1)(K-1)
Total	IJK

The Poisson–model (5.2) is a log–linear model according to the definition in section 3.1. In fact the logarithm of the point probability of an observed table $\{x_{ijk}, i=1,...I, j=1,...,J, k=1,...,K\}$ is

$$\ln f(x_{111},...,x_{IJK}) = \sum_{i=1}^{I} \sum_{j=1}^{J} \sum_{k=1}^{K} x_{ijk} \ln \lambda_{ijk}$$

$$- \sum_{i=1}^{I} \sum_{j=1}^{J} \sum_{k=1}^{K} \lambda_{ijk} - \sum_{i=1}^{I} \sum_{j=1}^{J} \sum_{k=1}^{K} \ln(x_{ijk}!).$$

The x_{ijk}'s are thus the sufficient statistics and the canonical parameters are $\tau_{ijk} = \ln \lambda_{ijk}$. Hence the ML–estimators for the λ_{ijk}'s are obtained through the likelihood equations as

$$x_{ijk} = E[X_{ijk}] = \lambda_{ijk}$$

as

(5.8) $$\hat{\lambda}_{ijk} = x_{ijk}, \quad i=1,...,I, \; j=1,...,J, \; k=1,...,K.$$

These trivial estimates are only of interest in the so–called **saturated model**, where none of the log–linear parameters vanish. For the saturated model the ML–estimates for the log–linear parameters are obtained from (5.8) though the reparameterization formulas, i.e. through the expression for the τ's in terms of the $\ln\lambda$'s. For example according to (5.7)

$$\hat{\tau}^{ABC}_{i \; j \; k} = l_{ijk} - \bar{l}_{ij.} - \bar{l}_{i.k} - \bar{l}_{.jk} + \bar{l}_{i..} + \bar{l}_{.j.} + \bar{l}_{..j} - \bar{l}_{...},$$

where $l_{ijk} = \ln x_{ijk}$, and similarly

$$\hat{\tau}^{AB}_{i \; j} = \bar{l}_{ij.} - \bar{l}_{i..} - \bar{l}_{.j.} + \bar{l}_{...}.$$

The importance of the log–linear parameterization (5.3), with the constraints (5.4) to (5.6), lies in the fact, that most hypotheses of interest in a three–way table can be formulated in terms of the log–linears parameters.

5.2. Hypotheses in a three–way table

In a three–way contingency table there are the following major types of hypotheses

$$H_1 : \tau^{ABC}_{ijk} = 0 \qquad\qquad \text{for all i, j and k}$$

$$H_2 : \tau^{ABC}_{ijk} = \tau^{AB}_{ij} = 0 \qquad\qquad - \quad - - \quad -$$

$$H_3 : \tau^{ABC}_{ijk} = \tau^{AB}_{ij} = \tau^{AC}_{ik} = 0 \qquad\qquad - \quad - - \quad -$$

$$H^{*}_4 : \tau^{ABC}_{ijk} = \tau^{AB}_{ij} = \tau^{AC}_{ik} = \tau^{A}_{i} = 0 \qquad\qquad - \quad - - \quad -$$

$$H_4 : \tau^{ABC}_{ijk} = \tau^{AB}_{ij} = \tau^{AC}_{ik} = \tau^{BC}_{jk} = 0 \qquad\qquad - \quad - - \quad -$$

$$H_5 : H_4 \text{ and } \tau^{A}_{i} = 0 \qquad\qquad \text{for all i}$$

$$H_6 : H_4 \text{ and } \tau^{A}_{i} = \tau^{B}_{j} = 0 \qquad\qquad \text{for all i and j}$$

$$H_7 : H_4 \text{ and } \tau^{A}_{i} = \tau^{B}_{j} = \tau^{C}_{k} = 0 \qquad\qquad \text{for all i, j and k}$$

From these the other hypotheses of interest are obtained through exchange of indices.

Hypotheses are often referred to as models. Thus H_2 can also be referred to as the log–linear model (5.3), but with no three–factor interactions and no two–factor interactions between variables A and B.

All hypotheses above except H_1 can be expressed in terms of independence, conditional independence and uniform distribution over categories. It is convenient to express independence between variables A and B on symbolic form as

$$A \oplus B,$$

conditional independence between A and B given variable C as

$$A \oplus B \,|\, C$$

and uniform distribution over the categories of variable A as

$$A = u.$$

Clusters of variables can be independent of a single variable or clusters of variables.

If e.g. A is independent of both B and C, it is expressed as

$$A \oplus B,C.$$

In order to illustrate the use of these symbols in contingency tables let A, B and C be the three categorical variables of the contingency table under model (5.1). This means that A is a categorical variable with observed value i, if a randomly chosen individual belongs to row i of the contingency table in fig.5.1. In the same way the observed value of variable B is j if a randomly chosen individual belongs to column j of the table and variable C has observed value k, if the individual belongs to layer k of the table. Since the marginal probability that variable A has observed value i under model (5.1) is $p_{i..}$ and correspondingly for B and C, the symbol $A \oplus B \oplus C$ is equivalent with

(5.9) $$P_{ijk} = p_{i..}p_{.j.}p_{..k}, \text{ for all i, j and k.}$$

By simple probability algebra it can be shown that $A \oplus B \mid C$ is equivalent with

(5.10) $$P_{ijk} = \frac{p_{i.k}p_{.jk}}{p_{..k}}, \text{ for all i, j and k.}$$

The symbol $A \oplus B,C$ is equivalent with

(5.11) $$P_{ijk} = p_{i..}p_{.jk},$$

and finally A=u is equivalent with

(5.12) $$p_{i..} = 1/I, \text{ for all i.}$$

Theorem 5.1.

The hypothesis H_2 to H_7 have the following interpretations in terms of independence, conditional independence and uniform distribution

$$H_2: A \oplus B \mid C$$

$$H_3: A \oplus B,C$$

$$H_4^*: A \oplus B,C; \; A=u$$

$$H_4: A \oplus B \oplus C$$

$$H_5: A \oplus B \oplus C; \; A=u$$

$$H_6: A \oplus B \oplus C; \; A=B=u$$

$$H_7: A \oplus B \oplus C; \; A=B=C=u$$

Proof:

Consider first H_2. Under H_2, $\tau_{ij}^{AB}=\tau_{ijk}^{ABC}=0$ and since $E[X_{ijk}]=np_{ijk}$.

$$p_{i.k} = \frac{1}{n} \exp\{\tau_0 + \tau_i^A + \tau_k^C + \tau_{ik}^{AC}\} \sum_{j=1}^{J} \exp\{\tau_j^B + \tau_{jk}^{BC}\},$$

$$p_{.jk} = \frac{1}{n} \exp\{\tau_0 + \tau_j^B + \tau_k^C + \tau_{jk}^{BC}\} \sum_{i=1}^{I} \exp\{\tau_i^A + \tau_{ik}^{AC}\}$$

and

$$p_{..k} = \frac{1}{n} \exp\{\tau_0 + \tau_k^C\} \sum_{i=1}^{I} \sum_{j=1}^{J} \exp\{\tau_i^A + \tau_j^B + \tau_{ik}^{AC} + \tau_{jk}^{BC}\}.$$

Multiplying these expressions yield

$$p_{i.k}p_{.jk}/p_{..k} = \frac{1}{n} \exp\{\tau_0 + \tau_i^A + \tau_j^B + \tau_k^C + \tau_{ik}^{AC} + \tau_{jk}^{BC}\},$$

which is equal to p_{ijk} under H_2, thus proving that (5.10) holds under H_2. On the other hand, if (5.10) is satisfied, then $\ln E[X_{ijk}] = \ln(np_{ijk})$ has the form

$$\ln(np_{ijk}) = \alpha_{ik} + \beta_{jk} + \gamma_k + \delta.$$

But it can be shown by easy algebra, that the fact that this expression does not contain terms with joint indices i and j combined with the constraints (5.4) and (5.5) entails that

$\tau^{ABC}_{ijk}=\tau^{AB}_{ij}=0$. Hence if (5.10) is satisfied, the hypothesis H_2 holds.

Under H_3, $\tau^{ABC}_{ijk} = \tau^{AB}_{ij} = \tau^{AC}_{ik}=0$, such that

$$p_{ij.} = \frac{1}{n} \exp\{\tau^A_i+\tau^B_j+\tau_0\} \sum_{k=1}^{K} \exp\{\tau^{BC}_{jk}+\tau^C_k\},$$

$$p_{i..} = \frac{1}{n} \exp\{\tau^A_i+\tau_0\} \sum_{j=1}^{J} \sum_{k=1}^{K} \exp\{\tau^{BC}_{jk}+\tau^C_k+\tau^B_j\}$$

and

$$p_{.j.} = \frac{1}{n} \exp\{\tau^B_j+\tau_0\} \sum_{i=1}^{I} \sum_{k=1}^{K} \exp\{\tau^{BC}_{jk}+\tau^A_i+\tau^C_k\}.$$

Hence

$$p_{ij.} = p_{i..}\,p_{.j.}/[\exp(\tau_0)\underset{i}{\Sigma}\exp(\tau^A_i) \underset{jk}{\Sigma\Sigma}\exp(\tau^{BC}_{jk}+\tau^B_j+\tau^C_k)]$$

But under H_3

$$p_{...} = \frac{1}{n}\underset{i}{\Sigma}\underset{j}{\Sigma}\underset{k}{\Sigma} \exp\{\tau_0+\tau^A_i+\tau^B_j+\tau^C_k+\tau^{BC}_{jk}\} = 1,$$

and it follows that

$$p_{ij.} = p_{i..}\,p_{.j.}.$$

The fact that H_3 holds thus imply that A and B are independent. That A and C are independent under H_3 is proved in a similar manner. If on the other hand A is independent of both B and C, or A $\oplus$ B,C, then $p_{ijk}=p_{i..}\,p_{.jk}$, which means that the logarithm of the mean value $E[X_{ijk}]=np_{ijk}$ has the form

$$\ln(np_{ijk}) = \alpha_{jk}+\beta_i+\gamma$$

From this form and the constraints (5.4) and (5.5) it follows by easy algebra that $\tau^{ABC}_{ijk}=\tau^{AB}_{ij}=\tau^{AC}_{ik}=0$ for all i, j and k. Hence A$\oplus$B,C implies that H_3 holds.

The remaining equivalences of the theorem are proved in similar ways. $\square$

It is worth noting that H_1 cannot be expressed in terms of independence, conditional independence or uniform distribution over categories. For three–ways tables this is the only such hypothesis. For higher order tables, to be dealt with in the next chapter, there are many such hypotheses.

The hypotheses H_2 to H_7 can be illustrated graphically, by representing each of the variables by a point or dot. If two variables in the log–linear parameterization (5.3) is connected through a non–zero interaction, then the dots representing the variables are connected by a line. If a variable is uniformly distributed over its categories, the dot is replace by an asterisk. The resulting graphs are for hypotheses H_1 to H_7 shown in fig. 5.2.

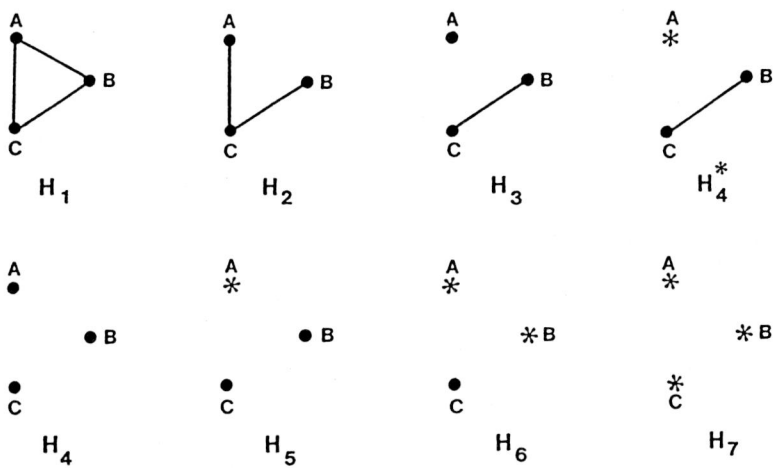

Fig. 5.2. Association graphs for hypotheses H_1 to H_7.

The graphs in fig. 5.2 are called **association graphs**. The connection between the graphs and the interpretation of the hypotheses in terms of independence and conditional independence has been explored by Darroch, Lauritzen and Speed (1980), by Goodman (1972), (1973) and by others. Two variables are independent if there is no route connecting the dots representing the variables on the graph, i. e. if the dots are not connected by lines either directly or via other points. Two variables A and B are conditionally independent given a third variable C, if the only route between A and B passes through C. Thus variables A and B are conditionally independent if their connecting line is discon–

nected when the dot representing the third variable is covered e.g. by a thumb. From the association graph it is thus easy to establish the interpretation of a given model for the data in terms of independence and/or conditional independence.

It should be noted that the association graphs for hypothesis H_1 and the saturated model are identical. This means that the fact that $\tau_{ijk}^{ABC}=0$ does not entail any interpretation in terms of independence and conditional independence. An intuitive reason for this is that, although the three variables do not interact jointly in terms of a three–factor interaction, each pair of variables interact.

The most important hypothesis in a three–way table is H_2, because the association between two variables can then be explained by their associations with the third variable. It is important to emphasize that H_2 does not not imply that A and B are independent. This means that an acceptance of H_2 based on the statistical analysis of a three–way contingency table, does not entail that an independence hypotheses will be accepted based on an analysis of the marginal two–way table of the two conditionally independent variables.

Under the saturated model the likelihood equations are

$$x_{ijk} = E[X_{ijk}], \quad i=1,...,I, \ j=1,...,J, \ k=1,...,K.$$

with solutions $\hat{\lambda}_{ijk}=x_{ijk}$ for the Poisson model and $\hat{p}_{ijk}=x_{ijk}/n$ for the multinomial model.

When some of the interactions are zero, the likelihood equations are replaced by equations of the same basic structure. Consider e.g. the equations under H_1. The log–likelihood function for the Poisson model (5.2) is under H_1, where $\tau_{ijk}^{ABC}=0$,

$$\ln f(x_{111},...x_{IJK}) = \sum_{i=1}^{I} \sum_{j=1}^{J} \sum_{k=1}^{K} x_{ijk}\left(\tau_0+\tau_i^A+\tau_j^B+\tau_k^C+\tau_{ij}^{AB}+\tau_{ik}^{AC}+\tau_{jk}^{BC}\right)$$

$$-\lambda_{...} - \sum_{i=1}^{I}\sum_{j=1}^{J}\sum_{k=1}^{K} \ln(x_{ijk}!)$$

$$= x_{...}\tau_0 + \sum_{i=1}^{I} x_{i..}\tau_i^A + \sum_{j=1}^{J} x_{.j.}\tau_j^B + \sum_{k=1}^{K} x_{..k}\tau_k^C +$$

$$+ \sum_{i=1}^{I}\sum_{j=1}^{J} x_{ij.}\tau_{ij}^{AB} + \sum_{i=1}^{I}\sum_{k=1}^{K} x_{i.k}\tau_{ik}^{AC} + \sum_{j=1}^{J}\sum_{k=1}^{K} x_{.jk}\tau_{jk}^{BC} - \lambda_{...} - \sum_{i=1}^{I}\sum_{j=1}^{J}\sum_{k=1}^{K} \ln(x_{ijk}!).$$

From the linear form of this expression it follows that the sufficient statistics for the interaction parameters and main effects in the model are the corresponding marginals of the contingency table. The likelihood equations are thus the following set of equations

$$(5.13) \qquad x_{...} = E[X_{...}]$$

$$(5.14) \qquad x_{i..} = E[X_{i..}], \ i=1,...,I-1,$$

$$(5.15) \qquad x_{.j.} = E[X_{.j.}], \ j=1,...,J-1,$$

$$(5.16) \qquad x_{..k} = E[X_{..k}], \ k=1,...,K-1.$$

$$(5.17) \qquad x_{ij.} = E[X_{ij.}], \ i=1,...,I-1, \ j=1,...,J-1$$

$$(5.18) \qquad x_{i.k} = E[X_{i.k}], \ i=1,...,I-1, \ k=1,...,K-1.$$

$$(5.19) \qquad x_{.jk} = E[X_{.jk}], \ j=1,...,J-1, \ k=1,...,K-1,$$

The number of equations would have exceeded the number of free parameter, if all equations up to indices i=I, j=J and k=K have been included. It is, however, easy to see from table 5.1 that the total number of equations in (5.13) to (5.19) is also the number of unconstrained log–linear parameters under H_1. Because, on the other hand, equations (5.13) to (5.16) are all obtainable from equations (5.17) to (5.19) if the indices i, j and k are allowed to run all the way to I, J and K, any solution to the equations

$$(5.20) \qquad x_{ij.} = E[X_{ij.}], \ i=1,...,I, \ j=1,...,J,$$

$$(5.21) \qquad x_{i.k} = E[X_{i.k}], \ i=1,...,I, \ k=1,...,K,$$

$$(5.22) \qquad\qquad x_{.jk} = E[X_{.jk}], \quad j=1,\dots,J, \ k=1,\dots,K.$$

will also be a solution to all the equations (5.13) to (5.19). Hence the ML–estimates are usually found by solving (5.20) to (5.22) rather than (5.13) to (5.19) although there are IJ+IK+JK equations in the latter system and only IJ+IK+JK−I−K−J+1 in the former.

This observation is general and shall be used extensively in the following. The rule is as follows: Under a given hypothesis (or model) the likelihood equations are obtained by equating the sufficient marginals for the interactions of highest order with their expectations. Sufficient marginals of lower orders are only set equal to their mean values, if such marginals can not be obtained by summation over an already included marginal. Thus for H_1 we do not need to include equations (5.13) to (5.16) because they can all be obtained by summation from equations (5.17), (5.18) and (5.19).

Once it has been established that the likelihood equations are obtained by equating observed and expected marginals, several consequences can be drawn. Firstly, a model or a hypothesis can be uniquely identified by its sufficient marginals. Thus the statement: "The model is log–linear with $\tau^{ABC}_{ijk}=0$ for all i, j and k", can equivalently be expressed as: "The model is log–linear with sufficient marginals $x_{ij.}$, $x_{.jk}$ and $x_{i.k}$". This last statement can even be abbreviated if the symbols AB, BC and AC are used for the marginals $x_{ij.}$, $x_{.jk}$ and $x_{i.k}$. Thus the model under H_1 is uniquely identified by the symbol

$$AB, \ BC, \ AC.$$

For computer programs this way of identifying models and hypotheses is very convenient and helpful.

Secondly the ML–estimates for the log–linear parameters need not in many cases be computed. The test quantities, necessary to test various hypotheses, depend only on the τ's through the expected numbers, and the estimated expected numbers for a given model can for many important models be expressed directly in terms of the **sufficient marginals**.

The sufficient marginals and the symbols for the hypotheses H_1 to H_7 are listed in table 5.2 together with their interpretations in terms of independence and conditional independence. Note the difference between the sufficient marginals for hypotheses H_3 and

H_4^*. Under hypothesis H_3 the marginal $x_{i..}$ must be added to $x_{.jk}$, since $x_{i..}$ cannot be derived by summation over the $x_{.jk}$'s. For H_4^*, however, $x_{.jk}$ alone is sufficient.

Table 5.2. Sufficient marginals, letter symbols and interpretations for the hypotheses of a three–way table.

Hypothesis	Sufficient marginals	Letter symbol	Interpretation
H_1	$x_{ij.}, x_{i.k}, x_{.jk}$	AB, AC, BC	—
H_2	$x_{i.k}, x_{.jk}$	AC, BC	$A \oplus B \mid C$
H_3	$x_{.jk}, x_{i..}$	BC, A	$A \oplus B, C$
H_4^*	$x_{.jk}$	BC	$A \oplus B, C$; A=u
H_4	$x_{i..}, x_{.j.}, x_{..k}$	A, B, C	$A \oplus B \oplus C$
H_5	$x_{.j.}, x_{..k}$	B, C	$A \oplus B \oplus C$; A=u
H_6	$x_{..k}$	C	$A \oplus B \oplus C$; A=B=u
H_7	$x_{...}$	—	A=B=C=u

As an example of the use of table 5.2 consider H_2. Under H_2, the complete set of likelihood equations is (5.13), (5.14), (5.15), (5.16), (5.18) and (5.19). But by the general rule it suffice to find solutions to (5.21) and (5.22) since all of the equations in (5.13) to (5.16) can be obtained by summations in (5.21) and (5.22).

The conditions for a unique solution of the likelihood equations are for the log–linear parameterization extremely simple. It can be shown that an observed set of marginals is on the boundary of the support if and only if one of the marginals appearing in likelihood equations are zero. So a necessary condition for the existence of ML–estimates for the parameters of a log–linear model is according to theorem 3.6A that none of the sufficient marginals are zero. If the table contains zero counts, the situation is more complicated. This situation is discussed in Haberman (1974), p.37–38 and appendix B. Tables with zero counts or marginals are called **incomplete tables.** We return to this subject in chapter 7.

5.3. Hypothesis testing

As mentioned above we may either speak of a hypothesis or of a model. Thus the hypothesis H_2 is equivalent with a log–linear model, where $\tau_{ijk}^{ABC}=0$ and $\tau_{ij}^{AB}=0$ for all i, j and k. When testing a given hypothesis it is important to state what alternative hypothesis, it is to be tested against. For multi–dimensional tables, where there are many potential hypotheses to consider, it is especially important how the testing is planned and carried out. For three–way tables the situation is still so relatively simple that common sense arguments are often enough to determine what hypotheses to test and in what order to test these hypotheses. In order to set the stage for some more general observations, consider, however, the hypothesis H_2, where the three–factor interactions as well as the two–factor interactions between variables A and B are zero.

In order to derive a test statistic for H_2 consider model (5.1), where the log–likelihood function is

$$\ln L = \ln \left[\begin{array}{c} n \\ x_{111}, \ldots x_{IJK} \end{array} \right] + \sum_{i=1}^{I} \sum_{j=1}^{J} \sum_{k=1}^{K} x_{ijk} \ln p_{ijk}.$$

Under the saturated model the ML–estimates are

$$\hat{p}_{ijk} = x_{ijk}/n,$$

while under H_2

$$\hat{p}_{ijk} = \exp(\hat{\tau}_0 + \hat{\tau}_i^A + \hat{\tau}_j^B + \hat{\tau}_k^C + \hat{\tau}_{ik}^{AC} + \hat{\tau}_{jk}^{BC}) = \hat{\mu}_{ijk}/n,$$

where the $\hat{\mu}_{ijk}$'s are the estimated expected numbers.

According to (3.39) the log–likelihood ratio for the hypothesis H_2 against the saturated model is then

(5.23)
$$Z(H_2) = -2\ln r = 2 \sum_{i=1}^{I} \sum_{j=1}^{J} \sum_{k=1}^{K} X_{ijk}(\ln X_{ijk} - \ln \hat{\mu}_{ijk}).$$

In the saturated model there are IJK parameters, while the number of unconstrained parameters under H_2 is

$$(I-1)(K-1)+(J-1)(K-1)+(I-1)+(J-1)+(K-1)+1 = K(I+J-1)$$

according to table 5.1. Hence $Z(H_2)$ given by (5.23) is approximately χ^2–distributed with

$$df = IJK-K(I+J-1) = IJK-IK-JK + K$$

degrees of freedom, according to theorem 3.11. The test statistic (5.23) is called **a good-ness of fit test statistic** for the model under H_2 as it is used to evaluate the fit of the model as contrasted to the saturated model. The test of H_2 is thus against the alternative that the p_{ij}'s are unrestricted. For later reference we state the result as

(5.24)
$$Z(H_2) = 2 \sum_{i=1}^{I} \sum_{j=1}^{J} \sum_{k=1}^{K} X_{ijk}[\ln X_{ijk}-\ln\hat{\mu}_{ijk}] \sim \chi^2(df(H_2)),$$

where

(5.25)
$$df(H_2) = (I-1)(J-1)(K-1)+(I-1)(J-1).$$

Here the degrees of freedom $df(H_2)$ for H_2 is written on the form which can be obtained directly from table 5.1.

One of the important features of the analysis of contingency tables based on a log–linear parameterization is that hypotheses H_1 to H_7 (without H_4^*, which we return to in a moment) are **hierarchical**, i.e. each subsequent hypothesis is more restrictive than the preceeding are. All interactions and main effects which are zero under H_1 are e.g. also zero under H_2. This means that any property true under H_1 will also be true under H_2. It also means that H_2 can be tested with H_1 as an alternative, and that a test of H_2 against the alternative H_1 is equivalent to a test of the hypothesis

$$H_2': \tau_{i\ j}^{AB} = 0, \ i=1,...,I, \ j=1,...,J.$$

These considerations show that the hypothesis of two–factor interactions between variables A and B being zero is tested under the hierarchical structure as a test of H_2 with H_1

as the alternative. The test statistic for this hypothesis can also be derived from theorem 3.11. Under H_1 the model is log–linear with canonical parameters τ_{ij}^{AB}, τ_{jk}^{BC}, τ_{ik}^{AC}, τ_i^A, τ_j^B, τ_k^C and τ_0. According to table 5.1, there are IJK–(I–1)(J–1)(K–1) unconstrained log–linear parameters in this model. According to table 5.1 there are IJK–(I–1)(J–1)(K–1) –(I–1)(J–1) unconstrained log–linear parameters under H_2. This means that when H_2 is tested with H_1 as the alternative, we are testing H_2 with (I–1)(J–1) parameters set to zero, namely the unconstrained τ_{ij}^{AB}'s.

The test statistic is again most easily derived under the model (5.1). The log–likelihood is under any hypotheses given as

$$\ln L = \ln \left[\begin{matrix} n \\ x_{111}, \ldots x_{IJK} \end{matrix} \right] + \sum_{i=1}^{I} \sum_{j=1}^{J} \sum_{k=1}^{K} x_{ijk} \ln[\mu_{ijk}/n],$$

where $\mu_{ijk} = E[X_{ijk}]$. Hence the log–likelihood ratio for testing H_2 against H_1 is equal to

(5.26)
$$Z(H_2|H_1) = 2 \sum_{i=1}^{I} \sum_{j=1}^{J} \sum_{k=1}^{K} X_{ijk} [\ln \tilde{\mu}_{ijk} - \ln \hat{\mu}_{ijk}],$$

where $\tilde{\mu}_{ijk} = E[X_{ijk}]$ with the ML–estimates for the log–linear parameters under H_1 inserted and $\hat{\mu}_{ijk} = E[X_{ijk}]$ with the ML–estimates for the log–linear parameters under H_2 inserted. According to theorem 3.11, $Z(H_2|H_1)$ is then approximately χ^2–distributed with

$$df(H_2|H_1) = (I–1)(J–1)$$

degrees of freedom, since of the unconstrained parameters under H_1, (I–1)(J–1) are set equal to zero under H_2 according to table 5.1. Thus in addition to (5.24), we have

(5.27)
$$Z(H_2|H_1) \sim \chi^2((I–1)(J–1)).$$

The test statistics for other hypotheses have the same forms as (5.24) and (5.26). We may collect the results in the following theorem.

Theorem 5.2

The goodness of fit test statistic for a hypothesis H expressable in terms of log–linear parameter is

$$(5.28) \qquad Z(H) = 2 \sum_{i=1}^{I} \sum_{j=1}^{J} \sum_{k=1}^{K} X_{ijk}[\ln X_{ijk} - \ln \hat{\mu}_{ijk}],$$

where $\hat{\mu}_{ijk} = E[X_{ijk}]$, estimated under H. The distribution of $Z(H)$ is approximately

$$(5.29) \qquad Z(H) \sim \chi^2(df(H)),$$

where $df(H)$ is the number of log–linear parameters set equal zero under H. The test statistic for H against an alternative hypothesis H_A, also expressable in terms of log–linear parameters, is

$$(5.30) \qquad Z(H|H_A) = 2 \sum_{i=1}^{I} \sum_{j=1}^{J} \sum_{k=1}^{K} X_{ijk}[\ln \tilde{\mu}_{ijk} - \ln \hat{\mu}_{ijk}],$$

where $\tilde{\mu}_{ijk} = E[X_{ijk}]$, estimated under H_A. The distribution of $Z(H|H_A)$ is approximately

$$(5.31) \qquad Z(H|H_A) \sim \chi^2(df(H|H_A)),$$

where $df(H|H_A)$ is the difference between the number of unconstrained parameters set equal to zero under H and the number set equal to zero under H_A. Finally

$$(5.32) \qquad Z(H|H_A) = Z(H) - Z(H_A).$$

Proof:

Consider in this case the model (5.2). The log–likelihood function is here equal to

(5.33)
$$\ln f(x_{111},...,x_{IJK} | \lambda_{111},...,\lambda_{IJK})$$

$$= \sum_{i=1}^{I} \sum_{j=1}^{J} \sum_{k=1}^{K} x_{ijk} \ln \lambda_{ijk} - \lambda_{...} - \sum_{i=1}^{I} \sum_{j=1}^{J} \sum_{k=1}^{K} \ln(x_{ijk}!)$$

The transformed likelihood ratio is then

$$z(H) = -2\ln r = -2\ln f(x_{111},...,x_{IJK} | \hat{\lambda}_{111},...,\hat{\lambda}_{IJK}) + 2\ln f(x_{111},...,x_{IJK} | \hat{\chi}_{111},...,\hat{\chi}_{ijk}),$$

where $\hat{\lambda}_{ijk}$ is the ML–estimate for λ_{ijk} under H and $\hat{\chi}_{ijk}$ is the ML– estimate for λ_{ijk} in the saturated model. Since $\hat{\chi}_{ijk} = x_{ijk}$

$$z(H) = 2[\sum_{i=1}^{I} \sum_{j=1}^{J} \sum_{k=1}^{K} x_{ijk} \ln \hat{\lambda}_{ijk} - \hat{\lambda}_{...}]$$

$$+ 2[\sum_{i=1}^{I} \sum_{j=1}^{J} \sum_{k=1}^{K} x_{ijk} \ln x_{ijk} - x_{...}].$$

Since, however, $x_{...} = E[X_{...}] = \lambda_{...}$, whatever equations (5.14) to (5.19) are included in a ML–estimation of the parameters and $\hat{\lambda}_{ijk} = \hat{\mu}_{ijk}$, (5.28) follows. The distributional result (5.29) is a direct application of theorem 3.11. The result (5.31) is also a consequence of theorem 3.11, and equation (5.32) follows directly from (5.28) and (5.30). □

For later reference table 5.3 summarizes the number of degrees of freedom for the approximate χ^2–distributions of the goodness of fit test statistics for the hypotheses H_1 to H_7. These numbers can be obtained from table 5.1 by adding the number of parameters set equal to zero under a particular hypothesis. The number of degrees of freedom not found in table 5.3 are obtained by interchanging the indices i, j and k.

It is time to say a few words about H_4^*, which is parallel to H_4 in the sense that $H_5 \subset H_4^* \subset H_3$. Thus the following test statistics make sense $Z(H_4|H_3)$, $Z(H_5|H_4)$, $Z(H_4^*|H_3)$ and $Z(H_5|H_4^*)$. On the other hand $Z(H_4|H_4^*)$ and $Z(H_4^*|H_4)$ does not make sense, since neither H_4 testet against H_4^* nor H_4^* tested against H_4 are tantamount to setting a set of log–linear parameters equal to zero.

Table 5.3. Degrees of freedom for the approximate χ^2–distribution of $Z(H)$.

Hypothesis H	Parameters set equal to zero	Number of degrees of freedom for $Z(H)$
H_1	$\tau^{ABC}_{ijk}=0$	$(I-1)(J-1)(K-1)$
H_2	$\tau^{ABC}_{ijk}=\tau^{AB}_{ij}=0$	$(I-1)(J-1)K$
H_3	$\tau^{ABC}_{ijk}=\tau^{AB}_{ij}=\tau^{AC}_{ik}=0$	$(I-1)(KJ-1)$
H_4^*	$\tau^{ABC}_{ijk}=\tau^{AB}_{ij}=\tau^{AC}_{ik}=\tau^{A}_{i}=0$	$KJ(I-1)$
H_4	$\tau^{ABC}_{ijk}=\tau^{AB}_{ij}=\tau^{AC}_{ik}=\tau^{BC}_{jk}=0$	$IKJ-I-J-K+2$
H_5	all above plus $\tau^{A}_{i}=0$	$IKJ-J-K+1$
H_6	$-\ -\ -\ \tau^{A}_{i}=\tau^{B}_{j}=0$	$IKJ-K$
H_7	$-\ -\ -\ \tau^{A}_{i}=\tau^{B}_{j}=\tau^{C}_{k}=0$	$IKJ-1$

Because of the constraints (5.4) to (5.6) the values of the main effects τ^{A}_{i} for variable A are influenced by the values of the interactions τ^{AB}_{ij} and τ^{AC}_{ik}, which both relate to variable A. This means that it is meaningsless to test the hypothesis $\tau^{A}_{i}=0$ in a model, where neither τ^{AB}_{ij} nor τ^{AC}_{ik} vanish.

The analysis of a given observed three–way contingency table can be carried out in many different ways. Since all tests are based on the $Z(H)$'s and their differences, one idea is to make a list of the observed value of $Z(H)$ for all possible hypotheses, i.e. not only H_1 to H_7 of table 5.3, but also the hypotheses obtained by exchange of indices, for example the hypotheses corresponding to the symbols (AB,BC), (AB,C) or (AC). From such a list one gets a first impression of what models are likely to fit the data. It is important to avoid tests at a given level for all the hypotheses in this list, since the hypotheses can not be tested independently. Even if all the test statistics are independent, the probability of rejecting at least one correct model is much larger than α, when all test are carried out at level α. To see this let $Z_1...,Z_m$ be m independent test statistics for $H_1,...,H_m$ with critical regions $Z_1 \geq c_1,...,Z_m \geq c_m$, and let

$$P(Z_j \geq c_j | H_j) = \alpha.$$

Then the probability of incorrectly rejecting at least one correct hypothesis is

$$P(Z_j \geq c_j \text{ for at least one } j | H_1,...,H_m)$$

$$= P(\underset{j}{U}\{Z_j \geq c_j\}) = 1 - P(\underset{j}{\cap}\{Z_j < c_j\}) = 1 - (1-\alpha)^m.$$

If thus $\alpha=0.05$ and m=8, we reject at least one correct hypothesis with probability 0.337.

There is a rich litterature on what to do in this situation. Generally the situation is known as a case of **multiple test procedures**. There are two very simple procedures which are easy to understand and apply. One procedure is known as the **Bonferroni procedure**. It is based on the simple fact that if

$$P(Z_j \geq c_j | H_j) = \alpha/m$$

for all j, i.e. if all tests are carried out at level α/m, then by the Bonferroni inequality

$$P(Z_j \geq c_j \text{ for at least one } j) \leq \underset{j}{\Sigma}P(Z_j \geq c_j | H_j) = m\alpha/m = \alpha,$$

whether $Z_1,..,Z_m$ are independent or not.

The Bonferroni procedure thus gives us a guarantee that the maximum overall level is less than or equal to α.

Holm (1979), see also Schaffer (1986) and Hommel (1988), has improved the Bonferroni procedure by introducing the sequential procedure described in fig. 5.3, where $Z_{(1)},\ldots,Z_{(m)}$ is ordered such that $P(Z_{(1)}\geq z_{(1)})\leq P(Z_{(2)}\geq z_{(2)})\leq\ldots\leq P(Z_{(m)}\geq z_{(m)})$. The **sequential Bonferroni procedure** will apart from very trivial cases lead to tests which are more powerful than the Bonferroni procedure and the increase in power can be substantial as shown by Holm (1979).

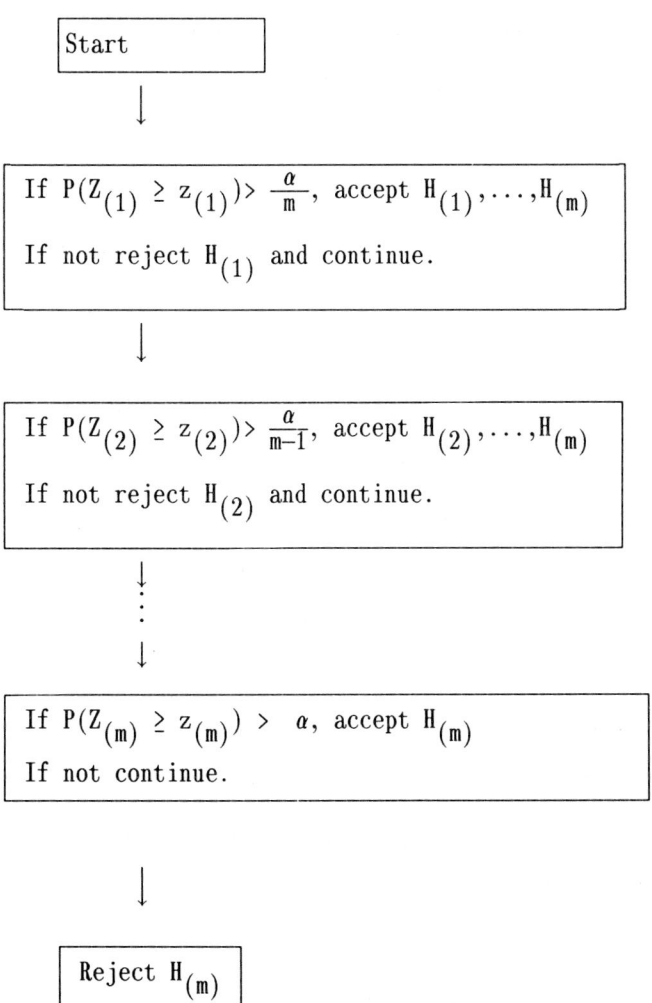

Fig. 5.3. The sequential Bonferroni procedure.

In some cases it is possible to determine a priori a hypothesis which is likely to hold. In such cases one may test this hypothesis directly through the test statistic (5.28). Suppose, for example, that we are fairly convinced that the three– factor interactions and the two—factor interactions between variables B and C are zero. We can then test

$$H_2: \tau_{ijk}^{ABC} = \tau_{jk}^{BC} = 0$$

by estimating the expected numbers under H_2, where the sufficient marginals are AB, AC, and compare the observed value of (5.28) with a χ^2–distribution with

$$df=(I-1)(J-1)(K-1)+(J-1)(K-1)$$

degrees of freedom.

It is, however, relatively seldom that a likely final hypothesis can be formulated directly. The best thing to do is then to set up a sequence of hierarchical hypotheses by determining which interactions are likely to vanish first. It may e.g. be more likely that B and C interact than A and C and more likely that A and C interact than A and B. The hypotheses H_1, H_2, H_3 and H_4 of table 5.2 should then be tested in that order. When a hierarchical order can be determined the relevant test statistics are (5.30). In the examples below it is shown how the testing is carried out and how different results are interpreted. Tables 5.5 and 5.9 thus shows the listing of z(H) for all possible hypotheses and tables 5.6 and 5.10 the test statistics hierarchically ordered.

Before we proceed to the examples, it remains to show how the expected numbers necessary for the test statistics (5.28) and (5.30) are computed. As mentioned above, it is often unnecessary to compute the ML–estimates of the log–linear parameters in order to obtain values of the expected numbers, As one example consider hypothesis H_4. According to theorem 5.1 and (5.9), we have under H_4

$$\mu_{ijk} = np_{ijk} = np_{i..}p_{.j.}p_{..k} = \mu_{i..}\mu_{.j.}\mu_{..k}/n^2.$$

Under H_4 the likelihood equations are, however,

$$x_{i..} = \mu_{i..}, \qquad\qquad i=1,...,I,$$

$$x_{.j.} = \mu_{.j.}, \qquad j=1,\ldots,J,$$

and

$$x_{..k} = \mu_{..k}, \qquad k=1,\ldots,K.$$

It follows that $\hat{\mu}_{ijk}$ under H_4 is given by

$$\hat{\mu}_{ijk} = x_{i..}x_{.j.}x_{..k}/n^2.$$

For the hypotheses H_2 to H_7 the expected numbers are listed in theorem 5.3. In order to cover all models, n is written as $x_{...}$.

Theorem 5.3.

The expected numbers μ_{ijk} are under hypotheses H_2 to H_7 estimated as

$$H_2: \hat{\mu}_{ijk} = x_{i.k}x_{.jk}/x_{..k}$$

$$H_3: \hat{\mu}_{ijk} = x_{.jk}x_{i..}/x_{...}$$

$$H_4: \hat{\mu}_{ijk} = x_{i..}x_{.j.}x_{..k}/x_{...}^2$$

$$H_5: \hat{\mu}_{ijk} = \frac{1}{I}x_{.j.}x_{..k}/x_{...}$$

$$H_6: \hat{\mu}_{ijk} = \frac{1}{IJ}x_{..k}$$

$$H_7: \hat{\mu}_{ijk} = \frac{n}{IJK}$$

Proof:

The expression for H_4 was proved above. For H_2, we get from (5.10)

$$\mu_{ijk} = np_{ijk} = (np_{i.k})(np_{.jk})/(np_{..k}) = \mu_{i.k}\mu_{.jk}/\mu_{..k},$$

for which the expression for H_2 follows when the sufficient marginals under H_2 are inserted in place of the μ's.

Under H_3, $\tau_{ijk}^{ABC} = \tau_{ij}^{AB} = \tau_{ik}^{AC} = 0$ for all i, j and k. But then

$$\mu_{.jk} = \exp(\tau_0 + \tau_j^B + \tau_k^C + \tau_{jk}^{BC})\sum_i \exp(\tau_i^A)$$

$$\mu_{i..} = \exp(\tau_0 + \tau_i^A)\sum_{jk}\sum \exp(\tau_j^B + \tau_k^C + \tau_{jk}^{BC})$$

and

$$\mu_{...} = e^{\tau_0}\sum_i \exp(\tau_i^A)\sum_{jk}\sum \exp(\tau_j^B + \tau_k^C + \tau_{jk}^{BC}),$$

such that

$$\mu_{ijk} = \mu_{.jk}\mu_{i..}/\mu_{...} \ .$$

Hence when the ML–estimates for the μ's are replaced by the sufficient marginals under H_3, the expression for H_3 follow. The expressions for H_5 is obtained by setting $x_{i..} = x_{...}/I$ in H_4, the expression for H_6 by setting $x_{.j.} = x_{...}/J$ and $x_{i..} = x_{...}/I$ in H_4 and finally the expression for H_7 by setting $x_{i..} = x_{...}/I$, $x_{.j.} = x_{...}/J$ and $x_{..k} = x_{...}/K$ in H_4. $\square$.

The necessary computations for solving the likelihood equations are carried out by means of standard statistical computer packages like SPSS, BMDF, GENSTAT or SAS.

From theorem 5.3 follows that there are explicit solutions to the likelihood equations for all models except H_1, where the likelihood equations are (5.20), (5.21) and (5.22). These equations are in almost all computer programs solved by the method of iterative proportional fitting, described in section 3.6. Let μ_{ijk}^o be initial estimates for the expected values. Then improved estimates μ_{ijk}^1, μ_{ijk}^2 and μ_{ijk}^3 in iterations 1, 2 and 3 are obtained as

$$\mu_{ijk}^1 = \mu_{ijk}^o \frac{x_{ij.}}{\mu_{ij.}^o}$$

$$\mu^2_{ijk} = \mu^1_{ijk} \frac{x_{i.k}}{\mu^1_{i.k}}$$

$$\mu^3_{ijk} = \mu^2_{ijk} \frac{x_{.jk}}{\mu^2_{.jk}} \; .$$

At iterations 1, 2 and 3 the μ's are thus adjusted to satisfy (5.20), (5.21) and (5.22) respectively. These three steps are repeated in iterations 4, 5 and 6 and so on. When the expected values do not change within the required accuracy in three consecutive iterations, the iterations are stopped and the ML–estimates found.

Example 5.1.

In 1968, 715 blue collar workers, selected from Danish Industry, were asked a number of questions concerning their job satisfaction. Some of these questions were summarized in a measure of job satisfaction. Based on similar questions the job satisfaction of the supervisors were measured. Also included in the investigation was an external evaluation of the quality of management for each factory. Table 5.4 shows the 715 workers distributed on the three variables

A: Own job satisfaction

B: Supervisors job satisfaction

C: Quality of management.

In table 5.5 the observed values of the goodness of fit test statistics (5.28) are shown for all hypotheses, which can be ordered hierarchically. The levels of significance $P(Z(H) \geq z(H))$ are computed in the approximating χ^2–distribution. In the table the hypotheses are identified by their sufficient marginals, as well as with the H–notations used earlier. An H in a parenthesis means, that the hypothesis is obtained from one of the hypotheses introduced in the start of section 5.2 by an interchange of indices.

Table 5.4. Own job satisfaction, supervisors job satisfaction and the quality of management for 715 blue collar workers in Denmark in 1968.

Quality of management	Supervisors job satisfaction	Own Job satisfaction	
		Low	High
Bad	Low	103	87
	High	32	42
Good	Low	59	109
	High	78	205

Source: Petersen (1968), table M/7.

Table 5.5. Observed values of the goodness of fit test statistic for all relevant hypotheses for the data in table 5.4.

Hypothesis	Sufficient marginals	$z(H)$	$df(H)$	Level of significance
H_1	AB, AC, BC	0.06	1	0.800
(H_2)	AB, AC	71.90	2	0.000
(H_2)	AB, BC	19.71	2	0.000
H_2	AC, BC	5.39	2	0.068
(H_3)	AB, C	102.11	3	0.000
(H_3)	AC, B	87.79	3	0.000
H_3	BC, A	35.60	3	0.000
(H_4^*)	AB	151.59	4	0.000
(H_4^*)	AC	87.79	4	0.000
H_4^*	BC	76.89	4	0.000
H_4	A, B, C	118.00	4	0.000
(H_5)	A, B	167.48	5	0.000
(H_5)	A, C	118.00	5	0.000
H_5	B, C	159.29	5	0.000
(H_6)	A	167.48	6	0.000
(H_6)	B	208.77	6	0.000
H_6	C	159.29	6	0.000
H_7	–	220.00	7	0.000

The first impression from table 5.5 is that the three–factor interaction vanish, while of the three two–factor interactions the one between A and B is most likely to be zero. If a set of hierarchical hypothesis were to be set a priori, it is likely that the sequence exhibited in table 5.6 would have been chosen. The observed test statistics in table 5.6 are obtained from the values in table 5.5 by substractions.

The most restrictive hypothesis, which can be accepted is thus H_2 with sufficient marginals AC and BC. The interpretation of the corresponding model is shown in fig. 5.4.

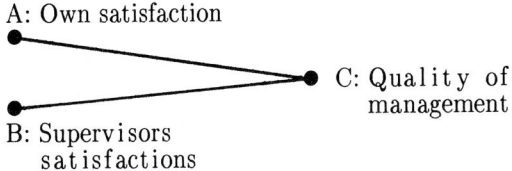

A: Own satisfaction

C: Quality of management

B: Supervisors satisfactions

Fig. 5.4. The interpretation of the least restrictive model fitted to the data of tabel 5.4.

Table 5.6. Observed test statistics under hierarchical testing for the data of table 5.4.

Hypothesis	Sufficient marginals	Interactions set to zero	$z(H\vert H_A)$	df	Level of significance
H_1	AB, AC, BC	$\tau_{ijk}^{ABC} = 0$	0.06	1	0.800
H_2	AC, BC	$\tau_{ij}^{AB} = 0$	5.33	1	0.022
H_3	BC, A	$\tau_{ik}^{AC} = 0$	30.21	1	0.000
H_4^*	BC	$\tau_{i}^{A} = 0$	41.29	1	0.000

The sequential Bonferroni procedure works well on table 5.5. In table 5.5 there are 18 hypotheses. Hence we start with the smallest level of significance and compare with $0.05/18=0.003$ for overall level $\alpha=0.05$. As can be seen the first 16 comparisons lead to rejection of all hypotheses except H_1 and H_2. The second largest level of significance is $P(Z(H_2)\geq 5.39)=0.068$. This is larger than $0.05/2$. Hence H_2 is accepted and consequently also H_1. It thus seems that we have reached a plausible result, which coincides with the conclusion reached above.

Verbally we may claim that the quality of management rather than the job satisfaction of the supervisor influence the job satisfaction of the worker. That the job satisfaction of the worker is only independent of the job satisfaction of the supervisor, conditionally on the level of management quality, can be seen by analysing the marginal table between variables A and B shown in table 5.7. A test of independence in this table, using the transformed log–likelihood ratio test statistics gives

$$z = 15.89, \text{df}=1,$$

such that the variables A and B are not independent, when variable C is not taken into consideration.

Table 5.7. Marginal contingency table for workers and supervisors job satisfactions.

Job satisfaction of supervisor	Job satisfaction of worker	
	Low	High
Low	162	196
High	110	247

Example 5.2.

The Swedish traffic authorities investigated in 1961 and 1962 on a trial basis the possible effects of speed limitations. In certain weeks a speed limit of 90 km/hour was enforced, while in other weeks no limits were enforced. Table 5.8 shows for two periods of the same length, one in 1961 and one in 1962, the observed number of killed person in traffic accidents on main roads and on secondary roads.

Table 5.8. Persons killed in traffic accidents in a periods of length 18 weeks in 1961 and 1962.

Year	Speed limit	Main roads	Secondary roads
1961	90 km/hour	8	42
	Free	57	106
1962	90 km/hour	11	37
	Free	45	69

Source: Unpublished data from the Swedish Traffic Authorities.

The goodness of fit test statistics for all models are shown in table 5.9. The notation is as in table 5.5, and the three variables are labeled

A: Year
B: Speed limit
C: Road type

Based on a priori knowledge it is most reasonable to expect that it is mainly road type that influences the effect of speed limits and that there is probably no major differences between the two years considered. For these reasons, the sequence of hierarchical tests shown in table 5.10 was chosen.

Table 5.9. Goodness of fit test statistic for all relevant models and the data in table 5.8.

Hypothesis	Sufficient marginals	$z(H)$	df	Level of significance
H_1	AB, AC, BC	0.19	1	0.660
(H_2)	AB, AC	11.36	2	0.003
(H_2)	AB, BC	1.34	2	0.513
H_2	AC, BC	2.44	2	0.295
(H_3)	AB, C	12.05	3	0.007
(H_3)	AC, B	13.16	3	0.004
H_3	BC, A	3.13	3	0.372
(H_4^*)	AB	60.27	4	0.000
(H_4^*)	AC	102.19	4	0.000
H_4^*	BC	10.09	4	0.039
H_4	A, B, C	13.85	4	0.008
(H_5)	A, B	62.06	5	0.000
(H_5)	A, C	102.88	5	0.000
H_5	B, C	20.81	5	0.001
(H_6)	A	151.09	6	0.000
(H_6)	B	69.02	6	0.000
H_6	C	109.83	6	0.000
H_7	–	158.04	7	0.000

The conclusion of the analysis is thus that there is an interaction between road type and speed limits, such that the drop in number of total traffic accidents, obvious throughout the table, depends on the road type considered. The trend is that more accidents are prevented on main roads, when a speed limit is enforced. It is natural in this case to test H_4^* rather than H_4 against H_3. We might namely be interested in testing that the accident level is the same in 1961 as in 1962 whether there is an interaction between road type and speed limitation or not. As it turns out H_4^* is rejected, so the least complicated model which fits the data in a satisfactory way, is the one corresponding to H_3 with interpretation illustrated in the association graph figure 5.5.

Table 5.10. A sequence of hierarchical tests for the data in table 5.8.

Hypothesis	Sufficient marginals	Interactions set to zero	$Z(H\|H_A)$	df	Level of significance
H_1	AB, AC, BC	$\tau_{ijk}^{ABC}=0$	0.19	1	0.660
(H_2)	AB, BC	$\tau_{ik}^{AC}=0$	1.15	1	0.270
H_3	BC, A	$\tau_{ij}^{AB}=0$	1.79	1	0.084
H_4^*	BC	$\tau_i^{A}=0$	6.96	1	0.009
H_5	B, C	$\tau_{jk}^{BC}=0$	10.72	1	0.001
(H_6)	B	$\tau_k^{C}=0$	48.21	1	0.000
H_7	—	$\tau_j^{B}=0$	89.02	1	0.000

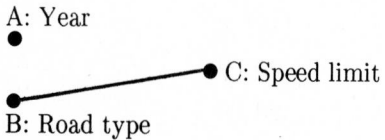

A: Year

C: Speed limit

B: Road type

Fig. 5.5. Association graph for the least complicated model, which describes the data in table 5.8.

In order to apply the sequential Bonferroni procedure on the data in this example, the levels of significance in order of magnitude and the corresponding limits $\alpha/(m+1-j)$ are for $\alpha=0.05$ shown in table 5.11. From this table it is seen that all hypotheses are rejected except those with sufficient marginals: (AB,AC,BC),(AB,BC),(AC,BC),(BC,A), and (BC).

Table 5.11. The hypotheses of table 5.9 ordered according to increasing level of significance and the adjusted levels $\alpha/(m+1-j)$, m=18.

Hypothesis		Sufficient marginals	Level of significance	$0.05/(m+1-j)$
j=1	H_7	–	0.000	0.003
2	H_6	C	0.000	0.003
3	(H_6)	B	0.000	0.003
4	(H_6)	A	0.000	0.003
5	(H_5)	A,C	0.000	0.004
6	(H_5)	A,B	0.000	0.004
7	(H_4^*)	AC	0.000	0.004
8	(H_4^*)	AB	0.000	0.005
9	H_5	B,C	0.001	0.005
10	(H_2)	AB,AC	0.003	0.006
11	(H_3)	AC,B	0.004	0.006
12	(H_3)	AB,C	0.007	0.007
13	H_4	A,B,C	0.008	0.008
14	H_4^*	BC	0.039	0.010
15	H_2	AC,BC	0.295	0.012
16	H_3	BC,A	0.372	0.017
17	(H_2)	AB,BC	0.513	0.025
18	H_1	AB,AC,BC	0.660	0.050

This result coincide with the result above except as regards H_4^*, which is accepted by a sequential Bonferroni procedure.

A uniform distribution over the categories of variable A means that there on the average were as many accidents in 1961 as in 1962.

The estimated log–linear parameters of the model BC,A are

$$\hat{\tau}^{BC}_{11} = \hat{\tau}^{BC}_{22} = -0.221$$

$$\hat{\tau}^{BC}_{12} = \hat{\tau}^{BC}_{21} = +0.221$$

$$\hat{\tau}^{A}_{1} = +0.137, \qquad\qquad \hat{\tau}^{A}_{2} = -0.137$$

$$\hat{\tau}^{B}_{1} = -0.619, \qquad\qquad \hat{\tau}^{B}_{2} = +0.619$$

$$\hat{\tau}^{C}_{1} = -0.491, \qquad\qquad \hat{\tau}^{C}_{2} = +0.491$$

$$\hat{\tau}_{0} = 3.533$$

These estimates can be derived directly from the likelihood equations, and it is instructive to do so. The likelihood equations corresponding to sufficient marginals BC are

$$x^{BC}_{.jk} = E[X_{.jk}]$$

or

$$19 = \exp(\hat{\tau}^{BC}_{11}+\hat{\tau}^{B}_{1}+\hat{\tau}^{C}_{1}+\hat{\tau}_{0}) \ \Sigma_{i} e^{\hat{\tau}^{A}_{i}}$$

$$79 = \exp(\hat{\tau}^{BC}_{12}+\hat{\tau}^{B}_{1}+\hat{\tau}^{C}_{2}+\hat{\tau}_{0}) \ \Sigma_{i} e^{\hat{\tau}^{A}_{i}}$$

$$102 = \exp(\hat{\tau}^{BC}_{21}+\hat{\tau}^{B}_{2}+\hat{\tau}^{C}_{1}+\hat{\tau}_{0}) \ \Sigma_{i} e^{\hat{\tau}^{A}_{i}}$$

$$175 = \exp(\hat{\tau}^{BC}_{22}+\hat{\tau}^{B}_{2}+\hat{\tau}^{C}_{2}+\hat{\tau}_{0}) \ \Sigma_{i} e^{\hat{\tau}^{A}_{i}} \ .$$

These equations imply that

$$\hat{\tau}^{BC}_{11} = \ln 19 - c - \hat{\tau}^{B}_{1} - \hat{\tau}^{C}_{1}$$

$$\hat{\tau}^{BC}_{12} = \ln 79 - c - \hat{\tau}^{B}_{1} - \hat{\tau}^{C}_{2}$$

$$\hat{\tau}^{BC}_{21} = \ln 102 - c - \hat{\tau}^{B}_{2} - \hat{\tau}^{C}_{1}$$

$$\hat{\tau}^{BC}_{22} = \ln 175 - c - \hat{\tau}^{B}_{2} - \hat{\tau}^{C}_{2},$$

where c is a common value. Since, however,

$$\hat{\tau}^{BC}_{11} + \hat{\tau}^{BC}_{12} = \hat{\tau}^{BC}_{11} + \hat{\tau}^{BC}_{21} = \hat{\tau}^{BC}_{12} + \hat{\tau}^{BC}_{22} = \hat{\tau}^{BC}_{21} + \hat{\tau}^{BC}_{22} = \hat{\tau}^{B}_{1} + \hat{\tau}^{B}_{2} = \hat{\tau}^{C}_{1} + \hat{\tau}^{C}_{2} = 0,$$

it follows that

$$c + \hat{\tau}^{B}_{1} = \tfrac{1}{2}(\ln 19 + \ln 79) = 3.657,$$

$$c + \hat{\tau}^{C}_{1} = \tfrac{1}{2}(\ln 19 + \ln 102) = 3.785$$

and from the similar equations involving $\hat{\tau}^{B}_{2}$ and $\hat{\tau}^{C}_{2}$

$$c = \tfrac{1}{4}(\ln 19 + \ln 79 + \ln 102 + \ln 175) = 4.276.$$

Hence

$$\hat{\tau}^{B}_{1} = -0.619$$
$$\hat{\tau}^{C}_{1} = -0.491$$

and consequently

$$\hat{\tau}^{BC}_{11} = -0.222.$$

The estimate for τ^{A}_{i} is obtained from the equations

$$x_{i..} = \exp(\hat{\tau}^{A}_{i} + \hat{\tau}_{0}) \underset{j\,k}{\Sigma\Sigma} \exp(\hat{\tau}^{BC}_{jk} + \hat{\tau}^{B}_{j} + \hat{\tau}^{C}_{k})$$

corresponding to sufficient marginals A, or

$$213 = \exp(\hat{\tau}_1^A + \hat{\tau}_0) \cdot 5.318$$

and

$$162 = \exp(\hat{\tau}_2^A + \hat{\tau}_0) \cdot 5.318$$

From $\hat{\tau}_1^A + \hat{\tau}_2^A = 0$ follows then than

$$\hat{\tau}_0 = \frac{1}{2}(\ln 213 + \ln 162) - \ln(5.318) = 3.553$$

such that

$$\hat{\tau}_1^A = 0.137.$$

The estimates of the two–factor interactions $\hat{\tau}_{jk}^{BC}$ show that the speed limitations as expected has had the largest effects on the main roads in Sweden since $\hat{\tau}_{11}^{BC} < 0$ indicates that there have been significantly fewer killed in the traffic than expected under a speed limitation on main roads. The positive value of $\hat{\tau}_1^A$, which is slightly significant, indicates that there as a whole has been slightly more killed persons in 1961 than in 1962. $\triangle$.

5.4. Decomposition of the test statistic

The additive form of the test statistic (5.28) allow us to decompose selected test statistics in order to further illustrate the interpretation of the models under consideration. As one example consider the test statistic for H_2 with

$$\hat{\mu}_{ijk} = x_{i.k} x_{.jk} / x_{..k}$$

according to theorem 5.3. Decomposed according to the value of k, $Z(H_2)$ can be written as

$$(5.34) \qquad Z(H_2) = 2 \sum_{i=1}^{I} \sum_{j=1}^{J} X_{ij1} \ln(X_{ij1} / \hat{\mu}_{ij1}) + \ldots + 2 \sum_{i=1}^{I} \sum_{j=1}^{J} x_{ijK} \ln(x_{ijK} / \hat{\mu}_{ijK}).$$

For a given k,

$$\hat{\mu}_{ijk} = x_{i.k} x_{.jk}/x_{..k}, \quad i=1,\ldots,I \quad j=1,\ldots,J,$$

are, however, the expected numbers under independence in the IJ cells in the k'th layer of the table. $Z(H_2)$ is thus the sum of K test statistics for independence in the conditional contingency tables given the levels of variable C. The terms in (5.34) are independent and each approximately χ^2–distributed with $(I-1)(J-1)$ degrees of freedom. Hence approximately

$$Z(H_2) \sim \chi^2(K(I-1)(J-1)),$$

in accordance with table 5.3. The decomposition of (5.34) shows that the test statistic for H_2 based on $K(I-1)(J-1)$ degrees of freedom, is composed of K test of independence between variables A and B each with $(I-1)(J-1)$ degrees of freedom.

This way of deriving $Z(H_2)$ does not reveal, however, the alternative decomposition

$$Z(H_2) = Z(H_1) + Z(H_2|H_1),$$

which correponds to successive tests of the H_1 and H_2. The number $K(I-1)(J-1)$ can thus be divided in two ways: Under the latter decomposition the degrees of freedom are divided as

$$K(I-1)(J-1)=(I-1)(J-1)(K-1)+(I-1)(J-1)$$

corresponding to a test of H_1 followed by a test of H_2 given H_1. Under the former decomposition the degrees of freedom are divided as

$$K(I-1)(J-1)=(I-1)(J-1)+\ldots+(I-1)(J-1)$$

corresponding to independence tests in each layer of the contingency table. Which one of the decomposition one choose to consider depends on what alternatives H_2 are compared to. The advantage of the decomposition (5.34) is, that if H_2 is rejected, the single terms in (5.34) may indicate that an independence hypothesis holds in some layers of the table.

Example 5.3.

In 1974 the Danish National Institute for Social Science Research investigated 1314 employee's who left their job during the second half of the year. The lay–offs were categorized according to three variables

A: Employment status on January 1st. 1975.

B: Cause of lay–off.

C: The length of present employment at time of lay–off.

Table 5.12 shows the three–way contingency table formed by these three categorical variables. The observed values of the test statistics in table 5.13 show that there does not seem to be any simple interpretation of the table in terms of independence, or conditional independence.

Table 5.12. Employment status on January 1st.1975, causes of lay–off and length of employment at time of lay–off for 1314 employee's who lost their job in the fall of 1974 in Denmark.

| Length of employment | Cause of lay–off | Employment status on January 1st, 1975 | |
		Got a new job	Still unemployed
Less than 1 month	Closure etc.	8	10
	Replacement	40	24
1–3 months	Closure etc.	35	42
	Replacement	85	42
3–12 months	Closure etc	70	86
	Replacement	181	41
1–2 years	Closure etc.	62	80
	Replacement	85	16
2–5 years	Closure etc.	56	67
	Replacement	118	27
More than 5 years	Closure etc	38	35
	Replacement	56	10

Source: Kjær (1978). Table 4.8.

Table 5.13. Selected observed test statistics for the data in table 5.12.

Model	Test Statistic	df	Significance probability
AB, AC, BC	9.01	5	0.108
AB, BC	24.63	10	0.006
AC, BC	165.92	6	0.000
AB, AC	64.62	10	0.000

The decomposition (5.34) with respect to variable B for the data in table 5.12 takes the form

$$24.63 = 1.44 + 23.19$$

The term 1.44 corresponds to a test of independence between variables A and C given level 1 of variable B and the term 23.19 corresponds to a test of independence between variables A and C given level 2 of variabel B. Both terms in (5.34) are in the present situation χ^2–distributed with five degrees of freedom. Hence only 23.19 is significant. Thus we cannot conclude that

$$A \oplus C | B,$$

but we may safely conclude that

$$A \oplus C | B(1),$$

i.e. A and B are independent given level 1 of variable B. The interpretation of the table is thus that there is a dependency between length of employment and the chance of getting a new job immediately, if the cause of the lay–off is a replacement. If the cause is a closure, this dependency is not manifest in the data. This conclusion can be sharpended, as we shall see by means of analysis of residuals in section 5.5. △.

5.5. Detection of model departures

There are many models of interest, which do not correspond to any of the hypothesis H_1 to H_7 or their equivalencies. It may thus be an important model to consider that variables A and B are independent for certain levels of variable C, although not for all.

As diagnostics for model departures two sets will be suggested here, standardized residuals and standardized estimates of log–linear parameters. As regards alternative diagnostics, cf. Kotze and Hawkins (1984). The residuals are defined as

$$d_{ijk} = x_{ijk} - \hat{\mu}_{ijk},$$

where $\hat{\mu}_{ijk}$ is $\mu_{ijk} = E[x_{ijk}]$ with the ML–estimates for the τ's under the given model inserted. The variance of d_{ijk} depends on which τ's are set equal to zero. In addition the τ's, which are non zero under the model, have to be replaced by estimates.

For a three–dimensional table, there exist simple approximations for the variance of d_{ijk} except for the model AB, BC, AC. Table 5.14 is due to Haberman (1978), p.231. It shows approximations to the variances of the residuals $x_{ijk} - \mu_{ijk}$ for all hypotheses except H_1. The standardized residuals are defined as

(5.35)
$$r_{ijk} = (x_{ijk} - \hat{\mu}_{ijk})/\hat{\sigma}_{ijk},$$

with $\hat{\sigma}^2_{ijk} = \hat{var}[x_{ijk} - \hat{\mu}_{ijk}]$.

Some computer programs provide exact or approximate expressions for (5.35) with $\hat{\sigma}_{ijk}$ estimated under the chosen model. Based on a table of the standardized residuals (5.35) it is possible to determine what cells of the table contribute significantly to the lack of fit.

Example 5.3. (Continued)

An analysis of the standardized residuals for the model AB,BC applied to the data in table 5.12, shown in table 5.15, reveals that the conclusion $A \oplus C | B(1)$ can be sharpended. We immediately notice that only the residuals for levels 1 and 2 of variable C have value significantly different from zero. This suggest that the model AB, BC may fit the data, if levels 1 and 2 of variable C, i.e. all employee's which have worked less than 3 months at the present employer, are omitted from the analysis. This can be illustrated by table 5.16, which shows the test statistics for the models of table 5.13 without levels 1 and 2 of variable C. It can thus be further concluded from the analysis of the data in table 5.12, that there is independence between the chance of getting a new job soon after the lay–off and the length of employment, if the employee has been employed more than 3 months at the time of lay–off. △

Table 5.14: Approximate estimated variances for the residuals .

Hypoth–esis	Sufficient marginal	$\widehat{\mathrm{var}}[X_{ijk}-\hat{\mu}_{ijk}]$
H_2	AC, BC	$\hat{\mu}_{ijk}(1-x_{i.k}/x_{..k})(1-x_{.jk}/x_{..k})$
H_3	BC, A	$\hat{\mu}_{ijk}(1-x_{.jk}/n)(1-x_{i..}/n)$
H_4^*	BC	$\hat{\mu}_{ijk}(1-x_{.jk}/n)(1-\frac{1}{I})$
H_4	A,B,C	$\hat{\mu}_{ijk}(1-x_{i..}x_{.j.}/n^2-x_{.j.}x_{..k}/n^2-x_{i..}x_{..k}/n^2+2x_{i..}x_{.j.}x_{..k}/n^3)$
H_5	B, C	$\hat{\mu}_{ijk}(1-\frac{1}{I}+\frac{1}{I}(1-x_{.j.}/n)(1-x_{..k}/n))$
H_6	C	$\hat{\mu}_{ijk}(1-\frac{1}{IJ})$
H_7	–	$\hat{\mu}_{ijk}(1-\frac{1}{IJK})$

Table 5.15. Standardized residuals for the model AB, BC and the data in table 5.12.

Length of employment	Course of lay–off	Employment status Got a new job	Still unemployed
Less than 1 month	Closure etc. Replacement	−0.11 −3.12	+0.11 +3.12
1–3 month	Closure etc. Replacement	0.04 −3.29	+0.04 +3.29
3–12 month	Closure etc Replacement	−0.23 +1.55	+0.23 −1.55
1–2 years	Closure etc. Replacement	−0.55 +1.63	+0.55 −1.63
2–5 years	Closure etc. Replacement	−0.04 +1.12	+0.04 −1.12
More than 5 years	Closure etc Replacement	+1.17 +1.42	−1.17 −1.42

Table 5.16. Test statistics for the models of table 5.13, if levels 1 and 2 of variable C are omitted.

Model	Test Statistic	df	Level of significance
AB, AC, BC	0.52	3	0.914
AB, BC	2.19	6	0.902
AC, BC	154.96	4	0.000
AB, AC	18.46	6	0.005

Instead of looking at the residuals, one may study the estimates of the log–linear parameters. According to theorem 5.1 two variables are independent, or conditionally independent, if all two–factor or higher order interactions involving the variables are zero. An inspection of the estimated interactions can thus reveal why an expected independence does not materialize. In order to obtain a correct evaluation of the significance of a given estimated interaction, the corresponding parameter estimate must be standardized. Consider thus e.g.

$$(5.36) \qquad \hat{\omega}_{ij}^{AB} = \hat{\tau}_{ij}^{AB} / \sqrt{\hat{\text{var}}[\hat{\tau}_{ij}^{AB}]} \; ,$$

where $\hat{\text{var}}[\hat{\tau}_{ij}^{AB}]$ is the variance of $\hat{\tau}_{ij}^{AB}$ with the estimated values of the τ's inserted. The exact expression for $\text{var}[\hat{\tau}_{ij}^{AB}]$ is rather complicated, but approximations are available, which are valid if the limiting χ^2–distribution of the test statistic is valid. One possibility is to apply theorem 3.7 according to which $\text{var}[\hat{\tau}_{ij}^{AB}]$ is a diagonal element in the matrix $n^{-1}\mathbf{M}$. Most computer programs provide both estimates of the log–linear parameters and standardized estimates (5.36) of the log–linear parameters. Since by theorem 3.7 the ML–estimates for the $\hat{\tau}_{ij}^{AB}$ are asymptotically normally distributed with mean 0 under the model,

$$(5.37) \qquad \hat{\omega}_{ij}^{AB} \sim N(0,1).$$

Hence the two–factor interaction τ_{ij}^{AB} may be said to contribute significantly to the lack of fit, if

(5.38) $$|\hat{\omega}_{ij}^{AB}| > 2,$$

since this event under the hypothesis $\tau_{ij}^{AB}=0$ has approximate probability 0.05. Great care should be exercised if this rule is applied too strictly. Firstly, the $\hat{\omega}_{ij}^{AB}$'s are not independent, such that the different standardized estimates cannot be evaluated independently of each other. Secondly, one in every 20 values of (5.38) should on the average exceed 2 even were the $\hat{\omega}_{ij}^{AB}$'s independent. None the less a table of the $\hat{\omega}_{ij}^{AB}$'s combined with the rule (5.38) is helpful to determine which of the τ_{ij}^{AB}'s are different from zero if the value of the test statistic makes it likely that not all are zero.

It should finally be noted that τ_{ij}^{AB} is an additive term in the exponent of the expression

$$\mu_{ijk} = E[x_{ijk}] = \exp\{...+\tau_{ij}^{AB}+...\},$$

such that the expected numbers in cell (i,j,k) is relatively larger when $\tau_{ij}^{AB}>0$ than when $\tau_{ij}^{AB}=0$, and relatively smaller when $\tau_{ij}^{AB}<0$. Hence a positive value of a two–factor interaction indicates a positive covariation between the two variables in question.

Example 5.4.

Consider the data in table 5.17 showing the non–response for a survey in Denmark in 1965.

Table 5.17. The non−response for a Danish survey in 1965 distributed according to sex and residence of the sampled person.

Residence	Response	Sex	
		Male	Female
Copenhagen	Yes	306	264
	No	49	76
Cities outside Copenhagen	Yes	609	627
	No	77	79
Countryside	Yes	978	947
	No	103	114

Source: Unpublished data from the Danish National Institute for Social Research.

Table 5.18 shows some of the possible models and their test statistics. The most restrictive model to fit the data thus seems to be BC with

A: Sex

B: Response or non−response

C: Residence

Table 5.18. Test statistics for selected hypotheses for the data in table 5.17.

Model	Test Statistic	df	Level of significance
AB, AC, BC	5.38	2	0.068
AB, BC	6.29	4	0.178
A, BC	10.27	5	0.068
BC	10.32	6	0.112
B, C	39.26	8	0.000

The standardized ML−estimates $\hat{\omega}^{BC}_{jk}$ for τ^{BC}_{jk} are

$\hat{\omega}^{BC}_{jk}$	k=1	2	3
j=1	−5.378	+2.101	+4.022
j=2	+5.378	−2.101	−4.022

All these values are significantly larger than 2, such that the conclusion, that the τ_{jk}^{BC}'s are not all zero, is confirmed. We note, however, that the great difference is between k=1 and k=2,3, or between the Capital, and the rest of Denmark.

As a final model, one may, therefore, consider a model, where variables B and C are independent if only persons outside Copenhagen are included.

The 2x2 contingency table formed by variables B and C, without level 1 for variable C is shown in table 5.19. The test statistic for independence in this table is

$$z = 0.52, \ \ df=1,$$

which is not significant. The hypothesis suggested by the standardized ML–estimates is confirmed and the rate of non–response is outside Copenhagen the same in the cities and in the countryside. $\triangle$.

Table 5.19. Response and non–response for the cities outside Copenhagen and the countryside.

Residence	Response	
	Yes	No
Cities outside Copenhagen	1236	1925
Countryside	156	217

5.6. Exercises

5.1. In 1973 a group of elderly in the city of Odense in Denmark were divided into groups. The persons in the E–groups were offered special help in their homes, while such help was not offered to the persons in the C–group. For each quarter in 1973, 1974, 1975 and 1976 was then reported how many of the elderly, who had to leave their homes to go to special care centers.

(a) Formulate a model for the data, which allows for a comparison of the E– and C–groups.

(b) Does the number who move to a care center from the two groups depend on the year or the season.

Year	Quarter	Number moved to care center from	
		E–group	C–group
1973	1	5	3
	2	3	3
	3	2	8
	4	7	6
1974	1	9	2
	2	9	20
	3	7	7
	4	8	7
1975	1	7	14
	2	8	12
	3	4	11
	4	5	6
1976	1	7	7
	2	8	8
	3	9	20
	4	6	6

(c) Does it matter for the analysis that the sizes of the E–group and the C–group are not reported.

5.2. The table below is from the Danish Welfare Study. It shows the number of persons in the sample, for which the social rank is reported, cross–classified as renters/owners and according to ownership of a freezer.

Social rank group	Renter or owner	Owns a freezer	
		Yes	No
I+II	Renter	304	38
	Owner	92	64
III	Renter	666	85
	Owner	174	113
IV	Renter	894	93
	Owner	379	321
V	Renter	720	84
	Owner	433	297

(a) Carry out two iterations of the marginal proportional fitting procedure, with all initial expected values equal to 297.31, for estimating the parameters of the model, where only the three–factor interaction is zero.

(b) Carry out a number of tests in order to find a reasonable model to fit the data.

(c) Give an interpretation of the model and estimate its parameter.

5.3. In 1975 the connection between high school average (K) and the performance after the first year of studying economics at the University of Copenhagen was investigated.

The table shows the number of economy students, who passed and failed after 1 year of study in 1971, 1972, 1973 and 1974 cross–classified with the high school average in six intervals.

High school average	Year	Passed	Failed
K < 7.0	71	8	24
	72	8	19
	73	7	20
	74	10	29
7.0 < K< 7.5	71	24	21
	72	18	17
	73	10	30
	74	21	30
7.5 < K< 8.0	71	13	16
	72	25	26
	73	23	24
	74	16	24
8.0 < K < 8.5	71	15	13
	72	9	22
	73	24	16
	74	17	15
8.5 < K< 9.0	71	15	15
	72	8	9
	73	16	10
	74	16	11
9.0 < K	71	14	8
	72	12	7
	73	6	10
	74	13	8

(a) Describe the data in table by a suitable log–linear model.

(b) Interprete the parameters of the chosen model.

(c) The investigators were especially interested in the connection between high school average and the chance of passing the examen. Which of the parameters cast light on this connection and in what way?

(d) It is common wisdom among students and teachers that the chance of passing after one year is 50%. Does the data support this claim? Can the 50% claim in any way be modified or improved?

5.4. In a study among persons between age 50 and age 66, the sample size was cross-classified according to type of dwelling, sex and marriage-status. The table shows the resulting contingency table.

Sex	Marriage status	Dwelling		
		Apartment	House	Farm
Male	Married	30	32	5
	Not married	64	229	14
Female	Married	68	41	5
	Not married	76	193	44

(a) A priori it is reasonable to believe that the strongest interactions are between dwelling and marriage status and the weekest between dwelling and sex. Test various models by a sequential test procedure.

(b) Use theorem 5.3 and table 5.14 to compute expected values and standardized residuals directly from the observed numbers.

(c) Study the residuals for the model A,BC, where A is sex, B marriage status and C dwelling. Do they reveal important information?

(d) Use the residuals for the independence model A,B,C to describe the way the variables interact.

5.5. In the data base of the Danish Welfare Study a special computer search was carried out to check how broken marriages or permanent relationships depended on sex and social rank. The table shows the resulting contingency table.

Sex	Social rank group	Marriage or permanent relationship broken	
		Yes	No
	I	14	102
	II	39	151
Men	III	42	292
	IV	79	293
	V	66	261
	I	12	25
	II	23	79
Women	III	37	151
	IV	102	557
	V	58	321

It is known a priori that there is a connection between sex and social rank. But it is of no interest in this connection.

(a) Is there a connection between broken marriage and one or both of the other two variables?

(b) Use residuals to describe the dependencies in the table.

(c) Compare the use of standardized residuals and standardized log–linear parameters as indicators of model departues for the present data set.

5.6. Reconsider the data in exercise 4.2.

(a) Analyse the table as a three–way contingency table.

(b) Compare the results from the analysis based on the three–way table with the conclusions drawn in exercise 4.2.

5.7. We return to the investigation concerning sport and youth in exercise 4.12. Among the questions in the survey was one concerning the time used to read about sport. The table below shows the sample cross–classified according to the answers to this question as well as sex and whether the students attend a high school or a vocational or commercial school.

Time spent on reading about sports (hours)	Sex	Attend School	
		High School	Vocational or Commercial
None	Boy	34	64
	Girl	49	135
0–1/2	Boy	29	61
	Girl	31	118
1/2 – 1	Boy	40	81
	Girl	37	142
1 – 2	Boy	37	65
	Girl	32	64
2 – 4	Boy	24	40
	Girl	11	37
4 – 6	Boy	6	15
	Girl	0	3
More than 6	Boy	3	7
	Girl	4	2

(a) Analyse the data by a log–linear model.

(b) Estimate the parameters of the most simple model which fits the data.

(c) Characterize the dependencies in the table based on the estimates from (b).

6. Multi–dimensional Contingency Tables

6.1. The log–linear model

In chapter 5, log–linear models for three–dimensional tables were treated in great details. Hence we shall not for higher order tables go into details with the parameterizations of the models or with the exact expressions for test quantities and their distributions. Besides for higher order tables the mathematical expressions quickly becomes large and cumbersome to write down.

An m–way contingency table can be written as follows

$$x_{i_1 i_2 \ldots i_m}, \quad i_1 = 1, \ldots, I_1, \; i_2 = 1, \ldots, I_2, \ldots, i_m = 1, \ldots, I_m,$$

where i_1 is the index of the first, i_2 the index of the second, and i_m the index of the last of the m categorical variables forming the table. When $X_{i_1 i_2 \ldots i_m}$ is the random variable corresponding to $x_{i_1 i_2 \ldots i_m}$, the contingency table can be parameterized through the

log–linear parameterization

(6.1)
$$\mu_{i_1 i_2 \ldots i_m} = E[X_{i_1 i_2 \ldots i_m}]$$

$$= \exp\{\tau^{AB\ldots S}_{i_1 i_2 \ldots i_m} + \ldots + \tau^{ABC}_{i_1 i_2 i_3} + \ldots + \tau^{AB}_{i_1 i_2} + \ldots + \tau^{A}_{i_1} + \ldots \tau^{S}_{i_m} + \tau_0\}.$$

It is difficult to keep track of all these parameters, but for concrete applications matters are much simplified by two facts. Firstly any model can be identified through its sufficient marginals, which also form the basis for the ML–estimation of the parameters. Secondly the actual estimates of the parameters are seldom needed in order to identify a model and give a valid interpretation of the model.

The τ–parameters in (6.1) with two or more indices are called **interactions**.

$\tau^A_{i_1},...,\tau^S_{i_m}$ are called **main effects** and τ_0 the **over–all effect**. If there are k indices for a log–linear parameter it is called a **k–factor interaction**. As a rule models which include non–null interactions of higher order than 4 are difficult to interprete and of limited practical use.

The main tools for statistical analyses of multiple contingency tables are test statistics of the general form (5.28) and (5.30). For multiple contingency tables the results in theorem 5.2 extend directly. Consider thus a hypothesis H, which consist of setting a number of log–linear parameters equal to zero. The test–statistic for H is

$$(6.2) \qquad Z(H) = 2 \sum_{i_1=1}^{I_1} \sum_{i_2=1}^{I_2} \cdots \sum_{i_m=1}^{I_m} X_{i_1 i_2 \cdots i_m} [\ln X_{i_1 i_2 \cdots i_m} - \ln \hat{\mu}_{i_1 i_2 \cdots i_m}],$$

where $\hat{\mu}_{i_1 i_2 \cdots i_m}$ is $E[X_{i_1 i_2 \cdots i_m}]$ with the non–null τ's under H replaced by their ML–estimates. It can be proved that Z(H) under H is approximately χ^2–distributed, i.e.

$$Z(H) \sim \chi^2 (df).$$

The degrees of freedom df for Z(H) are the number of τ's set equal to zero under H.

As in chapter 5 the fit of an observed table to a given model is measured by the observed value z(H) of (6.2). The hypothesis and the corresponding model are rejected if the value of z(H) is large. The level of significance

$$p = P(Z(H) \geq z(H))$$

can accordingly be approximately evaluated in a χ^2–distribution with df degrees of freedom.

It is often logical to test a hypothesis against an alternative hypothesis H_A, under which fewer interactions than under H are zero. In this way it can be tested whether the interactions assumed to be zero under H, but not under H_A, are in fact zero.

For higher order tables there are in general so many candidates for a reasonable model that a complete set of hierarchical hypothesis can not be set up beforehand. Special attention must, therefore, be devoted to setting up a reasonable strategy for testing the hypotheses of interest. We return to his problem in section 6.3.

6.2. Interpretation of log–linear models

For higher dimensional contingency tables, there are often many potential models which merit consideration. Hence it is important to have an easy way to identify a model. It turns out that also for higher order tables the sufficient marginals are convenient instruments for identifying models.

As noted in chapter 5, we do not consider models as valid, if an interaction between a given set of variables is zero, but there are non–null interactions between a larger set of variables, which include the given set. Models which are valid under this criterion are called **hierarchical models**. Consider e.g. for a four–way table the model with

$$\tau^{ABCD}_{ijkl} = \tau^{ABC}_{ijk} = \tau^{ABD}_{ijl} = \tau^{CD}_{kl} = 0, \text{ for all i,j,k l}$$

and with at least one non–zero interaction in all other sets of variables. This model is not hierarchical, since $\tau^{CD}_{kl}=0$ for all k and l, but at least one $\tau^{ACD}_{ikl} \neq 0$.

Consider now a four–way table formed by the four categorical variables A, B, C and D. Table 6.4 below shows an example of such a table. As for three–dimensional models a sufficient marginal corresponds to one of the equations in the minimal set of likelihood equations. Under the log–linear parameterization (6.1) all likelihood equations have the structural form of equating an observed marginal of the table with its mean value. As we saw in chapter 5 all equations corresponding to non–null interactions need not, however, be included. The general rule is, that an equation can be omitted if it can be obtained from another likelihood equation by summation. None of the equations in the minimal set of likelihood equations can thus be obtained by summation over any other equation in the set.

A concrete example illustrates the rule. Consider the hierarchical model for a four–way table characterized by

(6.3) $$\tau^{ABCD}_{i\,j\,k\,l} = \tau^{ABC}_{i\,j\,k} = \tau^{ABD}_{i\,j\,l} = 0 \quad \text{for all i,j,k,l.}$$

The minimal set of likelihood equations are then

(6.4) $$x_{i.kl} = E[X_{i.kl}] , \qquad \text{for all i, k and l}$$

(6.5) $$x_{.jkl} = E[X_{.jkl}] , \qquad \text{for all j, k and l}$$

and

(6.6) $$x_{ij..} = E[X_{ij..}] , \qquad \text{for all i and j.}$$

All other equations can be obtained from these by summation. Consider e.g. the equations corresponding to $\tau^{AC}_{i\,k}$, which are not assume to be zero. The likelihood equations for $\tau^{AC}_{i\,k}$ are

$$x_{i.k.} = E[X_{i.k.}] , \qquad \text{for all i and k.}$$

But these equations can be obtained from (6.4) by summation over l. As another example, the equations corresponding to τ^{D}_{l} are

$$x_{...l} = E[X_{...l}] , \qquad l=1,...,L.$$

But these equations can be obtained from (6.4) by summation over i and k or from (6.5) by summation over j and k.

Note that even though (6.6) corresponds to a two–factor interaction, it can not be derived from equations (6.4) and (6.5) by summation.

On symbolic form (6.4) is the sufficient marginal ACD, (6.5) is BCD and (6.6) is AB. Equations (6.4) to (6.6) and hence the model can thus be expressed through the symbol ACD,BCD,AB.

The complete set of marginals for the hypothesis (6.3) are characterized by ACD, BCD,AB,AC,AD,BC,BD,CD,A,B,C,D. The rule is thus that a sufficient marginal on symbolic form is omitted if it occurs as a combination in on of the earlier elements of the list. Thus AC and AD are omitted because they appears as combinations in ACD, BD and CD because they appear in BCD, etc.

The best way to illustrate the hierarchical hypotheses or models in a four–way table is to list the models, identified by their sufficient marginals, together with their interpretations in terms of independence, conditional independence and uniform distribution over categories. This is done in table 6.1 for a set of typical hierarchical models. All other hierarchical models can be obtained by exchange of variables. Several features of table 6.1 are important. Note firstly, that four models does not have an interpretation in terms of independence or conditional independence. These are models where all variables interact with each other, e.g. the saturated model and the model with sufficient marginals AB, AC, AD, BC, BD, CD. Note secondly that for any model involving higher order interactions, there exists a model in terms of only second order interactions, with the same interpretation. It is, thirdly, important to note that not all models in table 6.1 can be **hierarchically ordered** in one hierarchical order. The hypotheses marked by an asterix is a set of hierarchical ordered hypotheses, but there are may other such sets. The hierarchical order marked by an asterix corresponds to testing the following partial hypotheses successively

$$(1) \quad \tau_{1jkl}^{ABCD} = 0 \qquad (2) \quad \tau_{jkl}^{BCD} = 0 \qquad (3) \quad \tau_{ikl}^{ACD} = 0$$

$$(4) \quad \tau_{ijl}^{ABD} = 0 \qquad (5) \quad \tau_{ijk}^{ABC} = 0 \qquad (6) \quad \tau_{kl}^{CD} = 0$$

$$(7) \quad \tau_{jl}^{BD} = 0 \qquad (8) \quad \tau_{jk}^{BC} = 0 \qquad (9) \quad \tau_{il}^{AD} = 0$$

$$(10) \quad \tau_{ik}^{AC} = 0 \qquad (11) \quad \tau_{ij}^{AB} = 0 \qquad (12) \quad \tau_{l}^{D} = 0$$

$$(13) \quad \tau_{k}^{C} = 0 \qquad (14) \quad \tau_{j}^{B} = 0 \qquad (15) \quad \tau_{i}^{A} = 0$$

Table 6.1. All types of hierarchical models in a four–way table and their interpretations.

Model classification	Sequence of hierarchical models	Sufficient marginals	Interpretation	Association graph
	*	ABC,ABD,ACD,BCD	–	
	*	ABC,ABD,ACD	–	
	*	ABC,ABD,CD	–	
G,D		ABC,ABD	$C \otimes D \mid A,B$	
	*	ABC,AD,BD,CD	–	
		ABC,AD,BD	$C \otimes D \mid A,B$	
G,D		ABC,AD	$D \otimes B,C \mid A$	
G,D		ABC,D	$D \otimes B,C,A$	
G,D		ABC	$D \otimes B, C,A \& D=u$	
	*	AB,AC,AD,BC,BD,CD	–	
	*	AB,AC,AD,BC,BD	$C \otimes D \mid A,B$	
	*	AB,AC,AD,BC	$D \otimes B,C \mid A$	
G		AB,BC,CD,AD	$A \otimes C \mid B,D \& B \otimes D \mid A,C$	
		AB,AC,BC,D	$D \otimes B,C,A$	
		AB,AC,BC	$D \otimes B,C,A \& D=u$	
G,D	*	AB,AC,AD	$B \otimes C \otimes D \mid A$	
G,D		AB,BC,CD	$D \otimes A,B \mid C \& A \otimes C,D \mid B$	

Table 6.1. (Cont.): All types of hierarchical models in a four-way table and their interpretations.

Model classification	Sequence of hierarchical models	Sufficient marginals	Interpretation	Association diagram
G,D	*	AB,AC,D	$D \otimes A,B,C \ \& \ B \otimes C \mid A$	
G,D		AB,AC	$D \otimes A,B,C \ \& \ B \otimes C \mid A \ \& \ D=u$	
G,D		AB,CD	$A,B \otimes C,D$	
G,D	*	AB,C,D	$A,B \otimes C,D \ \& \ C \otimes D$	
G,D		AB,C	$A,B \otimes C,D \ \& \ C \otimes D \ \& \ D=u$	
G,D		AB	$A,B \otimes C,D \ \& \ C \otimes D \ \& \ C=D=u$	
G,D	*	A,B,C,D	$A \otimes B \otimes C \otimes D$	
G,D	*	A,B,C	$A \otimes B \otimes C \otimes D \ \& \ D=u$	
G,D	*	A,B	$A \otimes B \otimes C \otimes D \ \& \ C=D=u$	
G,D	*	A	$A \otimes B \otimes C \otimes D \ \& \ B=C=D=u$	
G,D	*	–	$A=B=C=D=u$	

Once the most restrictive model to fit the data in a satisfactory way has been identified by its sufficient marginals the interpretation of the model can thus be established. A helpful mean is the **association graph**, which as defined in section 5.2 is a graph with m dots representing the m variables, where two dots are connected by a line if and only if there exists a non–null interaction involving the two variables. From an association graph the interpretation of a model can be derived directly. The rule is that if the route between two variables, say from A to B, can be disconnected by covering the dot corresponding to one variable, say C, or the dots corresponding to several variables, say both C and

D, then A and B are conditionally independent given C in the first case and given both C and D in the second case. Two variables are independent unconditionally, if they are not connected at all in the graph. If a main effect, say τ_i^A, is 0 for all indices, it is shown by replacing the dot representing A with an asterisk. As an illustration consider the first non–trivial model in table 6.1, which has sufficient marginals ABC and ABD. For this model, the association graph is shown in fig.6.1.

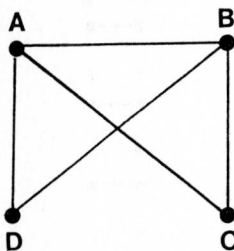

Fig. 6.1. Association graph for models ABC,ABD, ABC,AD,BD and

     AB,AC, AD,BC,BD.

On fig.6.1 the route from C to D is disconnected if the dots representing A and B are covered. Hence the interpretation

$$C \otimes D \,|\, A,B.$$

Note that the models ABC,AD,BD and AB,AC,AD,BC,BD also have the association graph shown in fig.6.1. As another example consider the two models ABC,AD and AB,AC,AD,BC which both have the association graph shown in fig. 6.2.

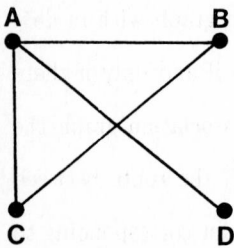

Fig. 6.2. Association graph for models ABC,AD and AB,AC,AD,BC.

On fig.6.2, the only routes from D to either C or B is through A, such that both models have the interpretation

$$D \otimes C,B \,|\, A.$$

The log–linear models for multi–way contingency tables can be classified into two partly overlapping classes of models, the **graphical models** introduced by Darroch, Lauritzen and Speed (1980) and the **decomposable models** introduced by Goodman (1968). In order to introduce the graphical models, we mention first that among all hierarchical models with the same interpretation, there is one for which the interpretation is equivalent to the model. This model is identified by the property that the set of sufficient marginals corresponds to the **cliques**, or the **maximal complete subsets** of the association graph. A complete subset is a set of neighbours in the graph which are all connected and a complete subset is maximal if it is not a true subset of another complete subset. Consider thus fig.6.1, where there are two cliques namely ABC and ABD. Hence model ABC, ABD is equivalent to the interpretation $C \otimes D \,|\, A,B$ in the sense that if $\tau^{ABCD}_{i\,j\,k\,l} = \tau^{ACD}_{i\,k\,l} = \tau^{BCD}_{j\,k\,l} = 0$ for all i,j,k and l then variable C is independent of variable D given the levels of both A and B and if C is independent of D given both A and B, then $\tau^{ABCD}_{i\,j\,k\,l} = \tau^{ACD}_{i\,k\,l} = \tau^{BCD}_{j\,k\,l} = 0$ for all i,j,k and l. The model AB,AC,AD,BC,BD, which differs from ABC,ABD in that in addition $\tau^{ABC}_{i\,j\,k} = 0$ and $\tau^{ABD}_{i\,j\,l} = 0$ for all i,j,k and l, also has interpretation $C \otimes D \,|\, A,B$ but in this case the model can not be derived from the interpretation, since the cliques are still ABC and ABD, and their corresponding interactions are not necessarily zero. Models for which the sufficient marginals corresponds to the cliques of the association graph are the graphical models. Darroch, Lauritzen and Speed (1980) showed that all decomposable models are graphical and that a graphical model is in addition decomposable if the association graph does not contain a **cycle** of length 4 or more. A cycle is a set of points, which are connected successively, but the diagonals are missing. The association graphs for models AB,AC,BD,CD and AB,BC,CD,AD, shown in fig.6.3, thus have cycles of length 4. For four–way tables the only graphical models, which are non–decomposable are AB,AC, BD,CD and its equivalents. Note on fig.6.3 the two ways a 4–cycle can manifest itself on an association graph.

Fig. 6.3. Association diagram for models AB,AC,BD,CD and AB,BC,CD,AD.

It is often the case that an association graph need to be rearranged in order to reveal its true nature.

Wermuth and Lauritzen (1983) contains a rather comprehensive treatment of association graphs for multiple contingency tables.

The importance of the concept of a decomposable model is partly due to the following theorem, first stated by Goodman (1968) and rigorously proved by Haberman (1974b), cf. also Andersen, A.H. (1974).

Theorem 6.1.

There is an explicit solution to the likelihood equations and the estimated expected numbers are direct functions of the sufficient marginals, if and only if the model is decomposable.

As an illustration reconsider the case of a three–way table. A typical set of hierarchical models, their interpretations and their association graphs are shown in table 6.2.

In table 6.2 a decomposable model is marked by a D and a graphical model by a G. The only non–graphical model is AB, BC, AC, which fails to be graphical, because A, B and C form a clique and $\tau^{ABC}_{ijk}=0$ for all i, j and k. All graphical model are decomposable. This means according to theorem 6.1, that the estimated expected numbers are direct functions of the sufficient marginals, except for the model AB,BC,AC. The exact expressions are listed in theorem 5.3.

Table 6.2. All typical hierarchical models in a three–way table and their interpretations.

Model classification	Sufficient marginals	Interpretation	Association graph
	AB, BC, AC	–	
G,D	AB, BC	$A \otimes C \mid B$	
G,D	AB, C	$A, B \otimes C$	
G,D	AB	$A, B \otimes C, C=u$	
G,D	A, B, C	$A \otimes B \otimes C$	
G,D	A, B	$A \otimes B, C=u$	
G,D	A	$B=C=u$	
G,D	–	$A=B=C=u$	

For four–way tables the graphical models are indicated by a G in table 6.1. The decomposable models are marked by a D. As noted above the only graphical model, which is not decomposable is AB,BC, CD,AD. For the decomposable models the exact expressions for the estimated expected numbers in terms of marginals are listed in table 6.3.

Tabel 6.3. Exact expression for the estimated expected numbers $\mu_{ijkl}=E[X_{ijkl}]$ for the decomposable models of a four–way table.

Sufficient marginals	Exact expression for $\hat{\mu}_{ijkl}$
ABC,ABD	$x_{ijk.}x_{ij.l}/x_{ij..}$
ABC,AD	$x_{ijk.}x_{i..l}/x_{i...}$
ABC,D	$x_{ijk.}x_{...l}/n$
ABC	$x_{ijk.}/L$
AB,AC,AD	$x_{ij..}x_{i.k.}x_{i..l}/x_{i...}^2$
AB,BC,CD	$x_{ij..}x_{.jk.}x_{..kl}/(x_{.j..}x_{..k.})$
AB,AC,D	$x_{ij..}x_{i.k.}x_{...l}/(nx_{i...})$
AB,AC	$x_{ij..}x_{i.k.}/(Lx_{i...})$
AB,CD	$x_{ij..}x_{..kl}/n$
AB,C,D	$x_{ij..}x_{..k.}x_{...l}/n^2$
AB,C	$x_{ij..}x_{..j.}/(nL)$
AB	$x_{ij..}/(KL)$
A,B,C,D	$x_{i...}x_{.j..}x_{..k.}x_{...l}/n^3$
A,B,C	$x_{i...}x_{.j..}x_{..k.}/(n^2L)$
A,B	$x_{i...}x_{.j..}/(nKL)$
A	$x_{i...}/(JKL)$

6.3. Search for a model

Since there are so many potential models in a higher order contingency table, it is important to have a good strategy for searching among the models. Goodness of fit test sta–

tistics are important tools for such a search, but they should not be used exclusively as formal tests of hypotheses. Rather should the level of significance, computed approximately from the appropriate χ^2–square approximation, serve as a guideline for the inclusion or exclusion of a set of interaction parameters. If formal tests of hypotheses are required, it is advisable to use multiple test procedures as for example the sequential Bonferroni procedure discussed in section 5.3. For four–dimensional tables, which form the basis for most examples below, the goodness of fit statistic for hypothesis H is given by

$$(6.7) \qquad Z(H) = 2 \sum_{i=1}^{I} \sum_{j=1}^{J} \sum_{k=1}^{K} \sum_{l=1}^{L} X_{ijkl}(\ln X_{ijkl} - \ln \hat{\mu}_{ijkl}),$$

where $\hat{\mu}_{ijkl}$ is $E[X_{ijkl}]$ with the ML–estimates for the non–null log–linear parameters under the model inserted. The distribution of $Z(H)$ can be approximated by

$$Z(H) \sim \chi^2(df(H)),$$

where $df(H)$ is the number of log–linear parameters set equal to zero under H. If H_A is an alternative hypothesis, where fewer log–linear parameters are set equal to zero, then

$$(6.8) \qquad Z(H|H_A) = 2 \sum_{i=1}^{I} \sum_{j=1}^{J} \sum_{k=1}^{K} \sum_{l=1}^{L} X_{ijkl}[\ln \tilde{\mu}_{ijkl} - \ln \hat{\mu}_{ijkl}] \sim \chi^2(df(H)-df(H_A)).$$

where $\tilde{\mu}_{ijkl}$ is $E[X_{ijkl}]$ with the ML–estimates of the non–null loglinear parameters under H_A inserted and $df(H_A)$ are the degrees of freedom for H_A. In the search for a model that fits the data both test statistics of the form (6.8) and of the form (6.7) are useful.

As mentioned in section 6.2, the model which includes all two–factor interactions and the saturated models have the same interpretation. Accordingly a search procedure, which often works, is to start by the model with fitted marginals AB, AC, AD, BC, BD, CD and test the fit of this model directly by means of (6.7). If the fit is satisfactory, one may then try sequentially through the use of (6.8) to exclude one or more of the two–factor interactions. If the model AB,AC,AD,BC,BD,CD does not fit the table, one has to go back to the saturated model and try to eliminate the four–factor and the three–factor interactions in turns. Unfortunately different search strategies may lead to different models, as the next example illustrates.

Example 6.1

In 1973 a retrospective study of cancer of the ovary was carrried out. Information was obtained from 299 women, who were operated for cancer of the ovary 10 years before. For these women the following four dichotomous variables were observed

A: Whether X–ray treatment was received or not.

B: Whether the women had survived the operation by 10 years or not.

C: Whether the operation was radical or limited.

D: Whether the cancer at the time of operation was in an early or in an advanced stage.

The observed number of women are shown in table 6.4.

Table 6.4. 299 women cross–classified according to four categorical variables.

D: Stage	C: Operation	B: Survival	A: X–ray No	Yes
Early	Radical	No	10	17
		Yes	41	64
	Limited	No	1	3
		Yes	13	9
Advanced	Radical	No	38	64
		Yes	6	11
	Limited	No	3	13
		Yes	1	5

Source: Obel (1975).

Our first search strategy is a stepwise exclusion of interactions, starting with the saturated model. Table 6.5 shows a number of models and the associated test statistics. The models in table 6.5 are selected partly based on prior beliefs on the part of the principal investigator and partly based on preliminary attempts with other models.

Table 6.5. Test statistics for a number of hierarchical models for the data of table 6.4.

Sufficient marginals	Test statistics $Z(H)$	df	Level of significance
ABC,ACD,ABD,BCD	0.60	1	0.438
ACD,ABD,BCD	1.23	2	0.540
ACD,BCD,AB	1.55	3	0.662
ACD,BC,BD,AB	1.93	4	0.748
AB,AC,AD,BC,CD,BD	7.17	5	0.208
AC,AD,BC,CD,BD	7.25	6	0.298
AD,BC,CD,BD	7.26	7	0.403
AD,BC,BD	7.86	8	0.447
AD,BD,C	9.39	9	0.402
BD,A,C	10.99	10	0.358
BD,C	28.99	11	0.002
BD,A	162.04	11	0.000
A,B,C,D	143.60	11	0.000

A satisfactory model is thus BD,A,C, showing that only the stage of the cancer influences the chance of survival, while neither the mode of operation nor treatment with X–rays seems to have an effect on the chance of survival.

A sequential procedure, which starts by testing the model AB, AC,AD,BC,CD,BD and then eliminate the two–factor interactions in turn would lead to the same model, as table 6.6 shows.

The test statistics and the degrees of freedom are obtained from table 6.5 in accordance with (6.8).

Table 6.6. Test statistics for a sequential testing of two–factor interactions being zero, starting with model AB,AC,AD,BC,BD,CD.

Sufficient marginals	Interactions set equal to zero	Test statistics $z(H)-z(H_A)$	df	Level of significance
AB,AC,AD,BC,CD,BD	All three– and four factors	7.17	5	0.208
AC,AD,BC,CD,BD	$\tau^{AB} = 0$	0.08	1	0.770
AC,AD,CD,BD	$\tau^{BC}_{jk} = 0$	0.01	1	0.933
AC,AD,BD	$\tau^{CD}_{kl} = 0$	0.60	1	0.530
AD,BD,C	$\tau^{AC}_{ik} = 0$	1.53	1	0.216
BD,A,C	$\tau^{AD}_{il} = 0$	1.60	1	0.206
A,B,C,D	$\tau^{BD}_{jl} = 0$	18.00	1	0.000

Suppose, however, that a sequential test procedure is chosen, starting with the saturated model. The test statistics corresponding to this procedure are shown in table 6.7. The very low level of significance for the step from model ACD,BC,BD,AB, to model AB,AC,AD,BC,BD,CD suggests that the three–factor interactions between ACD do not all vanish and the resulting model could have been ACD,BD. The main interpretation would still be that it is the stage of the cancer, which influences the chance of a survival by 10 years. But in addition one could conclude that the type of operation and the decision whether to use an X–ray treatment or not would depend on the stage of the cancer. The association graphs for the two models are shown in fig. 6.4.△.

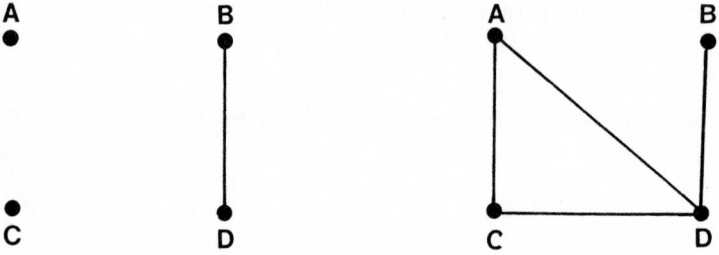

Fig. 6.4. Association graphs for two models fitted to the data in table 6.4.

Table 6.7. Sequential tests for a hierarchical ordered sequence of models for the data in table 6.4, starting with the saturated model

Sufficient marginals	Interactions set equal to zero	Test statistics $z(H)-z(H_A)$	df	Level of significance
ABC,ACD,ABD,BCD	$\tau^{ABCD}_{ijkl}=0$	0.60	1	0.438
ACD,ABD,BCD	$\tau^{ABC}_{ijk}=0$	0.63	1	0.429
ACD,BCD,AB	$\tau^{ABD}_{ijl}=0$	0.32	1	0.568
ACD,BC,BD,AB	$\tau^{BCD}_{jkl}=0$	0.38	1	0.538
AB,AC,AD,BC,BD,CD	$\tau^{ACD}_{ikl}=0$	5.24	1	0.002
ACD,BC,BD	$\tau^{AB}_{ij}=0$	0.09	1	0.743
ACD,BD	$\tau^{BC}_{jk}=0$	2.10	1	0.152
ACD,B	$\tau^{BD}_{jl}=0$	132.61	1	0.000

Example 6.2

The data for this example was collected in England in two periods from November 1969 to October 1971 and November 1971 to October 1973. The counts in table 6.8 are the numbers of traffic accidents involving trucks. In addition to the two periods, the accidents were classified according to three more categorical variables:

B: Whether the collision was in the back of the car or forward on the car. Forward includes the front and the sides.

C: Whether the truck was parked or not.

D: Light conditions with three levels: Day light, night on an illuminated road, night on a dark road.

The classification before and after November 1st. 1971 is called variable A.

The point of the study was, that a new compulsory safety measure for trucks was introduced in October 1971. The problem was, therefore, if the safety measure has had an effect on the number af accidents and on the point of collision on the truck.

For these data the strategy of starting with the model involving all two–factor interactions does not produce an immediate result. The observed test statistic for AB,AC,AD, BC,BD,CD is

$$z = 57.23, \ df = 9$$

with a level of significance below 0.0005. As an alternative we may consider the only model involving three–factor interactions, which has an interesting interpretation, namely ACD,BCD. This model has the association graph shown in fig. 6.5.

Table 6.8. Number of accidents involving trucks for two periods in England between 1969 and 1973, cross–classified according to three categorical variables.

D: Light Conditions	C:Parked	B: Collision	A: Periods Nov.69 to Oct.71	Nov.71 to Oct. 73
Daylight	Yes	Back	712	613
		Forward	192	179
	No	Back	2557	2373
		Forward	10749	9768
Night, illuminate street	Yes	Back	634	411
		Forward	95	55
	No	Back	325	283
		Forward	1256	987
Night, dark street	Yes	Back	345	179
		Forward	46	39
	No	Back	579	494
		Forward	1018	885

Source: Leaflet from Transport and Road Research Laboratory. Department of Environment. Crowthorne. Berkshire. UK. October 1976.

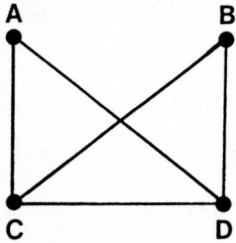

Fig. 6.5. Association graph for the model ACD,BCD.

The interpretation of model ACD,BCD is A $\otimes$ B|C,D, i.e. if differences in light conditions and whether the truck was parked or not, are taken into account, then the expected numbers of accidents in the back end and in the forward end of the truck was the same before and after the introduction of the safety measure. The test statistic for ACD, - BCD is

$$z = 6.86, \text{ df} = 6$$

with level of significance 0.334, so the model with the interpretation above can safely be accepted.

The introduction of the safety measure has not, accordingly, changed where on the trucks collisions take place. The analysis also shows that variables A and B are dependent through the common dependence on variables C and D. Hence one cannot make statements about A and B from a two–way table of A and B cross–classified, without loosing information contained in table 6.8.

If a model where all interactions involving variable A are zero fits the data the safety measure has had no effect whatsoever on the number truck accidents. The model concordant with ACD, BCD for which this is the case is the model with sufficient marginal BCD. The test statistic for model BCD against the alternative model ACD,BCD has observed test statistic

$$z(H|H_A) = 218.92, \text{ df} = 6$$

with level of significance p<0.0005 such that there is clearly a time effect in the data. A quick glance down table 6.8 shows that the number of accidents is in fact less in the second than in the first period for any combination of the three other variables.

Table 6.9 shows for the data of table 6.8 standardizes estimates of the three–factor interactions under the model ACD,BCD. For the interactions between ACD the figures in table 6.9 are thus

$$\hat{\omega}_{ikl} = \hat{\tau}^{ACD}_{ikl} / \sqrt{\widehat{\text{var}}[\hat{\tau}^{ACD}_{ikl}]} .$$

where $\widehat{\text{var}}[\hat{\tau}^{ACD}_{ikl}]$ are estimates for the variances of $\hat{\tau}^{ACD}_{ikl}$. In order to interprete the figures in table 6.8, one must keeep in mind that the expected number in a cell increases if a positive interaction is included and decrease if a negative interaction is included.

Table 6.9. Standardized estimates of three—factor interactions for the model ACD, BCD.

Variable D: Light condition	C: Parked	A: Periods Nov. 69 to Oct.71	Nov.71 to Oct. 73
Daylight	Yes	−3.811	+3.811
	No	+3.811	−3.811
Night, illuminated	Yes	−0.121	+0.121
	No	+0.121	−0.121
Night, dark	Yes	+2.976	−2.976
	No	−2.976	+2.976

Variable D: Light condition	C: Parked	B: Collisions Back	Forward
Daylight	Yes	−1.208	+1.208
	No	+1.208	−1.208
Night illuminated	Yes	+5.839	−5.839
	No	−5.839	+5.839
Night, dark	Yes	−4.274	+4.274
	No	+4.274	−4.274

In general three—factor interactions are difficult to interprete and this example is no exception. The upper half of table 6.9 shows that the combinations that increase the number of accidents significantly are daylight − parked − after November 1971, day—light − not parked − before November 1971, darkness − parked − before November 1971 and darkness − not parked − after November 1971. One way to describe this tendency is to say that the introduction of the safety measure seems to have limited the number on collisions on moving cars in daylight, while at night the safety measure has limited the number of collisions on parked cars. The lower part of table 6.9 shows that there is no significant dependency between where on the truck the collision happend and whether the truck moved or not, as long as we restrict attention to the daylight hours. However, after the safety measure was introduced parked cars on illuminated roads and moving cars on dark roads seem to be hit more often in the back while moving cars on illuminated roads and parked cars on dark streets are hit more often in the forward end.△.

Example 6.1 and 6.2 indicate how the search for a suitable model can be carried out for two concrete data sets. In recent years there have been an increasing interest in establishing general search procedures, which work well for a wide range of typical data sets. Almost all statistical analyses of contingency tables are to day carried out on computers. Search procedures which require little or no interference by the data analyst are, therefore, both appealing and potentially useful. Search procedures for contingency tables with many variables, which fast reach an acceptable model have been studied by Havranek (1984), Edwards and Havranek (1985), (1987) and Edwards and Kreiner (1983).

6.4 Diagnostics for model departures

The standardized residuals for a four–way table are defined as

$$(6.9) \qquad (x_{ijkl} - \hat{\mu}_{ijkl}) / \sqrt{\hat{var}[X_{ijkl} - \hat{\mu}_{ijkl}]},$$

where $\hat{\mu}_{ijkl}$ are the estimated expected numbers under the given model and $\hat{var}[X_{ijkl} - \hat{\mu}_{ijkl}]$ an estimate of the variance of $X_{ijkl} - \hat{\mu}_{ijkl}$. For all decomposable models for four–way tables, Haberman (1978), p.275 gives approximate expressions for the denominator in (6.9). These results are reproduced in table 6.10. For non–decomposable models the calculation of $var[X_{ijkl} - \hat{\mu}_{ijkl}]$ involves inversion of the information matrix according to the general formula (3.48).

Standardized residuals can be used as diagnostics for possisible model modifications, if an otherwise plausible model fails to fit the data satisfactorily, or to give a more differentiated picture for a model, which fits reasonable well.

Table 6.10. Approximations to the variances of the residuals for all decomposable models in a four-way table, excluding uniform distributions.

Sufficient marginals	Approximate variance of $X_{ijkl} - \hat{\mu}_{ijkl}$
ABC, ABD	$\hat{\mu}_{ijkl}\left(1-\dfrac{x_{ijk\cdot}}{x_{ij\cdot\cdot}}\right)\left(1-\dfrac{x_{ij\cdot l}}{x_{ij\cdot\cdot}}\right)$
ABC, AD	$\hat{\mu}_{ijkl}\left(1-\dfrac{x_{ijk\cdot}}{x_{i\cdot\cdot\cdot}}\right)\left(1-\dfrac{x_{i\cdot\cdot l}}{x_{i\cdot\cdot\cdot}}\right)$
ABC, D	$\hat{\mu}_{ijkl}\left(1-\dfrac{x_{ijk\cdot}}{n}\right)\left(1-\dfrac{x_{\cdot\cdot\cdot l}}{n}\right)$
AB, AC, AD	$\hat{\mu}_{ijkl}\left(1-\dfrac{x_{i\cdot k\cdot}\,x_{i\cdot\cdot l}}{x_{i\cdot\cdot}^{2}}-\dfrac{x_{ij\cdot\cdot}\,x_{i\cdot\cdot l}}{x_{i\cdot\cdot}^{2}}-\dfrac{x_{ij\cdot\cdot}\,x_{i\cdot k\cdot}}{x_{i\cdot\cdot}^{2}}+2\dfrac{x_{ij\cdot\cdot}\,x_{i\cdot k\cdot}\,x_{i\cdot\cdot l}}{x_{i\cdot\cdot}^{3}}\right)$
AB, BC, CD	$\hat{\mu}_{ijkl}\left(1-\dfrac{x_{i\cdot k\cdot}\,x_{\cdot\cdot kl}}{x_{\cdot j\cdot\cdot}\,x_{\cdot\cdot k\cdot}}-\dfrac{x_{ij\cdot\cdot}\,x_{\cdot\cdot kl}}{x_{\cdot j\cdot\cdot}\,x_{\cdot\cdot k\cdot}}-\dfrac{x_{ij\cdot\cdot}\,x_{\cdot jk\cdot}}{x_{\cdot j\cdot\cdot}\,x_{\cdot\cdot k\cdot}}\right.$ $\left.+\dfrac{x_{ij\cdot\cdot}\,x_{\cdot jk\cdot}\,x_{\cdot\cdot kl}}{x_{\cdot j\cdot\cdot}^{2}\,x_{\cdot\cdot k\cdot}}+\dfrac{x_{ij\cdot\cdot}\,x_{\cdot jk\cdot}\,x_{\cdot\cdot kl}}{x_{\cdot j\cdot\cdot}\,x_{\cdot\cdot k\cdot}^{2}}\right)$
AB, AC, D	$\hat{\mu}_{ijkl}\left(1-\dfrac{x_{i\cdot k\cdot}\,x_{\cdot\cdot\cdot l}}{n x_{i\cdot\cdot\cdot}}-\dfrac{x_{ij\cdot\cdot}\,x_{\cdot\cdot\cdot l}}{n x_{i\cdot\cdot\cdot}}-\dfrac{x_{ij\cdot\cdot}\,x_{i\cdot k\cdot}}{n x_{i\cdot\cdot\cdot}}\right.$ $\left.+\dfrac{x_{ij\cdot\cdot}\,x_{i\cdot k\cdot}\,x_{\cdot\cdot\cdot l}}{n x_{i\cdot\cdot\cdot}^{2}}+\dfrac{x_{ij\cdot\cdot}\,x_{i\cdot k\cdot}\,x_{\cdot\cdot\cdot l}}{n^{2}x_{i\cdot\cdot\cdot}}\right)$
AB, CD	$\hat{\mu}_{ijkl}\left(1-\dfrac{x_{ij\cdot\cdot}}{n}\right)\left(1-\dfrac{x_{\cdot\cdot kl}}{n}\right)$
AB, C, D	$\hat{\mu}_{ijkl}\left(1-\dfrac{x_{\cdot\cdot k\cdot}\,x_{\cdot\cdot\cdot l}}{n^{2}}-\dfrac{x_{ij\cdot\cdot}\,x_{\cdot\cdot\cdot l}}{n^{2}}-\dfrac{x_{ij\cdot\cdot}\,x_{\cdot\cdot k\cdot}}{n^{2}}+2\dfrac{x_{ij\cdot\cdot}\,x_{\cdot\cdot k\cdot}\,x_{\cdot\cdot\cdot l}}{n^{3}}\right)$
A, B, C, D	$\hat{\mu}_{ijkl}\left(1-\dfrac{x_{\cdot j\cdot\cdot}\,x_{\cdot\cdot k\cdot}\,x_{\cdot\cdot\cdot l}}{n^{3}}-\dfrac{x_{i\cdot\cdot\cdot}\,x_{\cdot\cdot k\cdot}\,x_{\cdot\cdot\cdot l}}{n^{3}}-\dfrac{x_{i\cdot\cdot\cdot}\,x_{\cdot j\cdot\cdot}\,x_{\cdot\cdot\cdot l}}{n^{3}}\right.$ $\left.-\dfrac{x_{i\cdot\cdot\cdot}\,x_{\cdot j\cdot\cdot}\,x_{\cdot\cdot k\cdot}}{n^{3}}+3\dfrac{x_{i\cdot\cdot\cdot}\,x_{\cdot j\cdot\cdot}\,x_{\cdot\cdot k\cdot}\,x_{\cdot\cdot\cdot l}}{n^{4}}\right)$

The standardized residuals are used to identify cells of the contingency table, where the fit between observed and expected numbers is particularly good or bad. Another diagnostics is known as **Cooks distance.** The precise definition of this measure is most easily explained within the framework of a regression model. It was in this connection the measure was introduced by Cook (1977). In chapters 8 and 9 it will be demonstrated, however, that a log–linear model for a contingency table can be formulated as a regression model, and in section 9.2, the exact definition of Cook's distance will be given. For the present application it suffice to say that Cook's distance is a measure of how much the log–linear parameters as a whole change if the observed number in a cell is removed from the data set. If Cook's distance for a cell is large, this cell thus havs a large influence on the values of the parameter estimates.

Example 6.2 (Cont.)

Table 6.11 show the standardized residuals and Cook's distances for the data in table 6.8 and the model ACD,BCD. Obviously it is for parked trucks at dark streets at night that the fit between model and data is the worst. Cook's distance, however, only attain large values for those trucks, which on dark streets are hit in the back. The two cell counts 345 and 179 thus have largest influence on the estimates of the log–linear parameters. △.

Table 6.11. Cook's distance (Cook) and standardized residuals (res) for the data in table 6.8 and the model ACD,BCD.

D: Light Conditions	C: Parked	B: Collision		A: Periods Nov.69 to Oct.71	A: Periods Nov.71 to Oct.73
Day-light	Yes	Back	Cook: res:	0.19 0.68	0.17 −0.68
		Forward	Cook: res:	0.05 −0.68	0.04 0.68
	No	Back	Cook: res:	0.03 −0.66	0.03 0.66
		Forward	Cook: res:	0.04 0.66	0.04 −0.66
Night, illuminate street	Yes	Back	Cook: res:	0.32 −0.63	0.24 0.63
		Forward	Cook: res:	0.04 0.63	0.02 −0.63
	No	Back	Cook: res:	0.13 −1.12	0.09 1.12
		Forward	Cook: res:	0.49 1.12	0.43 −1.12
Night, dark street	Yes	Back	Cook: res:	4.05 2.09	2.44 −2.09
		Forward	Cook: res:	0.57 −2.09	0.21 2.09
	No	Back	Cook: res:	0.01 0.24	0.01 −0.24
		Forward	Cook: res:	0.01 −0.24	0.01 0.24

6.5. Exercises

6.1. The Danish Institute for Building Research investigated in January 1983 the indoor climate in private homes in Denmark by interviewing a random sample of persons over 16 years of age. Among the variable measured for each person was

A: Smoking habits

B: Age

C: Sex

D: Frequency of headache

The resulting four–way contingency table is shown below.

Smoking habits	Age	Sex	Headache frequency	
			One or more times pr.week	Less than once a week
Smoker	Below 40	Male	11	142
		Female	45	83
	Above 40	Male	11	145
		Female	15	76
Non–smoker	Below 40	Male	8	117
		Female	29	89
	Above 40	Male	7	113
		Female	8	80

(a)   Find a suitable model to describe the data by means of a sequential test procedure.

(b)   Give an interpretation of the model arrived at in (a).

6.2. From the same investigation as in exercise 6.1 the table below shows the contingency table formed by the following variables, all categorized as binary.

A: The normal indoor temperature in the home.

B: Age

C: The presence of wall moisture or mould.

D: Dryness or irritation of the throat.

(a)   Compare the fit of the following two models

I: AD,BC,CD

II: AD,CD,B

both by testing the goodness of fit against the saturated model and by testing the fit of model II against model I as the alternative. Comment on the results.

Normal room temperature	Age	Moisture or mould	Irritation of throat	
			Yes	No
Under 23°	Below 40	Much	4	22
		Little or nothing	24	607
	Above 40	Much	3	12
		Little or nothing	52	684
Over 23°	Below 40	Much	3	6
		Little or nothing	20	219
	Above 40	Much	1	3
		Little or nothing	34	274

(b)   Draw association graphs for the two models in (a) and interprete them.

(c)   Show that both models are decomposable and check the computer program by calculating the expected values under both models directly.

6.3. From the Danish Welfare Study we consider the following four variables:

A: Daily alcohol consumption

B: Marriage status

C: Income

D: Urbanization

The categories of these variables follow from the four–way contingency table shown below.

(a)   The connection between alcohol consumption and the three other variables is of main concern. In addition Marriage status is assumed a priori to influence alcohol consumption the most and urbanization the least. Try accordingly to find a suitable model by eliminating the interactions in the following order BCD,ACD, ABD,ABC, CD,BD, CD,AD,AC,AB.

A: Daily alcohol-consumption	C: Income (1000 Dkr)	B: Marriage status	D:Urbanization Copen- hagen	Suburbian Copen- hagen	Three largest cities	Other Cities	Country side
Less than 1 unit	0–50	Widow	1	4	1	8	6
		Married	14	8	41	100	175
		Unmarried	6	1	2	6	9
	50–100	Widow	8	2	7	14	5
		Married	42	51	62	234	255
		Unmarried	7	5	9	20	27
	100–150	Widow	2	3	1	5	2
		Married	21	30	23	87	77
		Unmarried	3	2	1	12	4
	150–	Widow	42	29	17	95	46
		Married	24	30	50	167	232
		Unmarried	33	24	15	64	68
1 – 2 units	0–50	Widow	3	0	1	4	2
		Married	15	7	15	25	48
		Unmarried	2	3	9	9	7
	50–100	Widow	1	1	3	8	4
		Married	39	59	68	172	143
		Unmarried	12	3	11	20	23
	100–150	Widow	5	4	1	9	4
		Married	32	68	43	128	86
		Unmarried	6	10	5	21	15
	150–	Widow	26	34	14	48	24
		Married	43	76	70	198	136
		Unmarried	36	23	48	89	64
More than 2 units	0–50	Widow	2	0	2	1	0
		Married	1	2	2	7	7
		Unmarried	3	0	1	5	1
	50–100	Widow	3	0	2	1	3
		Married	14	21	14	38	35
		Unmarried	2	0	3	12	13
	100–150	Widow	2	1	1	1	0
		Married	20	31	10	36	21
		Unmarried	0	2	3	9	7
	150–	Widow	21	13	5	20	8
		Married	23	47	21	53	36
		Unmarried	38	20	13	39	26

Remark: 1 unit is approximately 1 bottle of beer or 2 cl. 40% alcohol.

(b) Discuss the use of direct model checks as opposed to sequential model checks.

(c) Estimate the two factor interactions, which are non–zero under the model and use them to describe the way the variables interact.

6.4. The Danish Institute for Traffic Safety Research collected the data in the table below of accidents involving a motor vehicle in order to study which factors influences the number of accidents.

A: General direction of the road	B: Number of parts involved	C: Driver in– fluenced by alcohol	D: Time of day		
			Morning	Afternoon	Evening
North– South	One	Yes	29	13	15
		No	6	3	11
	Several	Yes	294	120	53
		No	7	8	18
East– West	One	Yes	38	14	21
		No	4	7	16
	Several	Yes	206	102	52
		No	5	15	12

(a) Formulate a statistical model for the data in the table.

(b) The goodness of fit test statistic for the model with the four–factor and all three–factor interactions equal to zero has observed value 16.93 with 13 degrees of freedom. Explain how the degrees of freedom are calculated.

(c) Use the table below which shows a number of selected goodness of fit test statistic, all against the saturated model, to find a suitable model to describe the data.

(d) Draw the association graph and interprete the model.

Model	z(H)	degrees of freedom
AB,AC,AD,BC,BD,CD	16.93	13
AB,AC,AD,BC,BD	126.45	16
AB,AC,AD,BC,CD	113.97	16
AB,AC,BC,BD,CD	20.79	16
AB,AC,AD,BD,CD	50.34	14
AB,AD,BC,BD,CD	16.95	14
AC,AD,BC,BD,CD	21.47	14
AB,AD,BC,BD	126.86	17
AB,AD,BC,CD	114.00	17
AB,BC,BD,CD	21.71	17
AB,AD,BD,CD	50.68	15
AD,BC,BD,CD	21.80	15
AB,BC,BD	131.14	20
AB,BC,CD	119.73	20
AB,BD,CD	54.94	18
BC,BD,CD,A	29.54	18
BD,CD,A	62.78	19
BC,CD,A	127.56	21
BC,BD,A	138.97	21
CD,A,B	229.33	22
BD,A,C	240.74	22
A,B,C,D	407.30	25

6.5. Consider again the data in exercise 6.4.

(a) One candidate for a model is BC,BD,CD,A. Is this model decomposable? Why?

(b) Estimate the log–linear parameters of the model in (a) and use them to characterize the way variables B,C and D influences the number of accidents.

6.6. The Gallup Institute in Denmark interviewed in 1979 a random sample of 783 persons. Among the questions asked was one concerning attitude towards corporal punishment of children and one concerning memories of corporal punishment as a child. In the table below the answers are cross–classified with education and age.

(a) Formulate a log–linear model for the data.

(b) Find a reasonable simple model to fit the data.

(c) Interpret the resulting model.

(d) Use standardized residuals or standardized log–linear parameters to through light on the direction of model departures if any.

A: Attitude	B: Memory of punish-ment	C: Education (highest school level)	D: Age (years)		
			15–24	25–39	40–
No punish-ment of children	Yes	Elementary	1	3	20
		Secondary	2	8	4
		High	2	6	1
	No	Elementary	26	46	109
		Secondary	23	52	44
		High	26	24	13
Moderate punishment of children	Yes	Elementary	21	41	143
		Secondary	5	20	20
		High	1	4	8
	No	Elementary	93	119	324
		Secondary	45	84	56
		High	19	26	17

6.7. The four–way table below is from a 1971 investigation of satisfaction with housing conditions. The variables are

A: Type of housing

B: Influence on the management of the building

C: Degree of contact with other inhabitants

D: Satisfaction with housing conditions.

(a) Take the model with all 6 two–factor interactions included as base model and use a sequential procedure to find a reasonable model.

(b) Interpret the resulting model.

(c) Use the estimated log–linear parameters to describe the way the satisfactions depend on contact, influence and type of hopusing. Are there any surprises?

Type of housing	Influence	Contact	Satisfaction		
			Low	Medium	High
	Little	Little	21	21	28
		Much	14	19	37
Apartment building with 4 or more storeys	Some	Little	34	22	36
		Much	17	23	40
	Much	Little	10	11	36
		Much	3	5	23
	Little	Little	61	23	17
		Much	78	46	43
Apartment building, less than 4 storeys	Some	Little	43	35	40
		Much	48	45	86
	Much	Little	26	18	54
		Much	15	25	62
	Little	Little	13	9	10
		Much	20	23	20
Rented house, not detached	Some	Little	8	8	12
		Much	10	22	24
	Much	Little	6	7	9
		Much	7	10	21
	Little	Little	18	6	7
		Much	57	23	13
Rented house, detached or semi-detached	Some	Little	15	13	13
		Much	31	21	13
	Much	Little	7	5	11
		Much	5	6	13

6.8. A large scaled investigation of sportsactivities and attitudes among 16–19 year old high school students was carried out in Denmark in 1983 and 1985. One of the questions asked was concerning attitude to sport jointly with the other sex. The answers to this question is in the table below cross–classified with the sex of the student, the year of the interview and whether the student is in his or her first or third year in high school.

Sports joint with the other sex	Year	Grade	Sex	
			Boy	Girl
Very good idea	1983	First	31	103
		Third	23	61
	1985	First	41	77
		Third	31	52
Good idea	1983	First	51	67
		Third	39	72
	1985	First	67	80
		Third	31	70
Indifferent	1983	First	38	29
		Third	36	39
	1985	First	35	27
		Third	31	28
Bad idea	1983	First	10	15
		Third	15	16
	1985	First	12	10
		Third	4	4
Very bad idea	1983	First	4	2
		Third	2	3
	1985	First	2	3
		Third	7	3

(a) Explain how a log–linear model can throw light on what influences the view on joint sports.

(b) Choose a suitable model for the data and compute the standardized residuals for this model. Are there any cells of the contingency table, which deserve a comment?

(c) Choose a model which does not quite fit the data and use the standardized residuals to check for cells or variable levels, with may be omitted to improve the fit.

6.9. Consider a five–way contingency table formed by the variables A,B,C,D and E, and assume that a log–linear model with sufficient marginals AC,AD,BD,BE,CD fits the data.

(a) Draw an association graph for the model and give an interpretation in terms of conditional independencies.

(b)   Show that the model is not decomposable and find the decomposable model with the same association graph.

(c)   Specify the likelihood equations for the decomposable model.

(d)   Show that the expected values under the decomposable model are given by

$$\hat{\mu}_{ijkl,} = \frac{x_{i.kl.}\, x_{.j.l.}\, x_{.j..m}}{x_{...l.}\, x_{.j...}}$$

[Hint: Note that the expected values must have the form

$$\mu_{ijklm} = \varphi_{ikl}\, \alpha_{jl}\, \beta_{jm}.]$$

(e)   Use the likelihood equations and the constraints $\tau_{j.}^{BD} = \tau_{.1}^{BD} = 0$ to derive the estimate for τ_{j1}^{BD}.

7. Incomplete Tables, Separability and Collapsibility

7.1. Incomplete tables

An observed contingency table is incomplete if it contains zeros in certain cells. Such zeros are of two types, **random zeros** and **structural zeros**. A cell has a random zero, if the observed value in the cell is zero, but the expected value is positive. A cell has a structural zero if the expected number is zero, i.e. if it is known a priori that the cell will contain a zero. Random or structural zeros does not impaire the log–linear structure of a given model. It means, however, that certain log–linear parameters can not be estimated.

Consider e.g. a three–way table, where the log–likelihood has the form

(7.1)
$$\ln L = \sum_i \sum_j \sum_k \tau_{ijk}^{ABC} x_{ijk} + \sum_i \sum_j \tau_{ij}^{AB} x_{ij.} + ... + \tau_0 x_{...}$$

$$+\{\text{terms in } x\} + \{\text{terms in } \tau\}.$$

If certain x_{ijk}'s are zero then the corresponding three–factor interactions τ_{ijk}^{ABC} does not appear in the likelihood function and can not be estimated. In addition some of the lower order interactions may also vanish, namely if the corresponding marginal is zero. If e.g. $x_{111}=...=x_{11K}=0$ then $x_{11.}=0$ and τ_{11}^{AB} can not be estimated. For I=J=K=2 an example of a table , where τ_{111}^{ABC}, τ_{112}^{ABC} and τ_{11}^{AB} can not be estimated is shown in table 7.1.

Table 7.1. A 2x2x2 table with random zeros.

A:		B:		C:k=1	2
i=1		j=1		0	0
		2		24	33
2		1		41	32
		2		27	11

In a log–linear model the likelihood equations are derived by equating sufficient marginals with their mean values. Since zero marginals does not contribute to the likelihood equations, it follows that the set of likelihood equations will be the same whether the zeros of the table are random or structural. Thus log–linear parameters corresponding to zero marginals can not be estimated whether the zeros are random or structural. In a technical sense the estimation problem is, therefore, identical for cases with random zeros and cases with structural zeros.

There is still a difference between the two cases as regards the interpretation of the obtained estimates. If a cell count or a marginal is a structural zero, the corresponding log–linear parameter does not exist, since it does not enter any likelihood function. If a cell count or a marginal is a random zero, the corresponding parameter does exist, but it is not estimable based on the given data set.

As regards hypothesis testing the log–likelihood ratio test statistics for cases with random and structural zeros will be the same. Cells with structural or random zeros do not contribute to the log–likelihood ratio test statistic, since both the expected and the observed values are zero. It follows that the only problem connected with hypothesis testing in incomplete tables is how to count the degrees of freedom for the test statistic correctly. One may of course rely on the computer to do the count correctly, but it is important to be able to check the result. Hence we shall list a few important rules for counting degrees of freedom below.

The main rule is the following: Let H be hypothesis formulated in terms of log–linear parameters being zero and let $N_0, N_1(H)$ and $N_2(H)$ be defined as

N_0 = number of cells with a non–zero count

$N_1(H)$ = number of unconstrained log–linear parameters under H in a complete table

$N_2(H)$ = number of unconstrained log–linear parameters under H for which the corresponding sufficient marginal is zero.

The number of degrees of freedom for the log–likelihood ratio test statistic for testing H against the saturated model is then given by the formula

(7.2)
$$df(H) = N_0 - N_1(H) + N_2(H).$$

Under the saturated model the number of free log–linear parameters, which can be estimated, is equal to the number of cells N_0 with non–zero counts, because the likelihood equations in this case are

$$x_{ijk} = E[X_{ijk}], \quad \text{for all } i, j, k \text{ with } x_{ijk} > 0.$$

Under H not all free parameters can be estimated. Only those log– linear parameters, which corresponds to non–zero marginals have estimates. Hence the number of parameters that can be estimated under H is the number of unconstrained parameters $N_1(H)$ in a complete table, minus the number of zero–marginals $N_2(H)$ under H.

As an example consider again the 2x2x2 table in table 7.1 and the hypothesis

$$H: \tau^{ABC}_{ijk} = \tau^{AC}_{ik} = 0.$$

If it is ignored that the table is incomplete, the number of degrees of freedom for the log–likelihood ratio test statistic for H is

$$df(H) = 2.$$

The table has, however, only 6 non–zero cells. In a complete table the number of log–linear parameters for the model AB,BC is $(I-1)(J-1)+(J-1)(K-1)+(I-1)+(J-1)+(K-1)+1=6$. Under H there is, however, the zero marginal $x_{11.}=0$, such that τ^{AB}_{11} cannot be estimated. Hence the elements of (7.2) are

$$N_0 = 6$$
$$N_1(H) = 6$$
$$N_2(H) = 1$$

The correct number of degrees of freedom is thus

$$df(H) = 6-6+1 = 1.$$

The cells with zero counts are easy to count. The problem is those marginals which are zero because all cells adding up to it have zero counts. If all log–linear parameters are explicitly estimated for an incomplete table it is obvious which log–linear parameters have finite estimated values and contribute to the test statistic. But usually the estimates are not needed to derive the test statistic, and one has to be aware of zero marginals.

Haberman (1974b) has formulated a very simple rule that solve many problems. Assume that there is an index (i_0,j_0,k_0) with $x_{i_0 j_0 k_0} > 0$ such that for any other index (i,j,k), which coincide with (i_0,j_0,k_0) on one component, $x_{ijk} > 0$. Then $N_2(H)=0$. As an example of the application of this rule consider again table 7.1. For this table none of the indices satisfy the rule, because any index will have either $k=1$ in common with $(1,1,1)$ or $k=2$ in common with $(1,1,2)$. If, however, the zero in cell $(1,1,2)$ is changed to a positive number, the index $(2,2,2)$ satisfies the rule. Hence $N_2(H)=0$ in the new situation with $x_{112} > 0$. The rule does not imply that $N_2(H) > 0$ when the rule is not satisfied. Consider for example the hypothesis of no two or three factor interactions for table 7.1. Here $N_1(H)=4$ corresponding to $\tau_1^A, \tau_1^B, \tau_1^C$ and τ_0, but $N_2(H)=0$ since all marginals $x_{1..}$, $x_{.1.}$ and $x_{..1}$ are positive, and the degrees of freedom become $6-4=2$.

The methods developed above also apply to models, where the expected numbers in certain cells by assumption are equal to the observed numbers. Any cell, where the expected number by definition is equal to the observed number, does not contribute to the test statistic for a given hypothesis. It does not matter, therefore, what value we assign to such a cell. Also all estimates of parameters will be the same, whether we fit the expected number in such a cell by its observed number, or put both expected and observed numbers equal to zero.

These considerations imply that a model for an incomplete table can formally be written as

$$\ln\mu_{ijk} = \tau_{ijk}^{ABC} + \tau_{ij}^{AB} + ... + \tau_0, \qquad (i,j,k)\epsilon M_0,$$

where M_0 is the set of cells, which have either non–zero counts or for which by assumption $\mu_{ijk} = x_{ijk}$.

Much of the theory on incomplete tables was developed around 1970, cf. Bishop (1969), (1970), Fienberg (1970), (1972) and Haberman (1974b), chapter 7. A recent reference is Chen and Fienberg (1986).

7.2. Two–way tables and quasi–independence

For two–way tables the main hypothesis is the independence hypothesis

$$(7.3) \qquad \ln\mu_{ij} = \ln[E_{ij}] = \tau_i^A + \tau_j^B + \tau_0.$$

The hypothesis corresponding to (7.3) only implies independence between the two categorical variables if (7.3) holds for all i and j. But independence in a table with structural zeros imply that $\mu_{ij}=0$, which is in conflict with (7.3). Instead we consider the hypothesis of **quasi–independence**, defined as

$$(7.4) \qquad \ln\mu_{ij} = \tau_i^A + \tau_j^B + \tau_0 \quad , \quad (i,j)\epsilon M_0,$$

where M_0 is a subset of the cells, for example the cells with structural zeros.

No structure is assumed for the cells outside M_0.

The fact that the table is incomplete does not change the form of the likelihood equations, which are

$$x_{i.} = E[X_{i.}] \quad , i=1,...,I$$

and

$$x_{.j} = E[X_{.j}] \quad ,j=1,...,J \, ,$$

but the summations are only over pairs (i,j) in M_0. The test statistic for the model is

(7.5)
$$Z(H) = 2 \underset{(i,\,j)\epsilon M_0}{\Sigma\Sigma} X_{ij}[\ln X_{ij} - \ln\hat{\mu}_{ij}],$$

where

$$\hat{\mu}_{ij} = \exp(\hat{\tau}^A_i + \hat{\tau}^B_j + \hat{\tau}_0)$$

are the estimated expected values under H. In order to compute the degrees of freedom for the asymptotic χ^2–distribution of (7.5), we need the concept of an **inseparable two–way table.**

Definition 7.1.: A two–way contingency table $\{x_{ij};\ i=1,...,I,\ j=1,...,J\}$ **is inseparable under the quasi–independence hypothesis (7.4) if all cells can be connected without passing through a cell outside** M_0

The check of inseparability for a table requires a diagram like fig. 7.1. In (b) the table is inseparable as indicated by the drawn lines. In (a) the table can be separated in two subtables, such that a connecting line between the subtables must pass a cell outside M_0.

(a)   **(b)**

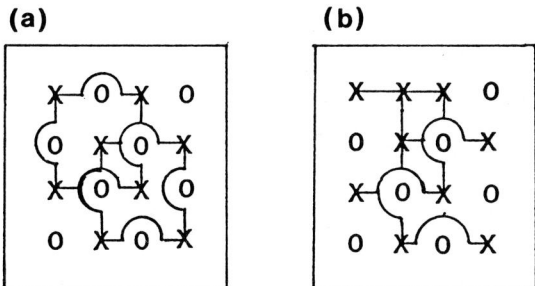

Fig. 7.1. Examples of a separable and an inseparable two–way table, where a zero denote a structural zero or a completely fittet cell.

For a complete table the degrees of freedom for (7.5) is $(I-1)(J-1)$. If all marginals $x_{i.}$ and $x_{.j}$ are null–zero and if the table is inseparable the degrees of freedom for an incomplete table are

$$df(H) = N_0(M) - I - J + 1,$$

where $N(M_0)$ is the number of elements in M_0.

Caussinus (1965) showed that if the table contains s inseparable subtables then the general formula for the degrees of freedom for the log–likelihood ratio test statistic for quasi–independence is

(7.6)
$$df(H) = N(M_0)-N_1-N_2+s,$$

where $N(M_0)$ is the number of elements in M_0, N_1 is the number of positive $x_{i.}$'s and N_2 is the number of positive $x_{.j}$'s. One intuitive explanation for the additive term s, which for inseparable tables has value 1, is that the model under H contains one main effect τ_0 for each of the inseparable subtables.

The hypotheses of uniform distribution over the row levels or over the column levels can under independence be tested for incomplete tables as well as for complete tables. Consider thus under H the hypothesis

$$H_2: \tau^B_j = 0 \text{ for all } j \epsilon M^B_0,$$

where $M^B_0 = \{j \,|\, (i,j) \epsilon M_0 \text{ for some } i\}$.

Since there are only N_2 positive $x_{.j}$'s, at most N_2-1 unconstrained τ^B_j's can be estimated. If, however, the table is separable into s separable subtables. there are s constraints among the τ^B_j's rather than 1. Hence the test statistic for H_2 has

$$df(H_2) = N_2 - s$$

degrees of freedom. Similarly for

$$H_1: \tau^A_i = 0 \text{ for all } i \epsilon M^A_0 ,$$

with $M^A_0 = \{i \,|\, (i,j) \epsilon M_0 \text{ for some } j\}$, the degrees of freedom are

$$df(H_1) = N_1 - s.$$

Example 7.1.

The Danish National Institute for Social Science Research conducted in 1982 a survey of

838 persons concerning their views on two retirement systems. One system called "early retirement" had been in effect for three years. The second called "partial pension" had just been introduced as a bill in the Danish Parliament. The interviewed persons were as regards the old system asked whether they regarded the system as a good system, a relatively good system or a bad system. For the new system, they were asked if we consider the proposal a good one, maybe a good one or a bad one. The responses are shown in table 7.2.

Table 7.2. 838 persons in 1982 cross–classified according to their views on two retirement systems.

Early retirement

Partial pension	Good system	Relatively good system	Bad system	No opinion
Good proposal	377	75	38	19
Maybe good proposal	92	25	15	8
Bad proposal	84	17	16	4
No opinion	34	17	6	11

Source: Olsen (1984). Table 3.12.

It is clear from the numbers that an independence hypothesis will almost surely be rejected. A more reasonable hypothesis could be quasi–independence with

$$\ln E[X_{ij}] = \tau_0 + \tau_i^A + \tau_j^B, \quad \text{for } i \neq j,$$

i.e. quasi–independence outside the diagonal. After removal of the diagonal, the table is still inseparable and all marginals still non–zero. The test statistic has, therefore, according to (7.6)

$$df = 12-4-4+1 = 5$$

degrees of freedom. The observed value of the test statistic is

$$z = 5.94, \quad df=5$$

with level of significance p=0.312. The hypothesis of quasi–independence can thus be accepted. The interpretation of this result is that the dependency between the views on the two retirement systems is due to the many cases where the views coincide. For the 409 persons, who do not have coinciding views, their opinion of the "early retirement" system is independent of their opinion of the "partial pension" system. The expected numbers under quasi–independence are shown in table 7.3. Note that the expected numbers in the diagonal are equal to the observed numbers.△.

Table 7.3. Expected numbers under quasi–independence for the data of table 7.2.

Early retirement

Partial pension	Good system	Relatively good system	Bad system	No opinion
Good proposal	(377)	74.8	38.7	18.5
Maybe good proposal	93.8	(25)	14.3	6.9
Bad proposal	76.8	22.6	(16)	5.6
No opinion	39.4	11.6	6.0	(11)

7.3 Higher order tables. Separability

For higher order tables the definition of an inseparable table is more complicated than definition 7.1, because the definition depends on the hypothesis to be tested.

Definition 7.2.: A table is inseparable if none of the categorical variables defining the table are contained in all the variable combinations that define the hypothesis and if the table is connected in the sense of definition 7.1.

This definition contains definition 7.1 since the variable combinations, which define (7.4), namely A and B, have no element in common.

A three dimensional table is thus inseparable under the hypotheses with sufficient marginals AB, BC, AC and A,B,C if the table is connected. Under the model AB, AC the table is, however, separable whether it is connected or not since A is contained in both sufficient marginals. The reason is that AB,AC is equivalent to conditional independence of B and C given A, in which case the analysis can be based on the I two–way tables between B and C for given levels of A. These tables are inseparable under the independence hypothesis if they satisfy definition 7.1.

In the following we shall often refer to the **generators** of a hypothesis. The set of generators is the set of sufficient marginals, which define the minimal set of likelihood equations. The generators are usual identified by the letter combinations corresponding to the variables. The generators for the hypothesis A⊗B|C or equivalently the model for which

$$\tau^{ABC}_{ijk} = \tau^{AB}_{ij} = 0, \quad \text{for all i, j and k}$$

in a three–way table are thus AC and BC.

In general if there is a variable combination, which belong to all generators of the hypothesis, then two cells with different indices on these variables must belong to different subtables, when the table is divided into inseparable subtables. As an example consider a 2x2x2 table with no zeros under the model AB, AC, BC. Since there is no common element to the generators and the table is connected, it is inseparable. Under the model AC,BC, however, C is a common element and the table can be divided into the two inseparable 2x2 tables of A cross–classified with B for levels C=1 and C=2, if these tables are inseparable.

The concept of separability is important in order to determine the number of degrees of freedom for the test statistic. The following result is often useful: If there are s inseparable subtables of a multi–way table and if the hypothesis H is generated by just two variable combinations A_1 and A_2, then the number of degrees of freedom for the log–likelihood ratio test statistic is given by

(7.7) $$df(H) = N_0 - N(A_1) - N(A_2) + s,$$

where N_0 is the number of cells with positive counts, $N(A_1)$ is the number of non–zero marginals involving all variables in A_1 and $N(A_2)$ the number of non–zero marginals involving all variables in A_2. If formula (7.7) does not apply the only method for determining the degrees of freedom is to count the unconstrained parameters without and under the model.

Example 7.2.

The data in table 7.4 is from the Danish Welfare study in 1976.

The Danish Welfare Study was based on a random sample of 5960 individuals. The non–response was 13.3%. In the obtained sample of size 5166, 3137 were classified according to social rank. Table 7.4 shows for each combination of social rank and socio–economic group in this subsample the number of respondents which frequently use medicin for headaches. Because there by definition is a certain overlapping between social rank and socio–economic group, 8 of the cells in table 7.4 are structural zeros. There are thus by definition no blue collar workers in social groups I–III and no self–employed and white collar workers in social group V. The reason for these structural zeros is that the social rank scale is primarily based on employment status and level of education. Thus all blue collar workers are either in group IV or group V, with the skilled workers in group IV.

Let the variables of table 7.4 be denoted

        A: Social rank.
        B: Socio–economic group.
        C: Headache medicin consumption.

Consider the following hypotheses involving three– and two–factor interactions:

$$H_1: \tau_{ijk}^{ABC} = 0, \qquad \text{all } i, j \text{ and } k$$

$$H_2: \tau^{ABC}_{ijk} = \tau^{AB}_{ij} = 0, \quad \text{all } i, j \text{ and } k$$

$$H_3: \tau^{ABC}_{ijk} = \tau^{AC}_{ik} = 0, \quad \text{all } i, j \text{ and } k$$

$$H_4: \tau^{ABC}_{ijk} = \tau^{BC}_{jk} = 0, \quad \text{all } i, j \text{ and } k$$

Table 7.4. Social rank, socio–economic group and use of headache medicin cross–classified for 3137 individuals in the Danish Welfare Survey.

Social rank group	Socio–economic group	Frequent use of headache medicin Yes	No
I+II	Self employed	5	77
	White collar workers	26	343
	Blue collar workers	0	0
III	Self employed	60	400
	White collar workers	42	377
	Blue collar workers	0	0
IV	Self employed	26	102
	White collar workers	102	685
	Blue collar workers	32	314
V	Self employed	0	0
	White collar workers	0	0
	Blue collar workers	160	839

Source: Hansen (1978): Table 6.A.50.

The test statistics and the degrees of freedom for these four hypotheses are shown in table 7.5 together with their level of significance.

Table 7.5. Test statistics, degrees of freedom and levels of significance for four hypotheses and the data in table 7.4.

Hypothesis	Sufficient marginals	Test statistic	Degrees of freedom	Level of significance
H_1	AB,AC,BC	1.74	2	0.419
H_2	AC,BC	354.90	4	0.000
H_3	AB,BC	24.63	5	0.001
H_4	AB,AC	8.57	4	0.073

From table 7.5 it is evident that a model only involving interactions between variables A and B and between variables A and C describes the data in a satisfactory way.

Further reductions in the model are not possible. Since the table is of dimension I=4, J=3, K=2, the degrees of freedom in table 7.5 can not be obtained from the formulae for the complete table. The degrees of freedom for H_1 in a complete table are thus $3 \cdot 2 \cdot 1 = 6$. It is easily seen that the structural zeros only affect the number of free parameters for the three–factor interactions τ_{ijk}^{ABC} and the two–factor interactions τ_{ij}^{AB}. As regards the three factor interactions, it is easy to verify, that if the constraints $\tau_{ij.}^{ABC} = \tau_{i.k}^{ABC} = \tau_{.jk}^{ABC} = 0$ are satisfied with summations over the non–empty cells of the table, then there are only two unconstrained parameters τ_{111}^{ABC} and τ_{211}^{ABC}. All three–factor interactions are shown in table 7.6.

Table 7.6. Three–factor interactions for the incomplete table 7.4 (with label ABC omitted).

		C:	
A:	B:	k=1	2
i=1	j=1	$+\tau_{111}$	$-\tau_{111}$
	2	$-\tau_{111}$	$+\tau_{111}$
	3	$-$	$-$
2	1	$+\tau_{211}$	$-\tau_{211}$
	2	$-\tau_{211}$	$+\tau_{211}$
	3	$-$	$-$
3	1	$-\tau_{111}-\tau_{211}$	$+\tau_{111}+\tau_{211}$
	2	$+\tau_{111}+\tau_{211}$	$-\tau_{111}-\tau_{211}$
	3	0	0
4	1	$-$	$-$
	2	$-$	$-$
	3	0	0

Table 7.7 shows that under the constraints $\tau_{i.}^{AB} = \tau_{.j}^{AB} = 0$, there are only two unconstrained interactions, involving A and B, namely τ_{11}^{AB} and τ_{21}^{AB}.

Table 7.7. Two–factor interactions between variables A and B for the incomplete table 7.4 (with label AB omitted).

B:

A:	j=1	2	3
i=1	τ_{11}	$-\tau_{11}$	–
2	τ_{21}	$-\tau_{21}$	–
3	$-\tau_{11}-\tau_{21}$	$\tau_{11}+\tau_{21}$	0
4	–	–	0

The degrees of freedom for H_1, where $\tau^{ABC}_{ijk}=0$, and for H_2, where $\tau^{ABC}_{ijk}=\tau^{AB}_{ij}=0$ follows immediatetly from tables 7.6 and 7.7. Note that the number of unconstrained three–factor interactions is less than $(I-1)(J-1)(K-1)=3\cdot2\cdot1=6$ due to the structural marginals $x_{ijk}=0$ in the saturated model. The number of unconstrained τ^{AB}_{ij}'s is less than $(I-1)(J-1)=3\cdot2=6$ because the marginals $x_{13.}$, $x_{23.}$, $x_{41.}$ and $x_{42.}$ are all structural zeros. For variable combinations AC and BC, there are no marginals, which are structural zeros. Hence the degrees of freedom for H_3 follow by adding $(I-1)(K-1)=3$ to the degrees of freedom for H_1 and for H_4 by adding $(J-1)(K-1)=2$ to the degrees of freedom for H_1.

As an exercise let us derive the degrees of freedom for H_1 from formula (7.2) and for H_4 from formula (7.7). The numbers in (7.2) for H_1 are

$N_0= 16$

$N_1(H_1) = 1+(I-1)+(J-1)+(K-1)+(I-1)(K-1)+(J-1)(K-1)+(I-1)(J-1)=18$

$N_2(H_1) = 4$

The last number is due to the fact that four marginals are zero for variable combination AB.

Under H_4, the table is separable since variable A is in both generators AB and AC. The interpretation of H_4 is B⊗C|A and variable A has four levels. Hence the table can be separated into four conditional tables, given the levels of variable A. These four subtables are inseparable. In order to apply formula (7.7) let A_1 be combination AC and A_2 combination AB. The number of non–zero marginals corresponding to τ^{AC}_{ik} is $I\cdot K=8$. Corresponding to τ^{AB}_{ij}, there are, however, 8 rather than $I\cdot J=12$ non–zero marginals, since

$$x_{13.} = x_{23.} = x_{41.} = x_{42.} = 0.$$

Hence the numbers in formula (7.8) become

$$N_0 = 16$$
$$N(A_1) = 8$$
$$N(A_2) = 8$$
$$s = 4,$$

such that the degrees of freedom for H_2 is in fact 4 as claimed in table 7.4.

The model AB,AC and its interpretation B⊗C|A shows that only social rank seems to influence the consumption of headache medicin. Surprisingly if the social rank is taken into account the use of headache medicin is independent of socio–economic group. The interactions between variables A and C are shown in table 7.8.

Table 7.8. Estimated interaction parameters between variables A and C.

Frequent use of headache medicin	Social rank			
	I+II	III	IV	V
Yes	−0.281	−0.007	+0.079	+0.194
No	+0.281	−0.007	−0.079	−0.194

The table shows that headache medicin is used more frequent by people with low social rank and less frequently by people with high social rank.△.

The concept of separability applies to complete as well as incomplete tables. For complete tables the only condition in definition 7.2 for the table to be inseparable is that the set of generators do not have a common element. For a three–way table this gives the classification in table 7.9 of separable and inseparable tables under H. If the generating class only have one element, separability has the special meaning that the corresponding marginal table is saturated. The test statistic is then trivial and separability means that the table can be separated in its cells.

Table 7.9. Separable and inseparable models in a three–way table.

Sufficient marginals	Separable /Inseparable
AB,AC,BC	Inseparable
AC,AB	Separable
AB,C	Inseparable
AB	Separable
A,B,C	Inseparable
A,B	Inseparable
C	Separable

The only non–trivial case of separability for a three–way table is AC, AB. In this case there are I inseparable sub–tables corresponding to the I levels of variable A. As already noted in section 5.4, the test statistic for conditional independence of C and B given A can be decomposed in I components according to the levels of variable A. This result has general validity. If the set of generators for a hypothesis has a common group of variables then the transformed log–likelihood ratio test statistic can be decomposed according to the combined levels of this group of variables. Such a decomposition correponds to the inseparable subtables of the original table.

Separability for a complete table is a graphical property if the model is decomposable. The rule is that a given decomposable model can be decomposed according to a given set of variables if on the association graph for the model all variables are fully connected to the set. Consider e.g. the association graph in fig. 7.2.

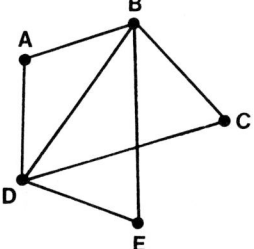

Fig. 7.2. Association graph of a five–dimensional contingency table.

The decomposable model corresponding to fig. 7.2 is

(7.8)   ABD, BCD, BDE.

The common group of variables for the generating set is BD, such that the table under this model can be decomposed according to the cells of the marginal table of B and D. This is in accordance with the interpretation

(7.9) $A \otimes C \otimes E \,|\, B,D.$

of the model. The common set BD and hence the relevant decomposition is easily identified on the association graph. If the model is not decomposable, these results do not hold. The model

$$AB,AD,BCD,BE,DE$$

thus have the same interpretation (7.9) as (7.8), but the intersection of these marginals is empty and no decomposition is possible.

Formula (7.7) is also true for complete tables. Consider for example in a complete three–way table the model AB,AC. Here $A_1 = AB$ and $A_2 = AC$,

$$s = I$$
$$N(A_1) = IJ$$
$$N(A_2) = IK$$

and the number of degrees of freedom for testing the model AB, AC against the saturated model is

$$IJK - IJ - IK + I = I(J-1)(K-1),$$

which is the correct number.

7.4. Collapsibility

For many tables the statistical analysis of a subset of variables can be based on a contingency table only involving the variables in the subset. This property is called **collapsibility** with respect to the subset, and we say that the table can be collapsed onto the marginal table of the variables in the subset. The most simple example of collapsibility is the model for independency in a two–way table. In a two–way table formed by variables A and B, the generators under the independence hypothesis are A and B. As we saw in section 4.5 hypotheses concerning τ_i^A can then be studied in the marginal table of $x_{1.},...,x_{I.}$ and hypotheses concerning τ_j^B in the marginal table of $x_{.1},...,x_{.J}$. The table can thus be collapsed onto the marginal table of A in order to study properties of variable A and onto the marginal table of B in order to study properties of variable B. There are several definitions of collapsibility stressing various aspects of a statistical analysis. It is most consequent, however, to say that a contingency table under a given model can be collapsed onto a set of variables $A_1,...,A_r$ if the expected values of the cell counts are the same when derived from the full table as when derived from the marginal table of $A_1,...,A_r$. If a table is collapsible onto $A_1,...,A_r$ according to this definition, it follows that ML–estimates for the log–linear parameters of the table and the transformed likelihood ratio test for the goodness of fit of the model will also be the same. To make the definition precise it is necessary to define how the equivalent of a given model is defined in the collapsed table. The rule is that if all variables not in the set $A_1,...,A_r$ are removed from the generators of the model, and the new set of generators is reduced to its minimal form, then these are the generators of the derived model. Consider e.g. the five–dimensional table formed by variables A,B,C,D and E, and assume that we under the model ABE, ACD,ADE want to collaps onto the marginal table formed by A,B and C. When D and E are removed, the new generators are AB,AC,A. Reduced to minimal form the generators for the model in the collapsed table are thus AB and AC.

Definition 7.3.: An m–way contingency table composed of variables $A_1,...,A_m$ can under the log–linear model H be collapsed onto the variables $A_1,...,A_r$, if the expected numbers estimated under H in the full model are the same as those estimated under the derived model in the marginal table of $A_1,...,A_r$.

There are, however, other definitions. Thus Wermuth (1987) discussed a definition, where a three–dimensional table is collapsible if certain statistical measures as odds ratios are preserved is when the table is collapsed onto a two–dimensional table.

A survey of different definitions of collapsibility and further results can be found in Kreiner (1987).

A not entirely trivial example is the three–way table formed by A,B and C with generators AC, BC. The expected numbers under the full model are

$$(7.10) \qquad \hat{\mu}_{ijk} = x_{i.k} x_{.jk} / x_{..k}$$

The derived model for the table collapsed onto the marginal table of B and C has generator BC. For the collapsed model the expected numbers are thus

$$\hat{\mu}_{.jk} = x_{.jk},$$

The same result is, however, obtained by direct summation in (7.10). Hence the table is collapsible onto the marginal table of B and C.

Collapsibility is a graphical property for graphical models. The following theorem is due to Asmussen and Edwards (1983). On the association graph of the model the **boundary** ∂G of a set of variables G is the set of variables A for which $A \notin G$ and A is connected with at least one variable in G. Consider for example fig. 7.2. Here the boundary of $G = \{E\}$ is $\partial G = \{B,D\}$ and the boundary of $G = \{B,D\}$ is $\partial G = \{A,C,E\}$.

Theorem 7.1.

A hierarchical model H for an m–way contingency table is collapsible onto $A_1, \ldots, A_r$ if and only if the boundary of every connected component G_i of $\{A_{r+1}, \ldots, A_m\}$, is contained in a generator of H.

It follows immediately from this theorem that a table is always collapsible onto any subset of its generators. It is also obvious, that it is very easy to check whether a table is collapsible under a given model. Consider the following four models for a four way table which have the graphical representations shown in fig. 7.3.

  (a) AB,AC,AD,BC,BD,CD

  (b) ABC,ABD,CD

  (c) ABC,AD

  (d) AB,AC,BC,AD

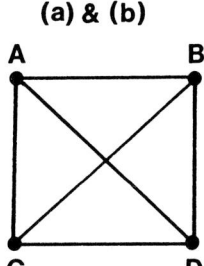
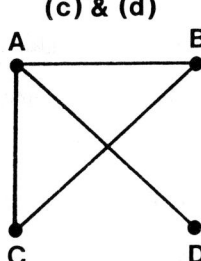

Fig. 7.3. Graphs of four hierarchical models for a four–way table.

For model (a) the table can not be collapsed onto A,B,C because the complement of $\{A,B,C\}$ is $\{D\}$, for which the boundary is $\partial\{D\}=\{A,B,C\}$. But ABC is not contained in a generator and the table cannot be collapsed onto D. In contrast (b) is collapsible onto A,B,C since now ABC is a generator of the model. Both model (a) and model (b) are collapsible onto A,B since $\partial\{C,D\}=\{A,B\}$ and AB is contained in a generator for both models. For models (c) and (d) the check of collapsibility onto A,B is slightly different, because C and D are disconnected in the set $\{C,D\}$. Hence the boundaries of both $\{C\}$ and $\{D\}$ should be checked. For (c) as well as (d), $\partial\{C\}=\{A,B\}$ and $\partial\{D\}=\{A\}$ are contained in generators of the model, such that the table is collapsible onto A,B. Consider next collapsibility onto C,D. The complement of $\{C,D\}$ is $\{A,B\}$ and $\partial\{A,B\}=\{C,D\}$, but CD is not contained in the generator of any of the models (c) and (d). The table is not accordingly collapsible onto the marginal table of A and B. The intuitive reason for this result is that there is a dependency between C and D through A and B, which will not reveal itself in the marginal table of just C and D.

The most famous example of lack of collapsibility is onto A,B in the three–way table with generators AC,BC. Here A and B are independent given C, but may well be, and often are, dependent in the marginal table of A and B.

Example 7.3.

Consider again the cancer survival data in table 6.3. For this data set a plausible model has generators

$$ACD, BD,$$

such that only the stage of the cancer (variable D) influences the chance of survival (variable B). The interactions between x–ray treatment (A), extensiveness of operation (C) and stage (D) reflects the fact that a decision on what treatment and kind of operation to use naturally will depend on the stage of the cancer. For this model we can without loss of information collaps onto the sub-tables of variables A,C,D and B,D, respectively. For A,C,D the complement is {B}, for which the boundary is $\partial\{B\}=\{D\}$ as illustrated in fig. 7.4.

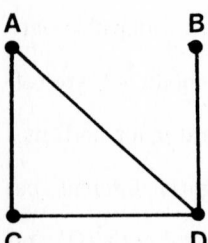

Fig. 7.4. The association graph for model ACD,BD.

Variable D is contained in both the generators, hence we can collaps onto the marginal table of A,C and D. As regards B,D, the complement is {A,C}, which is connected in fig. 7.4. The boundary of {A,C} is $\partial\{A,C\}=D$, which again is contained in both generators. Hence we can also collaps onto the marginal table of B and D. In the marginal table of A,C and D, the derived model has generator ACD, such that the model is saturated.

Hence the expected numbers are equal to the observed numbers. The marginal table of A,C and D is shown in table 7.10. The expected numbers for the four–dimensional table under model ACD,BD are shown in table 7.11. It is easy to verify that the marginals for variable combination A,C,D computed in table 7.11 are equal to the numbers in table 7.10, which is the condition for collapsibility.

Table 7.10. The marginal table for variables A,C and D and the data in table 6.1.

D: Stage	C:Operation	A: X–ray No	Yes
Early	Radical	51	81
	Limited	14	12
Advanced	Radical	44	75
	Limited	4	18

Table 7.11. The expected numbers for the data in table 6.1 under the model ACD, BD.

D: Stage	C: Operation	B: Survival	A: X–ray No	Yes
Early	Radical	No	10.0	15.9
		Yes	41.0	65.1
	Limited	No	2.7	2.4
		Yes	11.3	9.6
Advanced	Radical	No	36.8	62.8
		Yes	7.2	12.2
	Limited	No	3.3	15.1
		Yes	0.7	2.9

When collapsed onto the marginal table of variables B and D, the derived model has generator BD and is thus again saturated. The expected numbers are, therefore, the observed marginals shown in table 7.12.

Table 7.12. The marginal table for variables B and D and the data in table 6.1.

D: Stage	B: Survival No	Yes
Early	31	127
Advanced	118	23

Again it is easy to verify that the condition for collapsibility onto variables B an D is satisfied by checking that the marginals for the combination B,D obtained from table 7.11 are identical with the numbers in table 7.12.△.

7.5. Exercises

7.1. Consider the three–way table below, where an "x" denote a positive count and "0" a zero.

A	B	C k=1	2	3
i=1	j=1	x	0	x
	2	0	x	x
2	j=1	x	0	x
	2	x	x	x
3	j=1	x	0	0
	2	0	0	0
4	j=1	x	0	x
	2	x	x	x

(a) Count the degrees of freedom for the goodness of fit test statistics for the hypothesis with sufficient marginal

(1) AB,AC,BC
(2) AB,AC
(3) AC,BC
(4) AB,C

(b)   There are four unconstrained two–factor interactions between variables A and C, which can be estimated. If τ^{AC}_{11}, τ^{AC}_{12}, τ^{AC}_{21}, τ^{AC}_{22} are unconstrained, derived the remaining t^{AC}'s as functions of these four values.

(c)   Are matters simplified if the x in cell (3,1,1) is changed to a zero? How? Why?

7.2. Consider the very sparse three–way contingency table below, where an "x" means a positive cell count and "0" a structural zero.

(a)   Show that all three–factor interactions and all two–factor interactions between variables A and C are zero.

(b)   Compute the degrees of freedom for the hypothesis with sufficient marginals AB,C.

A	B		C	
		k=1	2	3
i=1	j=1	0	x	0
	2	0	x	0
	3	0	x	0
2	j=1	0	x	x
	2	0	0	x
	3	0	0	x
3	j=1	x	x	0
	2	x	x	0
	3	0	0	0

7.3. For the survival data in example 6.1, the contingency table was originally a five–way table, where also the pathology of the tumor was recorded. This 5–way table, shown below, is, however, incomplete.

(a)   Count the degrees of freedom for the goodness of fit test statistic for the model AB, AC,AD,AE,BC,BD,BE,CD,CE,DE.

(b)   Identify the three terms in formula (7.2) for question (a).

(c)   Specify for which interactions all terms can be estimated.

A:Stage	B.Type of cancer	C:X—ray	D:Pathology	E:Survival 10 year or less	More than 10 years
Limited	Extensive	No	Localized Spread	1 9	21 20
		Yes	Localized Spread	0 17	23 41
	Limited	No	Localized Spread	0 1	4 9
		Yes	Localized Spread	1 2	2 7
Advanced	Extensive	No	Localized Spread	1 37	3 3
		Yes	Localized Spread	1 63	4 7
	Limited	No	Localized Spread	0 3	0 1
		Yes	Localized Spread	0 13	1 4

7.4. Svalastoga (1959) collected data on the relationship between the social rank of father and son. Categorized in 5 social rank classes I to V, the data are shown in the table below.

Father social rank	Sons social rank				
	I	II	III	IV	V
I	18	17	16	4	2
II	24	105	109	59	21
III	23	84	289	217	95
IV	8	49	175	348	198
V	6	8	69	201	246

(a) Test the hypothesis of quasi—independence on these data.

(b) Does the analysis suggest other plausible models?

7.5. Consider a 5–way contingency table and assume that a log–linear model with sufficient marginals

AC,BC,CDE

fits the data.

(a)   Is this table separable when all cells have positive count? Why?

(b)   Describe the decomposition, which corresponds to the separability of the table.

7.6. Consider the 5–way formed by the variables A,B,C,D and E, and assume that the model with sufficient marginals

AC,BC,CDE

fits the data.

(a)   Is it possible to collaps the table onto:

(1) The two–way table formed by C and D.

(2) The two–way table formed by B and D.

(3) The marginal table of D.

(b)   For each of the three models in (a) derive the model corresponding to AC,BC,CDE.

(c)   The model AC,BC,CD,CE,DE has the same graf as AC,BC,CDE. Is the table collapsable onto the same subtables in (a) as AC,BC,CDE.

(d)   Find a two–way table for which the 5–way table is collapsable under AC,BC,CDE, but not under AC,BC,CD,CE,DE. Is there an intuitive explanation for this?

7.7. Consider a four–way table formed by A,B,C and D assume that the model with sufficient marginals AB,CD,B fits the data.

(a)   Determine for all three– and two–way subtables if the table is collapsable onto them.

(b)   Are there intuitive reasons for the results in (a)?

7.8. Reconsider the data in table 6.3.

(a)   Are the zeros in the table random or structural?

(b)   Clearly the zeros have an effect on what four–factor interactions can be estimated. A computer program prints out that the degrees of freedom for testing the model where all four–factor interactions are zero against the saturated model is 40. How is this number calculated?

(c)   Are the any three–factor interactions which cannot be estimated?

(d)   Compute the degrees of freedom for the hypothesis AB,AC,AD,BC,BD,CD.

8. The Logit Model

8.1. The logit model with binary explanatory variables

In chapters 4, 5 and 6 the categorical variables appeared in the model in a symmetrical way. In many situations, for example in examples 6.1 and 6.2 in chapter 6, one of the variable is of special interest. For the survival data in example 6.1, survival is the variable of special interest, and the problem is to study if the other three variables have influenced the chance of survival. Variable B in example 6.1 may, therefore, be called a **response variable** and variables A, C and D **explanatory variables**. This terminology is the same as the one used in regression analysis, and when survival is regarded as a response variable the data in example 6.1 can in fact be analysed by a regression model. In example 6.2 the position on the truck of the collision can be regarded as a response variable. We are here primarily interested in the effect of explanatory variable A, i.e. the introduction of the safety measure in November 1971, but have to take into account that the other explanatory variables, i.e. whether the truck was parked or not and what the light conditions were, may be of importance for the location of the collision. When the response variable is binary and the explanatory variables are categorical, the appropriate regression model is known as the **logit model**. More precisely the assumptions for a logit model are:

(a) The response variable is binary.

(b) The contingency table formed by the reponse variable and the explanatory variables can be described by a log–linear model.

Let the two levels of the response variable be denoted 1 and 2. The variation of the response is then conveniently described as a function of the explanatory variables through the conditional probability

$$p_{1|i_2\cdots i_m} = P\left\{\text{Variabel A at level 1}\,|\,\text{variables B,...,S at levels } i_2,\ldots,i_m\right\}$$

If $p_{i_1 i_2 \cdots i_m}$ is the probability under the multinomial model of observing an individual in cell $(i_1,\ldots,i_m)$ then

$$p_{1|i_2\cdots i_m} = p_{1i_2\cdots i_m}\big/\,(p_{1i_2\cdots i_m} + p_{2i_2\cdots i_m}).$$

The transformation

(8.1)
$$\text{logit}\,(p_{1|i_2\cdots i_m}) = \ln\frac{p_{1|i_2\cdots i_m}}{p_{2|i_2\cdots i_m}} = \ln\frac{p_{1i_2\cdots i_m}}{p_{2i_2\cdots i_m}}$$

is called the logit corresponding to the levels $i_2,\ldots,i_m$ of the explanatory variables.

In general the **logit–transformation** $y=\text{logit}(x)$ is defined as

$$y = \ln\frac{x}{1-x}.$$

Thus if $x=p_1/(p_1+p_2)$, we get $1-x=p_2/(p_1+p_2)$ and

$$y = \text{logit}(x) = \ln(p_1/p_2).$$

The logit–function is monotone with range $(-\infty,+\infty)$. Its graph is shown in fig. 8.1. High values of the logit correspond to x close to 1 while low values correspond to x close to 0. It follows that the logit (8.1) attains a high value if the observed levels $i_2,\ldots,i_m$ of the explanatory variables are most likely to occur together with the value 1 of the response variable, while the logit has a high negative value if the combination $i_2,\ldots i_m$ is most likely to occur together with the value 2 of the response variable. It follows that the logit is an important statistical tool for a study of the influence of the explanatory variables on the response variable.

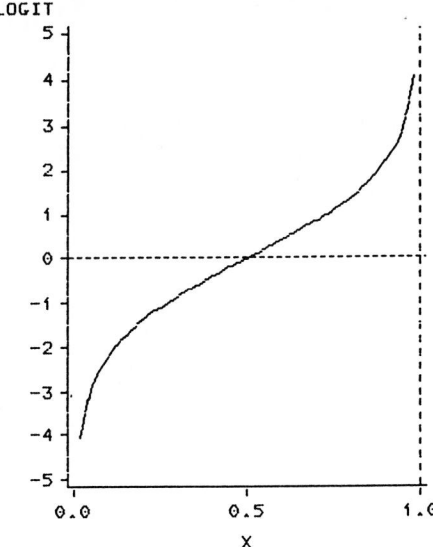

Fig. 8.1. The logit–transformation.

To avoid cumbersome expressions, we limit attention to the case with one response variable and three explanatory variables. Before formulating the logit–model in general as a regression model, consider the following special case:

Let a log–linear model be defined as

(8.2)
$$\ln\mu_{ijkl} = \ln(np_{ijkl})$$

$$= \tau_0 + \tau_i^A + \tau_j^B + \tau_k^C + \tau_l^D + \tau_{ij}^{AB}$$

$$+ \tau_{jk}^{BC} + \tau_{il}^{AD} + \tau_{jl}^{BD} + \tau_{ijl}^{ABD}.$$

The interpretation of this model is $C \otimes A,D \,|\, B$ with association graph shown in fig. 8.2

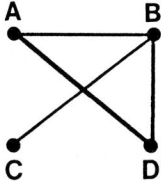

Fig. 8.2. Association graph for the model $C \otimes A,D \,|\, B$.

242

The logit with A as response variable is then

$$(8.3) \qquad \ln \frac{P_{1jkl}}{P_{2jkl}} = (\tau^A_1 - \tau^A_2) + (\tau^{AB}_{1j} - \tau^{AB}_{2j}) + (\tau^{AD}_{11} - \tau^{AD}_{21}) + (\tau^{ABD}_{1j1} - \tau^{ABD}_{2j1})$$

$$= 2[\tau^A_1 + \tau^{AB}_{1j} + \tau^{AD}_{11} + \tau^{ABD}_{1j1}],$$

since terms not containing A cancel out and

$$\tau^A_1 + \tau^A_2 = 0, \ \tau^{AB}_{1j} + \tau^{AB}_{2j} = 0, \ \text{etc.}$$

The logit thus has two properties:

(a) The logit only depends on the main effect for the response variable and on interactions between the response variable and the explanatory variables.

(b) The logit is twice the sum of those non−null interactions, which involve the response variable.

Property (a) means that if only the effects of the explanatory variables on the response are of interest, attention can be restricted to those interactions involving the response variable. This limit the number of models to include in the model search considerably. If the logit in (8.3) is denoted by g_{jkl}, then

$$g_{jkl} = 2[\tau^A_1 + \tau^{AB}_{1j} + \tau^{AD}_{11} + \tau^{ABD}_{1j1}].$$

It is convenient to introduce the parameters $\beta_0 = 2\tau^A_1$, $\beta^B_j = 2\tau^{AB}_{1j}$, $\beta^D_1 = 2\tau^{AD}_{11}$ and $\beta^{BD}_{j1} = 2\tau^{ABD}_{1j1}$. Then g_{jkl} has the linear form

$$(8.4) \qquad g_{jkl} = \beta_0 + \beta^B_j + \beta^D_1 + \beta^{BD}_{j1}.$$

The expression (8.4) shows that the model (8.2) can be formulated as the logit being a linear expression in parameters connected with explanatory variables B,C and D.

In the following the assigment of letters A, B, C and D to the variables is always such that A is the response variable.

In order to show that (8.4), and in general the logit model, can be expressed as a regression model, consider next the case, where the model only contain main effects and two–factor interactions involving the response variable, i.e. the model

(8.5) $$\ln(np_{ijkl}) = \tau_0 + \tau_i^A + \tau_{ij}^{AB} + \tau_{ik}^{AC} + \tau_{il}^{AD} + \{\text{terms in B, C and D}\}.$$

The logit of (8.5) is

(8.6) $$g_{jkl} = \ln \frac{p_{1jkl}}{p_{2jkl}} = 2\tau_1^A + 2\tau_{1j}^{AB} + 2\tau_{1k}^{AC} + 2\tau_{1l}^{AD}.$$

with the β–notation, introduced above, the logit can then be written

$$g_{jkl} = \beta_0 + \beta_j^B + \beta_k^C + \beta_l^D.$$

Since the explanatory variables B, C and D are assumed to be binary, $\tau_{12}^{AB} = -\tau_{11}^{AB}$, $\tau_{12}^{AC} = -\tau_{11}^{AC}$, and $\tau_{12}^{AD} = -\tau_{11}^{AD}$. Hence according to (8.6)

(8.7) $$g_{jkl} = \beta_0 + \beta^B z(j) + \beta^C z(k) + \beta^D z(l),$$

with $\beta^B = \beta_1^B$, $\beta^C = \beta_1^C$, $\beta^D = \beta_1^D$, $z(1)=1$ and $z(2)=-1$. The logit model is thus a multiple regression model, based on which the influence on the response variable of a given combination of levels of the explanatory variables can be studied. Consider e.g. two vectors

(a) $(j,k,l) = (1,1,2)$

and

(b) $(j,k,l) = (2,1,1)$

The values of the logit g_{jkl} for these combinations of j, k and l are

$$\text{(a)} \quad g_{112} = \beta_0 + \beta^B + \beta^C - \beta^D$$

and

$$\text{(b)} \quad g_{211} = \beta_0 - \beta^B + \beta^C + \beta^D.$$

Hence the effect on the response variable of a given set of observed values of the explanatory variables is fully described by the values of β's. Given the logit, g_{jkl}, the effect of the explanatory variables on the probability p_{1jkl} of observing the response variable at level 1 given the values i, j and k can be studied, since

$$(8.8) \qquad p_{1|jkl} = e^{g_{jkl}} / (1 + e^{g_{jkl}}).$$

From (8.8) follows that if for a given combination of j, k and l the probabilities of observing the response variable at levels i=1 and i=2 are equal, then $g_{jkl}=0$, while $g_{jkl}>0$ if $p_{1jkl} > p_{2jkl}$ and $g_{jkl}<0$ if $p_{1jkl}<p_{2jkl}$. The logit for given values of the explanatory variables thus measures the relative chance of observing response i=1 rather than i=2, and the larger the value of g_{jkl} the larger the chance of observing i=1.

One reason for chosing the logit rather than the conditional probabilities directly to measure the effect of the explanatory variables is that the logit has the complete real line as its range space. Had we instead chosen a model like

$$p_{1jkl} = \beta_0 + \beta^B z(j) + \beta^C z(k) + \beta^D z(l)$$

or

$$\ln p_{1jkl} = \beta_0 + \beta^B z(j) + \beta^C z(k) + \beta^D z(l)$$

certain combinations of j, k and l might for certain values of β^B, β^C or β^D bring us outside the range space of p_{1jkl} or $\ln p_{1jkl}$, while this can not happen for the logit model. One further desirable property of the logit transformation, not possessed for example by $g^*_{jkl} = \ln p_{1jkl}$, is symmetry in i=1 and i=2. In fact if the two values of i are interchanged, then g_{jkl} just changes sign, since

$$\ln \frac{p_{1jkl}}{p_{2jkl}} = -\ln \frac{p_{2jkl}}{p_{1jkl}},$$

and the model structure is unchanged.

Since the logit–model is merely a reformulation of the log–linear model for a multi–way contingency table, no new statistical tools need to be introduced. The analysis is, however, simplified by the fact that only interactions which involve the response variable have to be accounted for. This means that even for higher order tables, a simple strategy can be set up for the search for a satisfactory logit–model. In addition logit–models, which involves three–factor or higher order interactions between the response variable and two or more explanatory variables are difficult to give a simple interpretation. Hence a natural starting point for a model search is the model (8.5).

Assuming that β^D is most likely and β^C next most likely to be zero consider thus the hypotheses

$$H^{(1)}: \beta^D = 0$$
$$H^{(2)}: \beta^C = \beta^D = 0$$

and

$$H^{(3)}: \beta^B = \beta^C = \beta^D = 0.$$

Both the necessary test statistics for testing these hypotheses and the parameter estimates under the hypotheses are derived directly from the corresponding quantities for log–linear model of the four–way contingency table formed by the response variable A, and the explanatory variables B, C and D. The assumptions for the logit model (8.6) is that

$$\tau^{ABC}_{1jk} = \tau^{ABD}_{1jl} = \tau^{ACD}_{1kl} = \tau^{ABCD}_{1jkl} = 0.$$

Hence (8.6) is equivalent with the log–linear model with sufficient marginals

(8.9)   H: AB,AC,AD,BCD

Note here that nothing is assumed about the interactions between B, C, and D. Given this model the hypotheses $H^{(1)}$, $H^{(2)}$ and $H^{(3)}$ corresponds to the log–linear models

$$H^{(1)}: AB,AC,BCD$$
$$H^{(2)}: AB,BCD$$
$$H^{(3)}: A,BCD$$

It follows that $H^{(2)}$ can be tested against $H^{(1)}$ by means of the test statistic

(8.10) $$Z(H^{(2)}|H^{(1)}) = Z(H^{(2)})-Z(H^{(1)})$$

from chapter 6 applied to the four–way table formed by A,B,C and D. Similarly $H^{(3)}$ can be tested against $H^{(2)}$ by means of

(8.11) $$Z(H^{(3)}|H^{(2)}) = Z(H^{(3)}) - Z(H^{(2)}).$$

If the observed value of (8.10) is small then $H^{(2)}$: $\beta^C=\beta^D=0$ is accepted. If in addition the observed value of of (8.11) is small, then $H^{(3)}$:$\beta^B=\beta^C=\beta^D=0$ is accepted.

In order to test $H^{(1)}$:$\beta^D=0$, we can choose H given by (8.9) as alternative, such that $H^{(1)}$ is accepted if the observed value of

$$Z(H^{(1)}|H) = Z(H^{(1)}) - Z(H)$$

is small. Finally the logit model is accepted as an adequate description of the data if H tested against the saturated model is accepted.

The hypothesis

$$H^{(0)}: \beta^B = \beta^C = \beta^D = \beta_0 = 0,$$

is within the log–linear model framework equivalent to

$$H^{(0)}: BCD, A = u.$$

The ML–estimates for the β's are obtained from the ML–estimates of the τ's through the definition of the β's in terms of the τ's, i.e.

$$\hat{\beta}_0 = 2\hat{\tau}_1^A, \quad \hat{\beta}^B = 2\hat{\tau}_{11}^{AB}, \quad \hat{\beta}^C = 2\hat{\tau}_{11}^{AC}, \quad \hat{\beta}^D = 2\hat{\tau}_{11}^{AD}.$$

Example 8.1.

The data in table 8.1 is a cross–classification of the Danish Welfare Study according to the following four binary variables

A: Member of political party or not

B: Sex

C: Whether employed in the public or private sector

D: Whether living in Copenhagen or not.

For shortness the variables are denoted A: Membership, B: Sex, C: Sector and D: Urbanization. We shall analyse the data in table 8.1 by a logit–model with A as the response variable and B,C and D as explanatory variables. The observed goodness of fit test statistic for the model with sufficient marginals AB,AC,AD,BCD is

$$z = 0.93, \quad df=4,$$

leading to clear acceptance of the logit model

$$g_{jkl} = \beta_0 + \beta^B z(j) + \beta^C z(k) + \beta^D z(l).$$

Table 8.1. The Danish Welfare Study cross–classified according to membership of a political party, sex, employment sector and urbanization.

A: Member of party	B: Sex	C: Employment sector	D: Living in Copenhagen	
			Yes	No
Yes	Female	Public	12	31
		Private	5	20
	Male	Public	16	37
		Private	19	73
No	Female	Public	175	375
		Private	162	475
	Male	Public	111	266
		Private	241	906

Table 8.2 show the next steps in the analysis of the data by a logit model. The table display the log–linear models as identified by their sufficient marginals, the hypotheses in terms of the β's, the observed values of Z(H) for each hypotheses and the observed value of the test statistic $Z(H|H_A)$ for the hypothesis H given that all previous hypotheses have been accepted.

Table 8.2. Test statistics for an analysis of the data in table 8.1 by a logit–model, with p being the level of significance.

Log–linear model	Hypothesis	z(H)	df	p	$z(H|H_A)$	df	p
AB,AC,AD,BCD		0.64	4	0.958	0.64	4	0.958
AB,AC,BCD	$\beta^D=0$	0.87	5	0.973	0.22	1	0.637
AB,BCD	$\beta^D=\beta^C=0$	17.14	6	0.009	16.27	1	0.000
A,BCD	$\beta^D=\beta^C=\beta^B=0$	28.80	7	0.000	11.56	1	0.000

The most restrictive model to fit the data is AB,AC,BCD. The logit for this model is

$$g_{jkl} = \beta_0 + \beta^B z(j) + \beta^C(k).$$

The variables Sex and Sector thus influence the response variable, while Urbanization does not. Membership of a political party thus depends on sex and on whether the person is publicly or privately employed. Membership of a political party seems, however, to be as common in Copenhagen as outside Copenhagen. The estimates of the parameters for β_0, β^B and β^C in the final model are (with standard errors in parantheses):

$$\hat{\beta}_0 = 2\hat{\tau}_1^A = -2.560 \ (0.076)$$

$$\hat{\beta}^B = 2\hat{\tau}_{11}^{AB} = -0.324 \ (0.078)$$

$$\hat{\beta}^C = 2\hat{\tau}_{11}^{AC} = 0.302 \ (0.074).$$

The signs of the estimates for β^B and β^C show that membership of a political party is more frequent for employees in the public sector and more frequent among men.

For given values of the explanatory variables, the probability of membership can be estimated. Thus j=1, k=1 is a publicly employed woman and j=2, k=1 is a publicly employed man. For these combinations

$$\hat{g}_{11} = -2.582$$

and

$$\hat{g}_{21} = -1.934.$$

The corresponding probabilities of membership are

$$\hat{p}_{1|11} = 0.070$$

and

$$\hat{p}_{1|21} = 0.126.$$

The probability of a randomly chosen person being political active is thus in general small, but significantly lower for women than for men. △.

If the study is cross–sectional or prospective, i.e. if

$$E[X_{1jkl}] = np_{1jkl}$$

or

$$E[X_{1jkl}] = n_1 p_{1|jkl},$$

estimates of the probabilities $p_{1|jkl}$ of predicting i=1 given the values of the explanatory variables can be derived from the estimated logits through (8.8). This is not the case if the study is retrospective. In this case

$$E[X_{ijkl}] = n_i p_{jkl|i}, \quad i=1,2,$$

where n_i is the number of individuals selected with the response variable at level i and

$$p_{jkl|i} = P\{\text{variables B,C and D at levels j,k and l} | \text{variabel A at level i}\}.$$

Hence the logit is given by

$$g_{jkl} = \ln(\mu_{1jkl}) - \ln(\mu_{2jkl}) = \ln(n_1/n_2) + \ln(p_{jkl|1}/p_{jkl|2}).$$

But

$$p_{jkl|i} = p_{ijkl}/p_{i...},$$

such that

$$g_{jkl} = \ln(p_{1jkl}/p_{2jkl}) + \ln(n_1/n_2) + \ln(p_{2...}/p_{1...}).$$

The logit for a cross-sectional or prospective study is equal to the first term is this expression according to (8.3). The two last terms do not depend on j,k and l, however, such that an analysis by a logit-model for a retrospective study will thus only differ from an analysis based on a cross-sectional study or a prospective study by yielding a different parameter β_0. The evaluation of the influence of the explanatory variables can thus be summarized and interpreted in the same way. It is not, however, possible to estimate the probability $p_{1|jkl}$ from the logit, if $p_{1...}$ is unknown, since

$$p_{1|jkl} = e^{g_{jkl}} \frac{n_2}{n_1} \frac{p_{1...}}{1-p_{1...}} / [1 + e^{g_{jkl}} \frac{n_2}{n_1} \frac{p_{1...}}{1-p_{1...}}].$$

The only quantity, which is independent of $p_{1...}$ and thus can be estimated from the logit is the odds ratio

$$(8.12) \qquad \frac{p_{1|jkl} \cdot p_{2|j'k'l'}}{p_{2|jkl} \cdot p_{1|j'k'l'}} = \exp(g_{jkl} - g_{j'k'l'})$$

The odds–ratio (8.12) is the ratio between the odds of i=1 as compared with i=2 given the explanatory variables at levels j,k,l and at levels j',k',l'. The odds ratio (8.12) has the same interpretation as in section 4.2. The ratio is 1 if the explanatory variables do not influence the response variable, and the larger the value of the odds ratio, the more likely is the response i=1 given levels j,k,l of the explanatory variables than it is given levels j',k',l'.

Example 8.2.

In example 6.1 we studied a four–way table between binary variables. Let now the letters A to D be assigned as follows

A: Survival

B: Stage of cancer

C: Mode of operation

D: X–ray treatment

The data in table 6.4 can then be analysed by a logit–model with A as the response variable and stage of cancer, x–ray treatment and mode of operation as explanatory variables. The observed goodness of fit test statistic for the model with sufficient marginals AB, AC,AD,BCD is

$$z = 1.93, \ df = 4,$$

leading to a clear acceptance of the logit model

$$g_{jkl} = \beta_0 + \beta^B z(j) + \beta^C z(k) + \beta^D z(l).$$

Table 8.3 shows the next steps in the analysis of the survival data by a logit model. The table display the log–linear model as identified by its sufficient marginals, the hypoth–

eses in terms of the β's, the observed values of $Z(H)$ for each hypotheses and the observed value of the test statistic $Z(H|H_A)$ for the hypothesis H given that all previous hypotheses have been accepted.

Table 8.3. Test statistics for an analysis of the survival data by a logit–model.

Log–linear model	Hypothesis	z(H)	df	p	z(H|H$_A$)	df	p
AB,AC,AD,BCD		1.93	4	0.748	1.93	4	0.749
AB,AC,BCD	$\beta^D=0$	2.02	5	0.846	0.09	1	0.764
AB,BCD	$\beta^D=\beta^C=0$	4.12	6	0.661	2.10	1	0.147
A,BCD	$\beta^D=\beta^C=\beta^B=0$	136.73	7	0.000	132.61	1	0.000

The most restrictive model to fit the data is AB,BCD. The logit for this model is

$$g_{jkl}= \beta_0 + \beta^B z(j).$$

Only the stage of the cancer thus influences the chance of survival. The estimates of β^B and β_0 are

$$\hat{\beta}^B = 2\hat{\tau}^{AB}_{11} = -1.522$$

and

$$\hat{\beta}_0 = 2\,\hat{\tau}^A_1 = -0.112.$$

Hence

$$g_{jkl} = \begin{cases} -1.634 & \text{for } j=1 \\ +1.400 & \text{for } j=2 \end{cases}$$

The data in this example is based on a retrospective study, where $n_1=150$ were selected among those women, who survived and $n_2=149$ among those women, who did not. Hence it is not possible to estimate the probability of survival given the value of the explanatory variable "stage of cancer". In this case the only quantity, which can be estimated is the odds ratio (8.12) between the odds of survival for women with the cancer at early stage and the odds of survival for women with the cancer at an advanced stage. According to (8.12) with j=1 and j'=2 this odds ratio is

$$\frac{p_{1|1}}{p_{2|1}} \cdot \frac{p_{2|2}}{p_{1|2}} = 0.048 \ .$$

Since i=1 is not surviving and j=1 is cancer at an early stage according to table 6.4, the low value of the odds ratio indicate that the odds of not surviving against surviving is much lower with the cancer at an early stage, that at an advanced stage. The chances of survival are as expected much better with the cancer at an early stage. $\triangle$.

For log–linear models with interactions between the response variable and the explanatory variable of higher order than 2, as for example in (8.4), the logit model is formulated as follows. Let as before z(1)=1 and z(2)=−1, but let in addition

$$z(p,q) = \begin{cases} +1 \text{ for p=1, q=1} \\ -1 \text{ for p=1, q=2} \\ -1 \text{ for p=2, q=1} \\ +1 \text{ for p=2, q=2} \end{cases},$$

Then (8.4) can be written as

$$g_{jkl} = \beta_0 + \beta^B z(j) + \beta^D z(l) + \beta^{BD} z(j,l),$$

where $\beta^{BD} = \beta_{11}^{BD}$. That $\beta_{11}^{BD} = \beta^{BD} z(1,1)$ follows directly. Since $\beta_{12}^{BD} = -\beta_{11}^{B}$, because the variables are both binary, it follows that $\beta_{12}^{BD} = \beta^{BD} z(1,2)$. The remaining two cases are verified in the same manner. With score matrices constructed like z(j,l), j=1,..,J, l=1,...,L, the logit can thus be expressed as a regression model. This construction is only possible for combinations of binary variable.

8.2. The logit model with polytomous explanatory variables

The only assumption for the regression formulation (8.7) of the logit model is that the explanatory variables are dichotomous. The scoring $z(1)=1$ and $z(2)=-1$ is thus arbitrary. Suppose for example that $z(1)$ and $z(2)$ are defined as $z(1)=c$ and $z(2)=-c$ for arbitrary c. The value of g_{jkl} in (8.7) is then unchanged if the β's are redefined as $\beta^B=\beta^B/c$, $\beta^C=\beta^C/c$ and $\beta^D=\beta^D/c$. If one or more of the explanatory variables are polytomous the scoring is not, however, arbitrary.

Consider first a situation, where it is possible based on background information to assign scores to the J categories of explanatory variable B, while explanatory variables C and D are dichotomous. Variable B may for example be income with J income intervals. The category scores can then be chosen as the median incomes $z_{21},...,z_{2J}$ in the J intervals. A regression model analogous to (8.7) can now be formulated for the logits as

$$(8.13) \qquad g_{jkl} = \beta_0 + \beta^B z_{2j} + \beta^C z(k) + \beta^D z(l),$$

where as before $z(1)=1$ and $z(2)=-1$ for the binary explanatory variables, $\beta^C=2\tau^{AC}_{11}$ and $\beta^D=\tau^{AD}_{11}$.

The values $z_{21},...,z_{2J}$ are called **score values** for the categories. While β^C and β^D are direct functions of the τ's such that ML–estimates for these parameters are obtained from the ML–estimates for the τ's, this is not the case for β^B, when the categories are scored.

In order to derive the ML–estimate for β^B we need the likelihood function for the model (8.13). The logit is a parameter in the distribution of x_{1jkl} given $x_{1jkl}+x_{2jkl}=x_{\cdot jkl}$, which is a binomial distribution. The likelihood function is accordingly a product of binomial distributions, or

$$(8.14) \qquad L(\beta_0,\beta^B,\beta^C,\beta^D) = \prod_{j=1}^{J} \prod_{k=1}^{2} \prod_{l=1}^{2} \begin{bmatrix} x_{\cdot jkl} \\ x_{1jkl} \end{bmatrix} p_{1|jkl}^{x_{1jkl}} (1-p_{1|jkl})^{x_{2jkl}},$$

with

$$P_{1|jkl} = \frac{p_{1jkl}}{p_{1jkl}+p_{2jkl}} = e^{g_{jkl}}/(1+e^{g_{jkl}}).$$

Hence the log–likelihood function is

$$\ln L(\beta_0,\beta^B,\beta^C,\beta^D) = \underset{j\,k\,l}{\Sigma\Sigma\Sigma}[\beta_0+\beta^B z_{2j}+\beta^C z(k) + \beta^D z(l)]x_{1jkl}$$

$$- \underset{j\,k\,l}{\Sigma\Sigma\Sigma}x_{.jkl}\ln(1+e^{g_{jkl}}) + \{\text{terms in } x\} =$$

$$x_{1...}\beta_0+\beta^B\underset{j}{\Sigma}z_{2j}x_{1j..} + \beta^C(x_{1.1.}-x_{1.2.})+\beta^D(x_{1..1}-x_{1..2})$$

$$- \underset{j\,k\,l}{\Sigma\Sigma\Sigma}x_{.jkl}\ln(1+e^{g_{jkl}}) + \{\text{terms in } x\}.$$

Since the sums $x_{.jkl}$ in the logit model are regarded as fixed, the model is log–linear and the likelihood equations for the estimation of β_0 and β^B become

(8.15)
$$x_{1...} = \mu^*_{1...}$$

and

(8.16)
$$\overset{J}{\underset{j=1}{\Sigma}} z_{2j}x_{1j..} = \overset{J}{\underset{j=1}{\Sigma}} z_{2j}\mu^*_{1j..} \; ,$$

with $\mu^*_{1jkl}=E[X_{1jkl}|x_{.jkl}]=x_{.jkl}P_{1|jkl}$.

The likelihood equations for the estimation of β^C and β^D become

(8.17)
$$x_{1.1.}-x_{1.2.} = \mu^*_{1.1.}-\mu^*_{1.2.}$$

and

(8.18)
$$x_{1..1}-x_{1..2} = \mu^*_{1..1}-\mu^*_{1..2}$$

Due to (8.15), (8.17) and (8.18) can, however, be replaced by

(8.19) $$x_{1.1.} = \overset{*}{\mu}_{1.1.}$$

and

(8.20) $$x_{1..1} = \overset{*}{\mu}_{1..1}$$

Equations (8.15), (8.19) and (8.20) are satisfied if and only if the corresponding likelihood equations for a ML–estimation of the log–linear parameters τ_1^A, τ_{11}^{AC} and τ_{11}^{AD} are satisfied, provided that the likelihood equations for the estimation of τ_{jkl}^{BCD} are satisfied. In fact if

$$x_{.jkl} = np_{.jkl} = n(p_{1jkl} + p_{2jkl}),$$

(8.15) can be written as

(8.21) $$x_{1...} = \underset{j\,k\,l}{\Sigma\Sigma\Sigma} x_{.jkl} p_{1jkl} / (p_{1jkl} + p_{2jkl}) = \underset{j\,k\,l}{\Sigma\Sigma\Sigma} np_{1jkl} = np_{1...}$$

In the same way (8.19) and (8.20) are equivalent with

(8.22) $$x_{1.1.} = np_{1.1.}$$

and

(8.23) $$x_{1..1} = np_{1..1}$$

Equations (8.21), (8.22) and (8.23) are, however, the likelihood equations for the estimation of τ_1^A, τ_{11}^{AC} and τ_{11}^{AD}. Hence it is not necessary to develop new computer procedures to estimate β_0, β^C and β^D. From the ML–estimates for τ_1^A, τ_{11}^{AC} and τ_{11}^{AD}, β_0, β^C and β^D are derived as

$$\hat{\beta}_0 = 2\hat{\tau}_1^A,$$

$$\hat{\beta}^C = 2\hat{\tau}_{11}^{AC}$$

and
$$\hat{\beta}^D = 2\hat{\tau}^{AD}_{1\,1}.$$

Equation (8.16) is new, but is easily solved by numerical methods.

In a saturated model, where the conditional probabilities $p_{1|jkl}$ are unconstrained, the likelihood function is

$$(8.24) \qquad L = \prod_{j=1}^{J} \prod_{k=1}^{2} \prod_{l=1}^{2} \left[\frac{x_{.jkl}}{x_{1jkl}}\right]^{x_{1jkl}} p_{1|jkl}^{x_{1jkl}} (1-p_{1|jkl})^{x_{2jkl}}.$$

Hence the ML–estimate for $p_{1|jkl}$ in the saturated model is

$$(8.25) \qquad \tilde{p}_{1|jkl} = x_{1jkl}/x_{.jkl}.$$

Under the logit model, the ML–estimates of the conditional probabilities are

$$(8.26) \qquad \hat{p}_{1|jkl} = e^{\hat{g}_{jkl}}/(1+e^{\hat{g}_{jkl}})$$

with

$$\hat{g}_{jkl} = \hat{\beta}_0 + \hat{\beta}^B z_{2j} + \hat{\beta}^C z(k) + \hat{\beta}^D z(l).$$

From (8.24) follows that

$$(8.27) \qquad \ln L = \sum_{jkl}(x_{1jkl}\ln p_{1|jkl} + x_{2jkl}\ln(1-p_{1|jkl})) + \{\text{terms in } x\}.$$

From (8.26) follows further that

$$\ln\hat{p}_{1|jkl} = \hat{g}_{jkl} - \ln(1+e^{\hat{g}_{jkl}})$$

and

$$\ln(1-\hat{p}_{1|jkl}) = -\ln(1+e^{\hat{g}_{jkl}}).$$

Hence the transformed likelihood ratio test statistics for the fit of the model becomes

(8.28) $Z = -2\ln L(\hat{p}_{1|111},...,\hat{p}_{1|JKL}) + 2\ln L(\tilde{p}_{1|111},...,\tilde{p}_{1|JKL})$

$$= 2\sum_{jkl}\sum\sum[X_{1jkl}\ln X_{1jkl} + X_{2jkl}\ln X_{2jkl} - X_{.jkl}\ln X_{.jkl}]$$

$$- 2\sum_{jkl}\sum\sum[X_{1jkl}\hat{g}_{jkl} - X_{.jkl}\ln(1+e^{\hat{g}_{jkl}})].$$

According to the general result, theorem 3.9, Z is approximately χ^2–distributed with JKL–4 degrees of freedom, where K=L=2, since there are JKL unconstrained $p_{1|jkl}$'s and 4 β's. Hence the model is rejected if the level of significance

$$p = p(Z \ge z),$$

computed in a χ^2–distribution with JKL–4 degrees of freedom, is small.

The influence of the explanatory variables can be evaluated through a sequential test. The influence of variable D can thus be evaluated by testing the hypothesis

$$H_4: \beta^D = 0.$$

Under H_4 the logit is given by

$$g_{jkl} = \beta_0 + \beta^B z_{2j} + \beta^C z(k).$$

Let $\tilde{\beta}_0, \tilde{\beta}^B$ and $\tilde{\beta}^C$ be the ML–estimates for the parameters under H_4 and let

$$\tilde{p}_{1|jkl} = \exp(\tilde{\beta}_0 + \tilde{\beta}^B z_{2j} + \tilde{\beta}^C z(k))/[1 + \exp(\tilde{\beta}_0 + \tilde{\beta}^B z_{2j} + \tilde{\beta}^C z(k))].$$

Then the test statistic

(8.29) $Z_4 = 2\sum_{jkl}\sum\sum[X_{1jkl}\ln\hat{p}_{1|jkl} + X_{2jkl}\ln(1-\hat{p}_{1|jkl})]$

$$-2\sum_{jkl}\sum\sum[X_{1jkl}\ln\tilde{p}_{1|jkl} + X_{2jkl}\ln(1-\tilde{p}_{1|jkl})]$$

obtained as twice the difference between (8.27) with $p_{1|jkl}$ replaced by $\hat{p}_{1|jkl}$, and (8.27) with $p_{1|jkl}$ replaced by $\tilde{p}_{1|jkl}$ is approximately χ^2–distributed with one degree of freedom. A significant value of Z_4 indicates that H_4 can not be accepted such that variable D contributes to the description of the variation in the response variabel. If the observed value of Z_4 is non–significant, we may exclude variable D from the model without making the fit significantly worse.

If H_4 is accepted, one can go on to test

$$H_3: \beta^D = \beta^C = 0$$

sequentially through Z_3, which is equal to Z_4, when $\tilde{p}_{1|jkl}$ is the estimated value under H_3 and $\hat{p}_{1|jkl}$ the estimated value under H_4. Z_3 is also approximately χ^2–distributed with one degree of freedom.

In practice these sequential tests are easily obtained as differences from a list of (8.28) with $\hat{g}_{jkl}$ being the estimated logits under the model.

For models with a few explanatory varables it is often relatively easy to set up a reasonable sequence in which to check the inclusion or exclusion of the explanatory variables. If many explanatory variables are under consideration it is more difficult to plan a sequential procedure for testing the partial influence of the explanatory variables.

In many situations it is unrealistic to assume that the categories of an explanatory variable can be assigned score values. Even in this situation a regression model may be formulated. The method is related to what is often termed the **dummy variable method** in ordinary regression analysis.

Consider e.g. a model where B is polytomous, while C and D are dichotomous. The problem is then to formulate a regression model, where all the interactions τ_{1j}^{AB}, j=1,...,J between the response variable and explanatory variable B are identifiable in the model. This can be done through the J **dummy variables**

(8.30)
$$z_\nu(j) = \begin{cases} 1 & \text{if } j=\nu \\ 0 & \text{if } j \neq \nu \end{cases}, \; \nu, j=1,\ldots,J-1$$

and $z_\nu(J)=-1$ for $\nu=1,\ldots,J-1$.

The logit for the model is

(8.31)
$$g_{jkl} = 2\tau_1^A + 2\tau_{1j}^{AB} + 2\tau_{1k}^{AC} + 2\tau_{11}^{AD}.$$

Since C and D are binary, the last two terms can be written

$$2\tau_{1k}^{AC} = \beta^C z(k)$$

and

$$2\tau_{11}^{AD} = \beta^D z(l)$$

with $\beta^C=2\tau_{11}^{AC}$, $\beta^D=2\tau_{11}^{AD}$, $z(1)=1$ and $z(2)=-1$. The second term on the right hand side of (8.31) can be written as

(8.32)
$$2\tau_{1j}^{AB} = \sum_{\nu=1}^{J-1} \beta_\nu^B z_\nu(j),$$

with $\beta_\nu^B =2\tau_{1\nu}^{AB}$. For $j=1,\ldots,J-1$ this is obvious from (8.30). For $j=J$, (8.32) becomes

$$2\tau_{1J}^{AB} = -\sum_{\nu=1}^{J-1} \beta_\nu^B = -2\sum_{\nu=1}^{J-1} \tau_{1\nu}^{AB}$$

which is true because $\tau_1^{AB}=0$. Collecting the results, we have

(8.33)
$$g_{jkl} = \beta_0 + \sum_{\nu=1}^{J-1} \beta_\nu^B z_\nu(j) + \beta^C z(k) + \beta^D z(l),$$

where $\beta_0=2\tau_1^A$. This model has $3+(J-1)=J+2$ regression parameters $\beta_0, \beta^C, \beta^D$ and

$\beta^B_1,...,\beta^B_{J-1}$.

It should be kept in mind that (8.33) is merely a reformulation of (8.31). This means that the likelihood equations for the ML–estimation of $\tau^{AB}_{1\,1},...,\tau^{AB}_{1\,J}$ are the same as those for the estimation of $\beta^B_1,...,\beta^B_{J-1}$, with

$$\hat{\beta}^B_\nu = 2\hat{\tau}^{AB}_{1\,\nu} \ .$$

The goodness of fit test statistic for the logit model with dummy variable is given by (8.28) with

$$\hat{g}_{jkl} = \hat{\beta}_0 + \sum_{\nu=1}^{J-1}\hat{\beta}^B_\nu z_\nu(j) + \hat{\beta}^C z(k) + \hat{\beta}^D z(l).$$

Due to (8.30), $\hat{g}_{jkl}$ can also be written

$$\hat{g}_{jkl} = \hat{\beta}_0 + \hat{\beta}^B_j + \hat{\beta}^C z(k) + \hat{\beta}^D z(l) \ ,$$

where $\beta^B_J=-\beta^B_1-...-\beta^B_{J-1}$.

In (8.33) there are J+2 parameters to be estimated. Hence the goodness of fit test statistic has an approximate χ^2–distribution with JKL–J–2 degrees of freedom, where K=L=2. It should be noted that the introduction of dummy variables for the categories of a polytomous explanatory variable, does not change the model. It is merely a convenient way of introducing such a variable in the model for practical purposes. If for example the available computer programs require that a model is formulated as a regression model, (8.33) is the appropriate form. The dummy variable method on the other hand allow us to check the logit model without assuming score values for the categories of variable B.

Example 8.3.

The data in table 8.4 stems from an investigation of the Danish labour market in 1979. The table shows a sample of 398 persons with a university degree in three Danish counties distributed according to sex and employment status.

Consider a model, where employment is dichotomized by merging the categories full time and part time employed. The response variable A is then binary with categories: employed and unemployed, and the data can be analysed by a logit model. The explanatory variables are:

B: County.

C: Sex.

Table 8.4. A random sample of 398 persons in three Danish counties distributed according to sex and employment status.

A: Employment status

C: Sex	B: County	Unemployed	Part time employed	Full time employed
Male	Fyn	3	6	16
	Århus	43	24	44
	Copenhagen	16	25	43
Female	Fyn	3	2	4
	Århus	25	18	41
	Copenhagen	16	30	39

Source: Unpublished data from Statistics Denmark.

The logit model is accordingly

$$g_{jk} = 2\tau_1^A + 2\tau_{1j}^B + 2\tau_{1k}^C.$$

Since variable B is polytomous, we use dummy variables for B. The values $z_\nu(j)$ for the categories of B are assigned as follows

	$\nu=1$	2
j=1	1	0
2	0	1
3	−1	−1

The logit model with dummy variables then become

$$(8.34) \qquad g_{jk} = \ln \frac{\mu_{1jk}}{\mu_{2jk}} = \beta_0 + \sum_{\nu=1}^{2} \beta_\nu^B z_\nu(j) + \beta^C z(k),$$

where μ_{ijk} is the expected number in cell (ijk) with employment status i and sex k living in county j. The estimated parameters of this model are with standard errors in parantheses

$$\hat{\beta}_0 = -1.224 \ (0.173)$$

$$\hat{\beta}_1^B = -0.356 \ (0.313)$$

$$\hat{\beta}_2^B = +0.587 \ (0.192)$$

$$\hat{\beta}^C = -0.082 \ (0.117) \ .$$

These estimates seem to indicate, that sex does not contribute to the description of the response variable. The test statistics (8.28) for evaluating the fit of three logit models with various explanatory variables included are shown in table 8.5. Thus a model including both sex and county and a model only including county as explanatory variables both describes the data in a satisfactory way, while at least one of the explanatory variables are needed to explain the variation in employment. This is a clear indication that the contribution of sex is insignificant.

Given that a model with both explanatory variables included describes the data, the tests for $\beta_j^B = 0$ and $\beta^C = 0$ can be carried out as sequential tests. Thus the test statistic for $H_2 : \beta_j^C = 0$ given that (8.34) describes the data is obtained according to (8.29) by substraction in table 8.5 as

$$z_2 = 0.49.$$

Table 8.5. Test statistics for various logit–models applied to data in table 8.4.

Explanatory variables included	z	df	Level of significance	Sequential tests	df	Level of significa
Sex, County	3.11	2	0.211	–	–	–
County	3.60	3	0.309	0.49	1	0.488
None	17.03	5	0.004	13.43	1	0.000

The test statistic for $H_3 : \beta_j^B = 0$, $j=1,2$, given H_2, is

$$z_3 = 13.43.$$

These values and the corresponding degrees of freedom and levels of significance are shown in the last three columns of table 8.5. They confirm that the variation in employment status is adequately described by the variation over counties.

The three counties can be ordered in terms of urbanization, such that the county of Fyn is the least urbanized and the county of Copenhagen the most. Assume that this ordering can be represented in the model by scoring the counties Fyn: –1, Århus: 0 and Copenhagen: +1. A new logit model can then be formulated in accordance with (8.13), as

$$g_{jk} = \beta_0 + \beta^B z_{2j} + \beta^C z(k),$$

where $z_{21} = -1$, $z_{22} = 0$, $z_{23} = 1$, $z(1) = 1$ and $z(2) = -1$. The estimates of β_0, β^B and β^C with standard errors in parantheses are

$$\hat{\beta}_0 = -0.926 \ (0.127)$$

$$\hat{\beta}^B = 0.307 \ (0.180)$$

$$\hat{\beta}^C = -0.064 \ (0.116)$$

The test statistics for this model corresponding to table 8.5 are shown in table 8.6. It is obvious from these test statistics that a logit model with an equidistant scoring according to degree of urbanization does not fit the data. △.

Table 8.6. Test statistics for logit–models with an equidistant scoring of variable B applied to the data in table 8.4.

Explanatory variables included	Test statistic	df	Level of significance
Sex, County	13.53	3	0.004
County	13.84	4	0.008
None	17.03	5	0.004

8.3. Exercises

8.1. Formulate exercise 5.3 as a logit–model with interval mid–points of high school average and the year as explanatory variables and passed/failed as the response.

(a) Estimate the parameters of the logit–model and test which of the explanatory variables are needed to predict the result of the examen.

(b) Compare the result of the analysis by means of a log–linear model in excercise 5.3 and the analysis by means of a logit–model.

(c) Does the logit–model describe the data in a satisfactory way.

(d) Compute confidence limits for the probability of passing for a student with high school average 8.0.

8.2. Reconsider the data in exercise 5.2.

(a) Formulate a logit–model with ownership of a freezer as response variable and social rank and renter/owner status as explanatory variables.

(b) Use a dummy–variable for social rank group, estimate the parameters of the logit– model and compare with the log–linear parameters estimated in exercise 5.2.

(c) Assume now that the social rank categories have been scored I–II=2, III=3, IV=4. and V=5. Repeat the analysis and compare the regression parameters now obtained with those obtained in (b).

(d)   Test which of the explanatory variables are needed both with social rabl as a dummy variable and scored as in (c).

8.3. Reconsider the data in exercise 5.5.

(a)   Can the problem be formulated in terms of a logit–model?

(b)   Test whether a logit–model describe the data.

(c)   Which explanatory variables are needed.

(d)   Compare the predicted probability of a broken marriage for a man and a woman in social class I and for a man and a woman in social class V. Are these differences to be expected?

8.4. Suppose explanatory variable B is binary. If the categories are scored as

$$z(1) = 1$$
$$z(2) = -1,$$

then $\beta^B = 2\,\tau^{AB}_{11}$ and $\beta_0 = 2\tau^A_1$.

(a)   Show that if the binary categories of B are scored

$$z(1) = 1$$
$$z(2) = 0,$$

then

$$\beta^B = 4\tau^{AB}_{11}$$

and

$$\beta_0 = 2\tau^A_1 - 2\tau^{AB}_{11}.$$

(b)   Derive the corresponding formulas if the scoring is $z(1)=1$, $z(2)=2$.

8.5. Reconsider the problem and the data in exercise 6.1.

(a)   Check if a logit–model with headache frequency as response variable fits the data.

(b)   Which of the explanatory variable contribute to explain the variation in the response variable.

(c)   Check the formula

$$2\hat{\tau}^{AB}_{11} = \hat{\beta}^{B}$$

and the corresponding formulae and the estimates obtained in this exercise by a logit–model and by a log–linear model.

8.6. In the fall of 1974 the Danish National Institute for Social Research investigated the mobility of the Danish work force. The table below is from this investigation. The four recorded variables are Age, Change of Job, Plans to quit the job voluntarily and Sex.

(a)   Analyse the data by means of logit–model, where the age categories are score 22, 27 32,37 and 47, and Change of job is the response variable.

(b)   Compare the results in (a) with a logit–model, where age is represented by 4 dummy variables.

Age	Change of job	Plans to quit	Sex Men	Women
	Voluntarily	Yes	53	54
		No	70	66
19–24				
	Forced	Yes	31	6
		No	32	9
	Voluntarily	Yes	52	21
		No	38	39
25–29				
	Forced	Yes	18	2
		No	28	4
	Voluntarily	Yes	27	12
		No	19	15
30–34				
	Forced	Yes	11	2
		No	20	4
	Voluntarily	Yes	14	4
		No	11	5
35–39				
	Forced	Yes	6	1
		No	19	3
	Voluntarily	Yes	14	6
		No	21	12
40–				
	Forced	Yes	20	5
		No	35	7

(c)   Describe the way the explanatory variables in the final model influence whether the job change is forced or voluntary by computing the estimated probability that the change is voluntary for each combination of the observed combination of the explanatory variables.

8.7. Reconsider the data in exercise 6.2.

(a)   Let irritation of the throat be the response variable and analyse the data by a logit–model.

(b)   Compare the estimated parameters of the final logit–model and the estimated parameters of the log–linear model AC,CD,B. Are they directly comparable?

(c)   How do the explanatory variables influence the response variable.

8.8. Reconsider exercise 6.6.

(a)   Can the problem be formulated in terms of a logit model?

(b)   Analyse the data by a logit–model.

(c)   Compare with the analysis in exercise 6.6.

9. Logistic Regression Analysis

9.1. The logistic regression model

In chapter 8 the connection to log–linear models for contingency tables was stressed. The direct connection to regression analysis for continuous response variables will now be brought more clearly into focus. Assume as before that the response variable is binary and that it is observed together with p explanatory variables. For n cases the data will then consist of n vectors

$$(y_\nu, z_{1\nu}, \ldots, z_{p\nu}), \; \nu=1,\ldots,n$$

of jointly observed values of the response variable and the explanatory variables.

The logistic regression model then states that

$$Y_\nu = \begin{cases} 1 & \text{with probability } P_\nu \\ 0 & \text{with probability } 1-P_\nu \end{cases}$$

where

(9.1)
$$P_\nu = \exp(\beta_0 + \sum_{j=1}^{p} \beta_j z_{j\nu}) \; / \; [1+\exp(\beta_0 + \sum_{j=1}^{p} \beta_j z_{j\nu})]$$

or equivalently

(9.2)
$$\ln(P_\nu/(1-P_\nu)) = \beta_0 + \sum_{j=1}^{p} \beta_j z_{j\nu}.$$

It is assumed that the Y's are independent. If the Y's are not independent one has to take into account the dependencies. Suggestions for such models are due to Qu et al. (1987) and Conolly and Liang (1988). A very comprehensive survey of the history and methodology of logistic regression is due to Imrey, Koch and Stokes (1981).

The **logistic regression model** was introduced by Cox (1970) to describe the dependency of a binary variable on a set of continuous variables. The statistical model underlying logistic regression analysis was introduced by Berkson (1944), (1953) in connection with the analysis of so–called bio assays. The name of the model is derived from the fact that the logistic transformation of the probability of $Y_\nu = 1$ is linear. There are, however, no restrictions on the explanatory variables. All models treated in chapter 8 are thus special cases of (9.2), and we can mix binary, polytomous, whether scored or not, and continuous explanatory variables in the model.

The likelihood function for a set of n observations $y_1, \ldots, y_n$ is

$$L = P(Y_1 = y_1, \ldots, Y_n = y_n) = \prod_{\nu=1}^{n} P_\nu^{y_\nu}(1 - P_\nu)^{1 - y_\nu},$$

When the logistic expression (9.1) is inserted in L, the log–likelihood becomes

$$(9.3) \qquad \ln L(\beta_0, \beta_1, \ldots, \beta_p) = \sum_{\nu=1}^{n} y_\nu \left[\beta_0 + \sum_{j=1}^{p} \beta_j z_{j\nu} \right]$$

$$- \sum_{\nu=1}^{n} \ln[1 + \exp(\beta_0 + \sum_{j=1}^{p} \beta_j z_{j\nu})] = y_. \beta_0 + \sum_{j=1}^{p} \beta_j \sum_{\nu=1}^{n} y_\nu z_{j\nu} - nK(\beta_0, \beta_1, \ldots, \beta_p),$$

where $K(\beta_0, \beta_1, \ldots, \beta_p)$ does not depend on the observations $y_1, \ldots, y_n$. The model is accordingly log–linear with canonical parameters $\beta_0, \beta_1, \ldots, \beta_p$ and sufficient statistics $y_.$, $\Sigma y_\nu z_{1\nu}, \ldots, \Sigma y_\nu z_{p\nu}$. The dimension of the model is p+1 if the β's are unrestricted as is usually the case, and if there are no linear ties between the p+1 sufficient statistics. A typical example, where the log–linear model is of dimensional less than p+1, is if explanatory variables one and two are binary and $z_{2\nu}$ is −1 whenever $z_{1\nu}$ is 1 and $z_{2\nu}$ is 1 whenever $z_{1\nu}$ is −1. In this singular case $\Sigma y_\nu z_{1\nu} = -\Sigma y_\nu z_{2\nu}$ and there is no unique maximum for the likelihood function since $L(\beta_0, \beta_1, \ldots, \beta_p)$ is constant as long as $\beta_1 - \beta_2$ does not changes its value.

In the following it is assumed that (9.3) has dimension p+1. Since the sufficient sta-

tistics for $\beta_0, \beta_1, ..., \beta_p$ are $y_\cdot, \Sigma y_\nu z_{1\nu}, ..., \Sigma y_\nu z_{p\nu}$, the likelihood equations are

(9.4)
$$y_\cdot = E[Y_\cdot] = \sum_{\nu=1}^{n} P_\nu.$$

and

(9.5)
$$\sum_{\nu=1}^{n} y_\nu z_{j\nu} = \sum_{\nu=1}^{n} z_{j\nu} E[Y_\nu] = \sum_{\nu=1}^{n} z_{j\nu} P_\nu, \quad j=1,...,p.$$

Equations (9.4) and (9.5) have a unique set of solutions, which are the ML–estimates for $\beta_0, ..., \beta_p$ according to theorem 3.6 if the vector of lenght p+1 formed by the left hand sides of (9.4) and (9.5) corresponds to an interior point in the convex extension of the support.

The boundary of the convex extension of the support consist of piecewise–linear functions since the possible values of $t_0, ..., t_p$, where

$$t_0 = y_\cdot$$

and

$$t_j = \sum_{\nu=1}^{n} y_\nu z_{j\nu}, \quad j=1,...,p$$

are limited by linear inequalities in sums of the $z_{j\nu}$'s. When p>1 these inequalities are cumbersome to write down, but for p=1 it is straight forward. Assume that the z_{1j}'s are ordered such that $z_{1(1)} < z_{1(2)} \leq ... \leq z_{1(n)}$. Then since $y_\nu = 1$ or 0,

$$0 \leq t_0 \leq n$$

and

$$t_1 = 0 \text{ for } t_0 = 0,$$

$$z_{1(1)} + ... + z_{1(t_0)} \leq t_1 \leq z_{1(n-t_0+1)} + ... + z_{1(n)}, \quad \text{for } t_0 > 0,$$

forming a convex set in a (t_0, t_1)–space, and there is a unique solution to the likelihood equations if none of these inequalities are equalities.

Boundary points corresponds to infinitely large values of the estimates. Thus if the numerical procedure used to solve the likelihood equations converges to a finite vector then this vector will be the ML–estimates of the β's.

It can be helpful to distinguish between two cases:

Case A: There is one y–value for each combination of z's.

Applications of the asymptotic results in chapter 3 require for case A that n is reasonable large. Goodness of fit tests similar to the ones used in chapter 8 are not available, since the observed x–values for the Z–test statistic are either 1 or 0. It is, however, possible to compare different regression models and in this way evaluate the relative influence of different explanatory variables.

Case B: There are several observed y–values for each combination of the explanatory variables. In this case it is convenient to index the set of possible distinct combinations of the explanatory variables by the letter i=1,...,I.

Let

n_i = number of observations with values $z_{1i},...,z_{pi}$ of the explanatory variables

and

x_i = number of these n_i observations for which $y_\nu=1$.

By independence of the Y_ν's, the random variable X_i corresponding to x_i follows a binomial distributon with parameters n_i and P_i, where

$$(9.6) \qquad P_i = \exp(\beta_0 + \sum_{j=1}^{p} \beta_j z_{ji}) / [1 + \exp(\beta_0 + \sum_{j=1}^{p} \beta_j z_{ji})],$$

by virtue of (9.1).

Hence the likelihood function is

$$(9.7) \qquad L = \prod_{i=1}^{I} \begin{bmatrix} n_i \\ x_i \end{bmatrix} P_i^{x_i}(1-P_i)^{n_i-x_i}.$$

Due to (9.6) the log–likelihood function become

$$\ln L(\beta_0,\beta_1,...,\beta_p) = \beta_0 \sum_i x_i + \sum_{j=1}^{p} (\sum_i x_i z_{ji}) \beta_j - \sum_i n_i \ln(1+\exp(\beta_0 + \sum_j \beta_j z_{ji})) + \sum_i \ln \begin{bmatrix} n_i \\ x_i \end{bmatrix} .$$

Hence the sufficient statistics for case B are

$$t_0 = x_. = \sum_{i=1}^{I} x_i$$

and

$$t_j = \sum_{i=1}^{I} x_i z_{ji}, \quad j=1,...,p.$$

Since $E[X_i] = n p_i$, the likelihood equations thus become

(9.8)
$$x_. = \sum_i n_i P_i$$

and

(9.9)
$$\sum_{i=1}^{I} x_i z_{ji} = \sum_{i=1}^{I} n_i P_i z_{ji}, \quad j=1,...,p .$$

The boundaries of the convex extension of the support are again formed by linear inequalities, since the x_i's are positive and bounded by 0 and n_i. For the case $p=1$, the lower boundary of the convex extension of the support has the form

(9.10)
$$\begin{cases} z_{1(1)} t_0 \text{ for } 0 \le t_0 \le n_{(1)} \\ z_{1(1)} n_{(1)} + (t_0 - n_{(1)}) z_{1(2)} \text{ for } n_{(1)} \le t_0 \le n_{(1)} + n_{(2)} \\ ... \\ z_{1(1)} n_{(1)} + ... + z_{1(I-1)} n_{(I-1)} + (t_0 - n_{(I-1)} - ... - n_{(1)}) z_{1(I)} \\ \quad \text{ for } t_0 \ge n_{(1)} + ... + n_{(I-1)} \end{cases}$$

when the z_{1i}'s are ordered so that $z_{1(1)} \le z_{1(2)} \le ... \le z_{1(I)}$ and the n_i's accordingly. The upper limit has the form

$$(9.11) \quad \begin{cases} z_{1(I)}t_0 \text{ for } 0 \le t_0 \le n_{(I)} \\ z_{1(I)}n_{(I)} + z_{1(I-1)}(t_0 - n_{(I)}) \text{ for } n_{(I)} \le t_0 \le n_{(I)} + n_{(I-1)} \\ \cdots \\ z_{1(I)}n_{(I)} + \cdots + z_{1(2)}n_{(2)} + (t_0 - n_{(I)} - \cdots - n_{(2)})z_{1(1)} \text{ for } t_0 \ge n_{(I)} + \cdots n_{(2)} \end{cases}$$

Thus a unique set of ML–estimates for the β's exists if none of these inequalities are equalities.

Instead of drawing the convex extension of the support and checking if the observed point is on the boundary, one can, following an important result by Albert and Anderson (1984), check the existence and uniqueness of the ML–estimate by looking into the formation of the observations in the $(p+1)$–dimensional plane spanned by the z's. It was proved by Albert and Anderson (1984), that there exists a unique, finite solution to the likelihood equations if and only if it is **not** possible to find a vector $(\alpha_0,...,\alpha_p)$ with $(\alpha_1,...,\alpha_p) \ne (0,...,0)$ such that

$$(9.12) \qquad \alpha_0 + \alpha_1 z_{1i} + ... + \alpha_p z_{pi} \ge 0$$

whenever $x_i = n_i$,

$$(9.13) \qquad \alpha_0 + \alpha_1 z_{1i} + ... + \alpha_p z_{pi} = 0$$

whenever $0 < x_i < n_i$.

and

$$(9.14) \qquad \alpha_0 + \alpha_1 z_{1i} + ... + \alpha_p z_{pi} \le 0$$

whenever $x_i = 0$.

This result means that a set of observations $x_1,...,x_I$ corresponding to I distinct vectors of explanatory variables $(z_{11},..,z_{p1}),...,(z_{1I},...,z_{pI})$ is extreme in the sense that a unique and finite ML–estimate $(\hat{\beta}_0,...,\hat{\beta}_p)$ does not exists, if and only if the inequalities (9.12) to (9.14) are satisfied. The advantage of this result is that one can check directly in a p–dimensional diagram if the given observations form an extreme combination.

Consider thus the cases $p=1$ and $p=2$. For $p=1$ the inequalities (9.12) to (9.14) means that there is a point α on the real line such that, either

$$z_{1i} \geq \alpha, \text{ whenever } x_i = n_i$$

$$z_{1i} \leq \alpha, \text{ whenever } x_i = 0$$

and

$$z_{1i} = \alpha, \text{ if } 0 \langle x_i \langle n_i,$$

or the opposite inequalities hold. This simply means that only one of the z's, say z_{01}, have x_i different from 0 or n_i and for those z's which are smaller than z_{01}, either $x_i = 0$ for all i or $x_i = n$ for all i, and for those z's which are larger than z_{01}, either $x_i = n_i$ for all i or $x_i = 0$ for all i. Albert and Anderson (1984) termed this type of configuration **quasi–complete separation** of the data points.

For p=2, the equalities (9.12) to (9.14) means that when satisfied there is a line

$$(9.15) \qquad\qquad z_{2i} = \alpha + \beta z_{1i}$$

in the (z_{1i}, z_{2i})–plane which separates those vectors (z_{1i}, z_{2i}) for which $x_i = n_i$ from those for which $x_i = 0$. One the line (9.15) x_i may be less than n_i and larger than 0.

For a general p, a configuration of x's is extreme, if there exist a (p–1)–dimensional plane which separates those vectors $(z_{i1}, ..., z_{ip})$ for which $x_i = 0$ from those for which $x_i = n_i$.

In practice the test for an extreme configuration runs as follows: Let the rows of the matrix Z_1 be those

$$(1, z_{i1}, ..., z_{ip})$$

for which $0 < x_i < n_i$. Then the rank of Z_1 must be less than or equal to p, in order for (9.13) to be satisfied.

This is obviously a necessary condition for the configuration to be extreme and it will, in most cases, suffice to check this condition. Thus if the rank of Z_1 is larger than p, the configuration is not extreme, and in the normal case, where there are more than p vectors $(z_{i1},...,z_{ip})$ with $0 \langle x_i \langle n_i$, it often suffice to check this condition.

Based on the likelihood function (9.7) it is possible to derive a goodness of fit test for the regression model. If the P_i's are unconstrained, the ML–estimates are

$$\tilde{P}_i = x_i/n_i.$$

The transformed likelihood ratio test statistic for comparing a model, where P_i has the parametric form (9.6) with a model, where P_i is unconstrained is, therefore,

(9.16)
$$Z_0 = 2 \sum_{i=1}^{I} [X_i \ln\frac{X_i}{n_i} + (n_i - X_i)\ln(\frac{n_i - X_i}{n_i})]$$

$$-2 \sum_{i=1}^{I} [X_i \ln\hat{P}_i + (n_i - X_i)\ln(1 - \hat{P}_i)]$$

with

$$\hat{P}_i = \exp(\hat{\beta}_0 + \sum_{j=1}^{p} \hat{\beta}_j z_{ji}) / [1 + \exp(\hat{\beta}_0 + \sum_{j=1}^{p} \hat{\beta}_j z_{ji})] .$$

The test statistic Z_0 is thus a goodness of fit test statistic for the regression model. If the n_i's are reasonable large, Z_0 follows an approximate χ^2–distribution with $I-(p+1)$ degrees of freedom, since there are I unconstrained P_i's and $p+1$ β's. The regression model is rejected as a description of the data if the level of significance

$$p = P(Z_0 \geq z_0)$$

computed in a χ^2–distribution with $I-p-1$ degrees of freedom is sufficiently small.

As an alternative to (9.16) one may use the Pearson test statistic

$$Q_0 = \sum_i (X_i - n\hat{P}_i)^2/(n\hat{P}_i) + \sum_i (n - X_i - n(1 - \hat{P}_i))^2/(n(1 - \hat{P}_i)),$$

which is easily rewritten as

(9.17)
$$Q_0 = \sum_i (x_i - n\hat{P}_i)^2 / (n\hat{P}_i(1-\hat{P}_i)).$$

As for Z_0, Q_0 is approximately χ^2–distributed with $I-p-1$ degrees of freedom.

Assuming that the explanatory variables are ordered according to relative influence on the response variable, a reasonable first step is to test the hypothesis

$$H_p: \beta_p = 0,$$

i.e whether the p'th variable is without influence. Let $\tilde{\beta}_0, \tilde{\beta}_1,...,\tilde{\beta}_{p-1}$ be the ML–estimates under H_p, i.e. let

$$L(\tilde{\beta}_0, \tilde{\beta}_1,...,\tilde{\beta}_{p-1},0) = \max L(\beta_0, \beta_1,...,\beta_{p-1},0).$$

Then $\tilde{\beta}_0, \tilde{\beta}_1,...,\tilde{\beta}_{p-1}$ are the solutions to the equations

$$y_. = \sum_{\nu=1}^{n} P_\nu,$$

$$\sum_{\nu=1}^{n} y_\nu z_{j\nu} = \sum_{\nu=1}^{n} z_{j\nu} P_\nu, \qquad j=1,...,p-1,$$

where

$$P_\nu = \exp(\beta_0 + \sum_{j=1}^{p-1} \beta_j z_{j\nu}) \,/\, [1+\exp(\beta_0 + \sum_{j=1}^{p-1} \beta_j z_{j\nu})]$$

in case A, and

$$x_. = \sum_i n_i P_i,$$

$$\sum_i x_i z_{ji} = \sum_i n_i P_i z_{ji}, \quad j=1,...,p-1$$

where

$$P_i = \exp(\beta_0 + \sum_{j=1}^{p-1} \beta_j z_{ji}) / [1+\exp(\beta_0 + \sum_{j=1}^{p-1} \hat{\beta}_j z_{ji})]$$

in case B.

The transformed likelihood ratio test statistic for testing H_p against the model with all explanatory variables included is

$$(9.18) \qquad Z_p = -2\ln L(\tilde{\beta}_0, \tilde{\beta}_1, \ldots, \tilde{\beta}_{p-1}, 0) + 2\ln L(\hat{\beta}_0, \hat{\beta}_1, \ldots, \hat{\beta}_p).$$

According to theorem 3.18, Z_p is approximately χ^2–distributed with one degree of freedom. If the observed value of Z_p is found to be non–significant a logistic regression model with $\beta_p = 0$ describes the data as well as a model with all p explanatory variables included. The next step is then to test whether β_{p-1} can be set to zero without any loss in goodness of fit between data and model, and so on.

Sometimes there are strong prior reasons to believe that several of the β's are zero. Suppose for example that prior beliefs suggest the hypothesis

$$H_{(r)}: \beta_{p-r+1} = \ldots = \beta_p = 0 \ .$$

Under $H_{(r)}$ a model with only the first p–r explanatory variables included describes the data as well as a model including all p explanatory variables. This hypothesis can be tested by

$$(9.19) \qquad Z_{(r)} = -2\ln L(\tilde{\beta}_0, \tilde{\beta}_1, \ldots, \tilde{\beta}_{p-r}, 0, \ldots, 0) + 2\ln L(\hat{\beta}_0, \hat{\beta}_1, \ldots, \hat{\beta}_p)$$

which is approximately χ^2–distributed with r degrees of freedom.

The ML–estimates $\tilde{\beta}_1, \ldots, \tilde{\beta}_{p-r}$, constrained by $H_{(r)}$, satisfy

$$L(\tilde{\beta}_0, \tilde{\beta}_1, \ldots, \tilde{\beta}_{p-r}, 0, \ldots, 0) = \max L(\beta_0, \beta_1, \ldots, \beta_{p-r}, 0, \ldots, 0)$$

Since all test statistics are differences between the values of $-2\ln L$ with various arguments, it is a reasonable starting point for an analysis to list the value of $-2\ln L$ for all models.

Example 9.1

The Institute for Building Research in Denmark made in 1970 an investigation of the indoor climate in Danish Schools. The students in a number of school classes were asked whether they felt that the indoor climate at the moment was pleasant or not so pleasant. Simultaneously a number of objective measurements were taken of the actual climate in the class room, namely

T: Temperature

M: Degree of moisture

C: Amount of carbondioxide in the air

F: Amount of fresh air

D: Amount of dust

V: Degree of ventilation

Table 9.1 shows the number of students in the class, the number of yes–answers to the question above and the values of the 6 indoor climate indicators.

Table 9.1. Number of students claiming that the indoor climate is pleasant together with 6 objective indoor climate indicators for three Danish school classes observed on differents days.

Class	Date	Number of students in the class	Number of yes answers	T	M	C	F	D	V
7A	4/3	19	18	22.0	30	0.09	8.5	0.20	230
7A	3/3	20	6	21.5	25	0.11	6.1	0.08	230
7A	2/3	19	4	21.5	25	0.11	4.8	0.06	230
8A	4/3	18	13	18.5	25	0.09	9.2	0.07	236
8A	3/3	14	12	20.0	25	0.05	8.7	0.08	236
8A	2/3	18	4	20.0	25	0.11	5.2	0.12	236
8A	18/3	17	14	20.5	30	0.08	13.1	0.09	249
8A	17/3	19	18	21.0	30	0.08	12.5	0.06	249
9B	18/3	16	9	21.5	30	0.09	8.7	0.07	215
9B	17/3	18	8	21.0	30	0.07	9.3	0.09	215

Source: Unpublished data from the Danish Institute for Building Research.

These numbers can be analysed by a logistic regression model with p=6 and a yes–answer as response variable. The estimates and their standard errors are shown in table 9.2.

Table 9.2. ML–estimates and their standard errors for a logistic regression model applied to the data in table 9.1.

Variable	Parameter	Estimate	Standard error
Intercept	β_0	4.656	9.064
T:Temperature	β_1	1.320	0.342
M:Moisture	β_2	−1.141	0.29
C:Carbondioxide	β_3	20.29	17.7
F:Fresh air	β_4	1.449	0.33
D:Dust	β_5	25.305	7.56
V:Ventilation	β_6	−0.0705	0.036

The form of the convex extension of the support can be illustrated by this example if we assume that the only explanatory variable is F. Table 9.3 show the values of z_{4i}, x_i, n_i and the cumulative sums necessary to establish the lower and upper limits (9.10) and (9.11) for t_1 given the value of t_0, after the observations have been rearranged according to z–values and i=5 and 9 with the same z–value have been merged.

Tabel 9.3. Computations of the cumulative sums (9.10) and (9.11) for the data in table 9.1 and explanatory variable F.

z_{4i}	x_i	n_i	$z_{4i}x_i$	$z_{4i}n_i$	Sum from below	Sum from above
4.8	4	19	19.2	91.2		1522.5
5.2	4	18	20.8	93.6	184.8	1431.3
6.1	16	20	97.6	122.0	306.8	1337.7
8.5	18	19	153.0	161.5	468.3	1215.7
8.7	21	30	182.7	261.0	729.3	1054.2
9.2	13	18	119.6	165.6	894.9	793.2
9.3	8	18	74.4	167.4	1062.3	627.6
12.5	18	19	225.0	237.5	1299.8	460.2
13.1	14	17	183.4	222.7	1522.5	
Sum	116	178	1075.7	1522.5		

Since $t_0 = 116$ is between $19+18+20+19+30=106$ and $106+18=124$ when we sum from below and between $17+19+18+18+30=102$ and $102+19=121$, when we sum from above, the relevant inequalities emerging from (9.10) and (9.11) are

$$729.3 + 10 \cdot 9.2 = 821.3 \le t_1$$

and

$$1054.2 + 14 \cdot 8.5 = 1163.2 \ge t_1.$$

The value 1057.7 of t_1 does not attain any of these limits, such that the likelihood equations have a unique solution in this case. The convex set enveloped by inequalities (9.10) and (9.11) is shown in fig. 9.1. The observed point is marked by a dot.

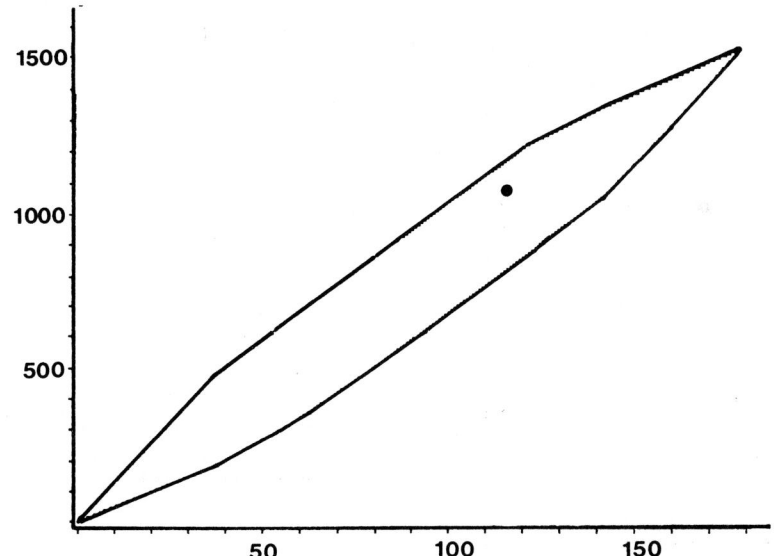

Figur 9.1. The boundary of the support for the indoor climate data, when F is the only explanatory variable.

Suppose next that there are two explanatory variables in the model, namely T: Temperature and M: Moisture. For this case it is somewhat more complicated to derive the limits of the convex extension of the support. To give an impression of the form of its shape, we have constructed the boundaries for the support for 6 values of t_0. The results are shown in fig.9.2.

The most interesting features on figur 9.2 are the three points marked by $+$, Δ and $*$. The point $+$ corresponding to the observed value $(t_1, t_2) = (2413.0, 3235)$ is well within the boundaries. The point $*$ with $(t_1, t_2) = (2472.5, 3235)$ is on the boundary, because no combination of x's can give a t_1-value higher than 2472.5. Table 9.4 column 4 show one combination of x's that leads to $t_1 = 2472.5$.

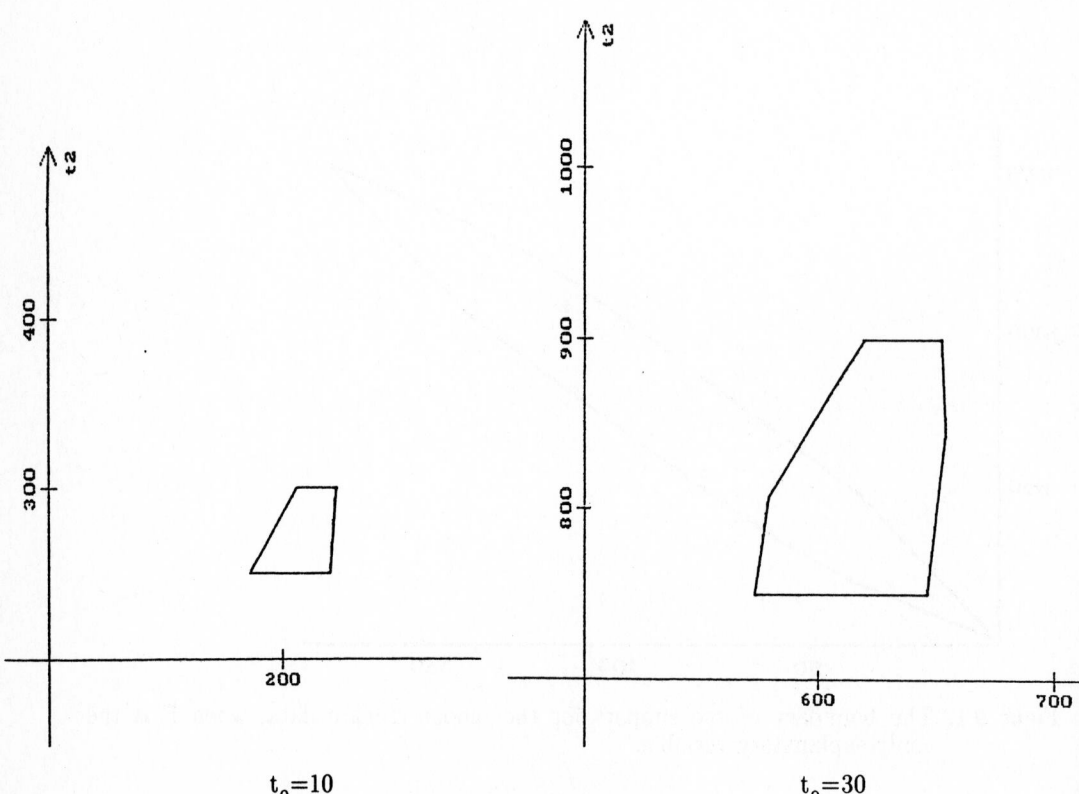

$t_0 = 10$

$t_0 = 30$

Figur 9.2. The boundaries of the support for $t_0 = 10, 30$

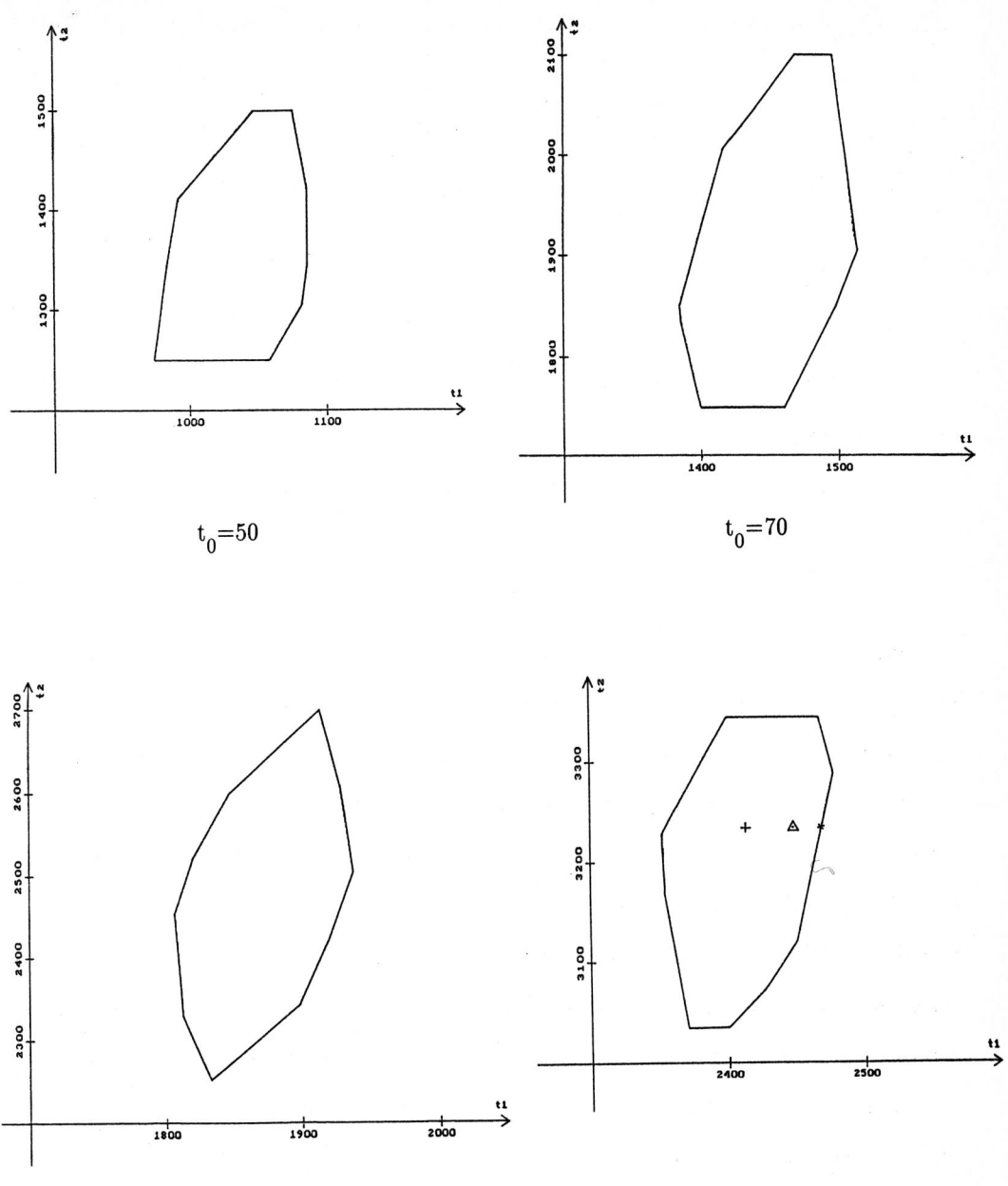

Figur 9.2. The boundaries of the support for t_0=50,70,90,116.

Table 9.4. Observed x–values with $t_2=3235$ and t_1 equal to either (a) 2472.5 or
(b) 2449.5.

n_i	z_{1i}	z_{2i}	x_i (a)	(b)
19	22.0	30	19	19
20	21.5	25	20	20
19	21.5	25	19	19
18	18.5	25	0	10
14	20.0	25	5	0
18	20.0	25	5	0
17	20.5	30	0	16
19	21.0	30	16	0
16	21.5	30	16	16
18	21.0	30	16	16
$t_1=$			2472.5	2449.5
$t_2=$			3235	3235

What characterizes (a) is, that the x's are placed where both z_{1i} and z_{21} have high values and no redistribution of the x's can make the value of t_1 higher given the value 3235 of t_2. Column (b) shows on the other hand a distribution of the x's for which $t_1=2449.5$ and its obvious that the value of t_1 can be increased. The important point to stress is that there is no simple rule by which to determine if (t_1,t_2) is on the boundary. Even for t_0 fixed at 116, t_1 does not attain its highest value, which is 2480.0, for the boundary point (2472.5, 3235). It is neither a criterion that an observed value of x_i is zero, since the point (2449.5, 3235), marked by Δ on fig. 9.2, which is in the interior, is generated by a set of x's containing more zeros than the boundary point (2472.5, 3235).

In order to check whether the configuration with all six explanatory variables inclu-ded is extreme, we form the matrix Z_1, which, since none of the x_i's are equal to 0 or n_i, becomes

$$Z_1 = \begin{bmatrix} 1 & 22.0 & 30 & 0.09 & 8.5 & 0.20 & 230 \\ 1 & 21.5 & 25 & 0.11 & 6.1 & 0.08 & 230 \\ 1 & 21.5 & 25 & 0.11 & 4.8 & 0.06 & 230 \\ 1 & 18.5 & 25 & 0.09 & 9.2 & 0.07 & 236 \\ 1 & 20.0 & 25 & 0.05 & 8.7 & 0.08 & 236 \\ 1 & 20.0 & 25 & 0.11 & 5.2 & 0.12 & 236 \\ 1 & 20.5 & 30 & 0.08 & 13.1 & 0.09 & 249 \\ 1 & 21.0 & 30 & 0.08 & 12.5 & 0.06 & 249 \\ 1 & 21.5 & 30 & 0.09 & 8.7 & 0.07 & 215 \\ 1 & 21.0 & 30 & 0.07 & 9.3 & 0.09 & 215 \end{bmatrix}$$

It can be checked by numerical methods that the rank of Z_1 is 7 and it follows that the configuration is not extreme, such that a unique set of ML–estimates in fact exists.

Fortunately boundary points in general corresponds to solutions of the likelihood equations, where one or more $\hat{\beta}$ is infinitely large. Hence a boundary point will in almost all cases reveal itself by a numerical estimation procedure, which diverges.

Since $\hat{\beta}_3$ and $\hat{\beta}_6$ are less than twice their standard errors, table 9.2 seems to indicate that the amount of carbondioxide and the degree of ventilation may be omitted in the model as explanatory variables. The tests for these hypotheses can be derived from table 9.5, which shows the estimated values of $-2\ln L$ under various models.

Table 9.5. The transformed likelihood ratio and the observed values of the likelihood ratio test statistic for various models applied to the data in table 9.1.

Variables included	$-2\ln L$	z	df	Level of significance
Saturated	169.58[1]			
TMCFDV	176.05	6.47	3	0.091
TMFDV	177.29	1.25	1	0.264
TMFD	179.87	2.58	1	0.108
TMF	190.72	10.85	1	0.001

Note: 1) equal to the first term in (9.16).

The conclusion of the analysis is that temperature, degree of moisture, the amount of fresh air and the amount of dust all makes significant contributions to the description of the indoor climate as perceived by the students. △.

For the logistic regression model (9.6), (9.7) it is assumed that the x_i's follow independent binomial distributions. This assumption is not satisfied, if the observations are a result of a clustered or stratified sampling design. How the statistical procedures are adjusted to account for such situations are discussed in Roberts, Rao and Kumar (1987).

In other situations model departures are due to over dispersion, i.e. the variance of Y_ν can not be described by the binomial variance

$$\text{var}[Y_\nu] = P_\nu(1-P_\nu),$$

whatever the choice of the P_ν's. In such situations Efron (1986) has suggested to base the analysis on the so–called double exponential family, introduced by Diaconis and Efron (1985).

9.2. Regression diagnostics

In section 3.7 it was shown that the likelihood equations for a generalized linear model are equivalent with the normal equations (3.64) for a weighted least square solution based on the adjusted y–values (3.65). Since the y–values are readjusted in each iteration of the estimation procedure, the method is widely known as an **iterative weighted least square method**. We now specify the method in details for the logistic regression model (9.6), (9.7) for I independent binomials. The sufficient statistics are

$$t_j = \sum_i x_i z_{ji} \, , \ j=0,...,p$$

and the likelihood equations are

$$E[T_j] = \sum_i n_i P_i z_{ji} \, ,$$

with $z_{0i}=1$, $i=1,...,I$.

Hence in the general linear model

(9.20) $$K'(\tau_i) = n_i P_i$$

with

$$\tau_i = \ln \frac{P_i}{1-P_i}$$

and P_i given by (9.6).

Accordingly

$$(9.21) \qquad K''(\tau_i) = n_i \frac{\partial P_i}{\partial \tau_i} = n(\frac{\partial \tau_i}{\partial P_i})^{-1} = n_i P_i(1-P_i)$$

and the normal equations (3.57) become

$$(9.22) \qquad \sum_i z_{ji} x_{i0} w_{i0} = \sum_{l=1}^{p} \beta_l \sum_i z_{li} z_{ji} w_{i0} \quad j=0,...,p.$$

with $w_{i0}=n_i P_{i0}(1-P_{i0})$ and the adjusted x'values x_{i0} defined as

$$(9.23) \qquad x_{i0} = \frac{x_i - n_i P_{i0}}{w_{i0}} + \sum_{l=1}^{p} \beta_{l0} z_{li}.$$

The iterative weighted least squares procedure thus prescribes in each iteration to compute the adjusted x–values (9.23) and then obtain least squares estimates of the β's using the w_{i0}'s as weights. On matrix form (9.22) become

$$(9.24) \qquad ZW_0 x_0' = ZW_0 Z' \beta,$$

where Z has elements $z_{ji}, j=0,...,p, i=1,...,I$ x_0 is a vector with elements $x_{i0}, i=1,...,I$ and W_0 a diagonal matrix with $w_{10},...,w_{I0}$ in the diagonal.

As shown by Pregiborn (1981) the fact that the regression parameters are obtained by a least square method allows for the definition of a number of diagnostics similar to those uses in modern regression analysis, cf. also McCullogh and Nelder (1983), chapter 4 and Jennings (1986).

Residuals for the logistic regression model can be defined in various ways. Most commonly used are the squared roots of the individual terms in (9.17), i.e.

$$(9.25) \qquad q_i = (x_i - n_i \hat{P}_i)/\sqrt{n\hat{P}_i(1-\hat{P}_i)}.$$

Pregiborn (1981) and McCullogh and Nelder (1983) also consider square roots of the individual terms in (9.16) defined as

$$(9.26) \qquad d_i = \pm \sqrt{2} \{ x_i [\ln x_i - \ln(n_i \hat{P}_i)] + (n_i - x_i)[\ln(n_i - x_i) - \ln(n_i(1-\hat{P}_i))] \}^{\frac{1}{2}},$$

where "+" is used, when $x_i > n_i P_i$ and "−", when $x_i < n_i \hat{P}_i$. Both q_i and d_i are useful as measures of the influence of the i'te term on the value of the test statistics. In order to evaluate their relative influence, it is, however, important that both q_i and z_i are properly scaled.

As regards q_i, we can use theorem 3.15 to derive the variance of $X_i - n\hat{P}_i$. The derivatives of P_i and $(1-P_i)$ with respect to β_j are

$$\partial P_i / \partial \beta_j = z_{ji} P_i(1-P_i)$$

and

$$\partial (1-P_i)/\partial \beta_j = -z_{ji} P_i(1-P_i),$$

which also covers the case j=0, if z_{0i} is set equal to 1. It then follows from (3.48) that

$$(9.27) \qquad \text{var}[X_i - n_i \hat{P}_i] = n_i P_i(1-P_i - \sum_{j=0}^{p} \sum_{k=0}^{p} \frac{n_i}{n} z_{ji} z_{ki} P_i(1-P_i)^2 m^{jk})$$

where m^{jk} are the inverse elements of the square matrix $\mathbf{M}$ of dimension p+1 with elements

$$m_{jk} = \sum_{i=1}^{I} \frac{n_i}{n} z_{ji} z_{ki} (P_i(1-P_i)^2 + P_i^2(1-P_i)) = \sum_{i=1}^{I} \frac{n_i}{n} z_{ji} z_{ki} P_i(1-P_i)$$

The matrix $\mathbf{M}$ can be written as

$$\mathbf{M} = \frac{1}{n} \mathbf{ZWZ'},$$

where $\mathbf{W}$ is diagonal with diagonal elements $n_i P_i(1-P_i)$.

If, therefore, z_i is the i'th column of $\mathbf{Z}$, the standardized residual based on q_i become

(9.28) $$r_i = (x_i - n_i \hat{P}_i)/\hat{\sigma}_i,$$

where

$$\hat{\sigma}_i^2 = n_i P_i(1-P_i)(1 - \frac{1}{n} n_i P_i(1-P_i) z_i' \mathbf{M}^{-1} z_i).$$

The standardized residual r_i is approximately distributed as a standard normal deviate.

Since approximately

$$(x_i - nP_i)/\sqrt{n_i P_i(1-P_i)} \simeq \sqrt{2}\sqrt{x_i \ln(\frac{x_i}{n_i P_i}) + (n_i - x_i)\ln(\frac{n_i - x_i}{n_i(1-P_i)})}$$

the variance of d_i in (9.26) can be estimated by

$$\tau_i^2 = 1 - \frac{1}{n} n_i P_i(1-P_i) z_i' \mathbf{M}^{-1} z_i = 1 - h_{ii},$$

such that the standardized residual based on d_i is

(9.29) $$r_i^* = d_i/\sqrt{1 - h_{ii}}.$$

Both r_i and r_i^* are useful as diagnostics for detecting model deviations in logistic regression analysis. Their role as diagnostics are primarily to indicate which of the observed responses contribute most to a significant test statistics.

Note that with $w_i = n_i P_i(1-P_i)$, (9.27) can be written

$$\text{var}[X_i - n\hat{P}_i] = w_i(1 - \frac{1}{n} z_i'(\mathbf{ZWZ'})^{-1}z_i w_i))$$

which is the diagonal element in the matrix

(9.30)
$$\mathbf{W(I - H)},$$

where

$$\mathbf{H} = \frac{1}{n} \mathbf{WZ'(ZWZ')}^{-1}\mathbf{ZW} .$$

The matrix $\mathbf{H}$ is known in applied regression analysis as the **hat–matrix**. As in ordinary regression analysis, the elements of the hat–matrix are the necessary adjustments of the residuals as compared to the estimated variance

$$\text{var}[X_i] = n\hat{P}_i(1 - \hat{P}_i).$$

The diagonal elements

(9.31)
$$h_{ii} = \frac{1}{n} w_i z_i'(\mathbf{ZWZ'})^{-1}z_i w_i$$

of the hat–matrix are also often used as diagnostics. A large value of h_{ii} indicate a data point, where the variance of the residual deviate more from the binomial variance than for other data point. Such a data point can, therefore, influence the evaluation of the fit of the model in an uncharacteristic way.

While the standardized residuals (9.28) or (9.29) tell us whether an observed number x_i differ significantly from the expected number $n_i P_i$ under the logistic regression model, the term h_{ii} is an indicator of how much the variance of the residual is influenced by an observation x_i.

In applied regression analysis a third widely used indicator is **Cooks distance**. An observation x_i is regarded as influencing the estimation of the parameters $\beta_0, \beta_1, ..., \beta_p$ if the estimates change significantly when x_i is removed from the dataset. Cook (1977) suggested to compare the original β-estimates with a new set of β-estimates, say $(\hat{\beta}_0^{(i)},...,\hat{\beta}_p^{(i)})$ obtained after x_i has been removed from the dataset. As a measure for the change from $\hat{\beta}=(\hat{\beta}_0,...,\hat{\beta}_p)$ to $\hat{\beta}^{(i)}=(\hat{\beta}_0^{(i)},...,\hat{\beta}_p^{(i)})$, Cook suggested to consider

$$D_i = \frac{1}{p}(\hat{\beta}-\hat{\beta}^{(i)})'V_\beta^{-1}(\hat{\beta}-\hat{\beta}^{(i)}),$$

where V_β is the variance matrix of the $\hat{\beta}$'s. In a logistic regression model the variance matrix of $\hat{\beta}$ is

$$\text{var}[\hat{\beta}] = (Z\ Z')^{-1},$$

hence Cook's distance becomes

$$D_i = \frac{1}{p}(\hat{\beta}-\hat{\beta}^{(i)})'ZWZ'(\hat{\beta}-\hat{\beta}^{(i)}).$$

Pregiborn (1981) showed that D_i can be approximated by

$$D_i = \frac{1}{p}(x_i-n_i\hat{P}_i)^2 h_{ii}/[n_i\hat{P}_i(1-\hat{P}_i)(1-h_{ii})^2].$$

or due to (9.28)

$$D_i = \frac{1}{p}r_i^2 h_{ii}/(1-h_{ii}).$$

Whether the Cooks distance differ from r_i^2/p thus depends on the value of h_{ii}. For $h_{ii}=\frac{1}{2}$, $D_i = \frac{1}{p}r_i^2$.

An observation, which influence the β-estimates more than the other observations is often referred to as an **outlier**. The easiest way to spot outliers is on plots of the residuals against the explanatory variables. Some outliers will not reveal themselves in this way, however, if they are due to a combination of values of two explanatory variables.

If a residual analysis shows that a logistic regression model does not fit the data, it is a possibility that a transformation of the response variables or of the logit

$$\ln(P_{\nu}/(1-P_{\nu})) \, ,$$

will yield more satisfactory residual diagrams. One attractive candidate for a transformation of the logit is the power transformation due to Box and Cox (1964).

With λ being a parameter to be estimated, the Box–Cox tranformation is given by

$$g_{\lambda}\{\ln(P_{\nu}/(1-P_{\nu}))\} = \begin{cases} \ln(P_{\nu}/(1-P_{\nu})) & \text{for } \lambda=0 \\ \lambda^{-1}\left[\left[\dfrac{P_{\nu}}{1-P_{\nu}}\right]^{\lambda} -1\right] & \text{for } \lambda \neq 0 \end{cases}$$

A discussion of the application of the Box–Cox transformation can be found in Guerro and Johnson (1982). Kay and Little (1987) discuss the possibility of transforming the explanatory variables to obtain a better fit.

Example 9.1. (Continued)

The final model to be accepted for the indoor climate data was

$$\ln \frac{P_i}{1-P_i} = \beta_0 + \beta_1 z_{1i} + \beta_2 z_{2i} + \beta_4 z_{4i} + \beta_5 z_{5i},$$

where T=temperature, M=moisture, F=fresh air and D=dust contributed to the description of the variation in the perceived indoor climate. The ML–estimates of $\beta_0, \beta_1, \beta_2, \beta_4$ and β_5 under the model are

$$\hat{\beta}_0 = \quad -11.16 \ (4.08)$$
$$\hat{\beta}_1 = \quad 1.042 \ (0.284)$$
$$\hat{\beta}_2 = \quad -0.703 \ (0.168)$$
$$\hat{\beta}_4 = \quad 0.950 \ (0.171)$$
$$\hat{\beta}_5 = \quad 17.53 \ (6.08)$$

with standard errors in parantheses.

Table 9.6 shows the logits $\ln(x_i/(n_i-x_i))$, the diagonals h_{ii} of the hat matrix, the standardized residuals r_i, the alternative residuals r_i^* and Cook's distance D_i.

Table 9.6. Logits, diagonals of the hat matrix, residuals and Cook's distances for the data of table 9.1 with Temperature, Moisture, Fresh Air, and Dust as explanatory variables.

Case	Logit	h_{ii}	Residuals r_i	r_i^*	Cook's distance D_i
1	2.89	0.80	1.37	1.49	1.83
2	1.39	0.54	1.37	1.42	0.54
3	-1.32	0.55	-1.64	-1.70	0.83
4	0.96	0.76	1.72	1.76	2.31
5	1.79	0.24	0.02	0.02	0.00
6	-1.25	0.63	-1.19	-1.22	0.59
7	1.54	0.20	-3.12	-2.35	0.60
8	2.89	0.34	0.39	0.41	0.02
9	0.25	0.50	1.71	1.69	0.73
10	-0.22	0.45	-0.78	-0.78	0.12

The residuals in table 9.6 show that only case number 7 has a significantly large value. Nevertheless, case 7 does not have an especially large value of Cook's distance. Hence case 7 can be expected to contribute to the value of the test statistic, but does not influence the values of the parameter estimates more than other cases.

Cases 1 and 4 have, however, relatively high values of Cook's distance. Accordingly these two cases can be expected to influence the parameter estimates. That cases 1 and 4 indeed correspond to outlying observations can be seen directly from the residual plots fig. 9.3, where the residuals r_i are plotted against the four included explanatory variables. None of the four plots in fig. 9.3 show systematic patterns apart from the large negative residuals for case 7. Two points differ, however, substantially from the other points. On the plot for Temperature the point for case 4 with Temperature $=18$ is clearly below the rest of the points, while on the plot for Dust the point for case 1 with Dust=0.20 is clearly above the other points. The two large values of Cook's distance thus correspond to out-

294

liers. In order to demonstrate the relative influence of cases 1, 4 and 7 on the parameter estimates and the test statistics, logistic regression analyses were carried out with these three cases excluded in turns. Table 9.7 show the results.

Table 9.7. Regression analyses with case 1,4 and 7 respectively excluded.

	All cases	Without Case 1	Without Case 4	Without Case 7
Intercept	−11.16	−6.98	−20.92	−14.20
Temperature	1.04	0.83	1.50	1.28
Moisture	−0.70	−0.62	−0.70	−0.85
Fresh Air	0.95	0.84	0.94	1.21
Dust	17.53	2.96	19.79	19.96
z	10.3	8.36	7.33	3.69
df	5	4	4	4
p	0.067	0.079	0.120	0.449

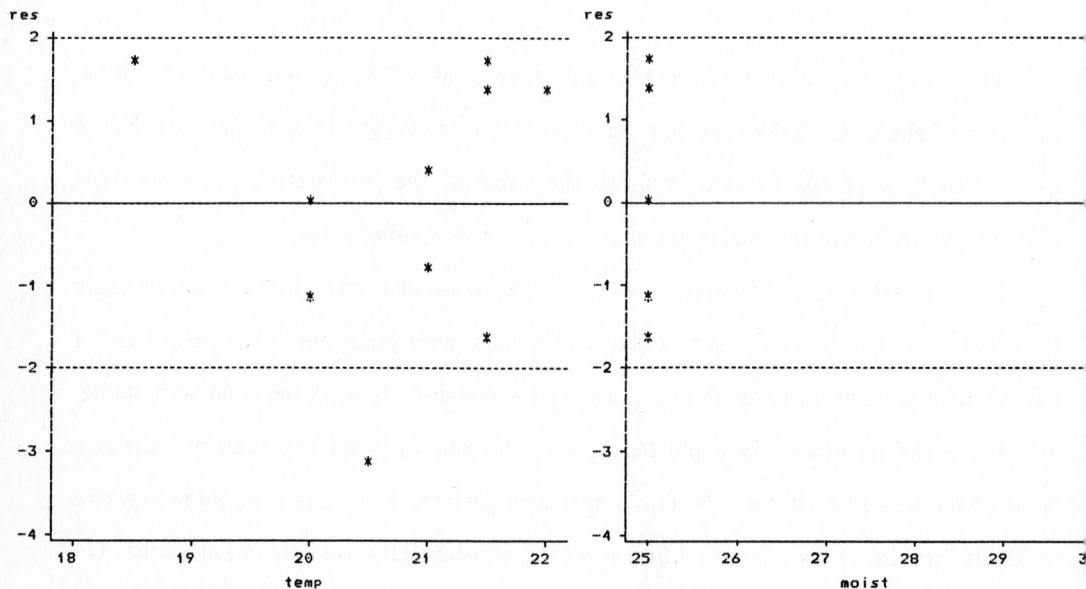

Fig. 9.3. Residuals plotted against the explanatory variables Temperature and Moisture.

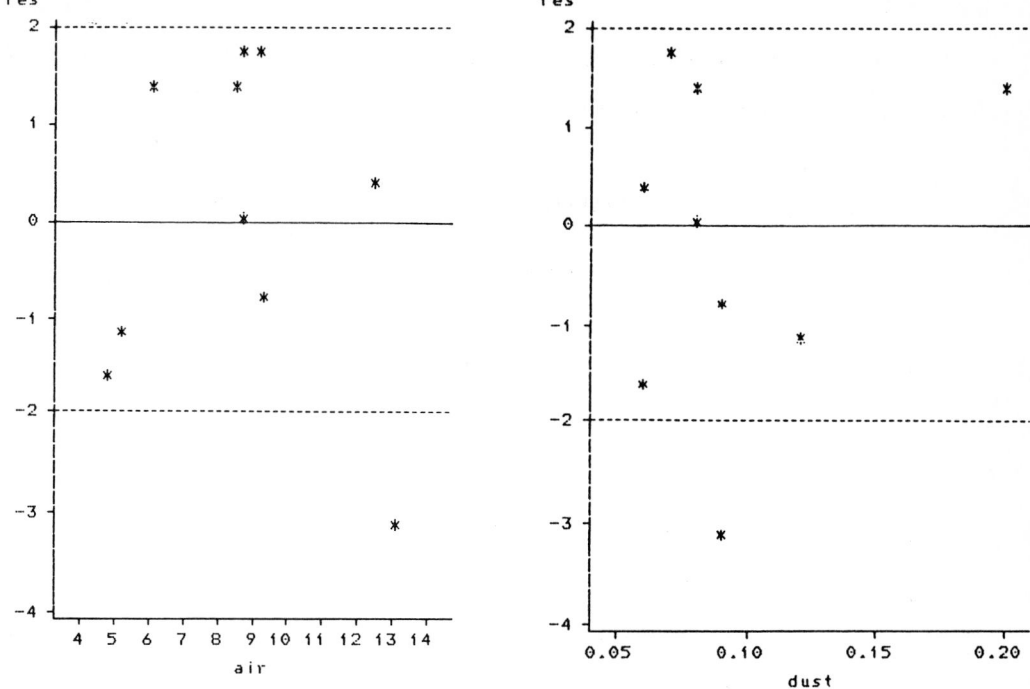

Figur 9.3. Residuals plotted against the explantory variables Fresh Air and Dust.

As expected exclusion of case 1 change the β–estimate for Dust dramatically and the exclusion of case 4 change the β–estimate for Temperature by 50%. The exclusion of case 7 only change the β–estimates in a moderate manner, but the value of the test statistic is decreased by more than 60%, thus dramatically changing the fit of the model.

The results in table 9.7 confirm the usefulness of both the standardized residuals and of Cook's distance as regression diagnostics.△.

Example 9.2

In the Danish Welfare Study from 1974, a sample of 2827 employed persons were cross--
classified according to following 4 variables

Employment: 1=Private employment

                   0=Public employment

Sex: 1=Male

                  2=Female

Social rank: 2=Rank group I and II

                  3=Rank group III

                  4=Rank group IV

                  5=Rank group V

Urbanization: 1=Copenhagen

                  2=Suburbs of Copenhagen

                  3=Cities

                  4=Countryside

We regard Sex, Social rank and Urbanization as explanatory and want to study whether the percentage of persons, which are privately employed, depends on these variables. Table 9.8 shows the data with x_i being the number of privately employed persons and n_i the total number of employed persons for combination (z_{i1}, z_{i2}, z_{i3}) of the explanatory variables.

Table 9.8. Number of privately employed, x_i, among all employed persons, n_i, in a 1974 Danish sample, for each combination of sex, social rank and urbanization.

Case	x_i	n_i	Sex z_{1i}	Social rank z_{2i}	Urbanization z_{3i}
1	7	9	1.00	2.00	3.00
2	27	34	1.00	2.00	4.00
3	12	15	1.00	2.00	2.00
4	7	12	1.00	2.00	1.00
5	20	21	1.00	2.00	5.00
6	12	23	1.00	3.00	3.00
7	41	69	1.00	3.00	4.00
8	17	33	1.00	3.00	2.00
9	13	23	1.00	3.00	1.00
10	24	36	1.00	3.00	5.00
11	24	58	1.00	4.00	3.00
12	64	162	1.00	4.00	4.00
13	49	93	1.00	4.00	2.00
14	41	82	1.00	4.00	1.00
15	48	102	1.00	4.00	5.00
16	21	49	1.00	5.00	3.00
17	48	167	1.00	5.00	4.00
18	23	44	1.00	5.00	2.00
19	18	40	1.00	5.00	1.00
20	58	142	1.00	5.00	5.00
21	21	36	2.00	2.00	3.00
22	42	87	2.00	2.00	4.00
23	34	62	2.00	2.00	2.00
24	14	25	2.00	2.00	1.00
25	24	40	2.00	2.00	5.00
26	6	24	2.00	3.00	3.00
27	30	98	2.00	3.00	4.00
28	12	46	2.00	3.00	2.00
29	7	21	2.00	3.00	1.00
30	10	41	2.00	3.00	5.00
31	21	78	2.00	4.00	3.00
32	52	226	2.00	4.00	4.00
33	20	75	2.00	4.00	2.00
34	23	68	2.00	4.00	1.00
35	36	162	2.00	4.00	5.00
36	7	51	2.00	5.00	3.00
37	20	164	2.00	5.00	4.00
38	9	42	2.00	5.00	2.00
39	8	33	2.00	5.00	1.00
40	23	234	2.00	5.00	5.00
Totals	993	2827	1412.00	3619.00	3351.00

Source: Data obtained from the databased of the Danish Welfare Study. Cf.Hansen (1978).

Table 9.9 summarizes a logistic regression analysis of these data.

Table 9.9. Parameter estimates and test statistics for a logistic regression analysis of the data in table 9.8.

Variable	Parameter	Standard error
Intercept	3.586	(0.251)
Sex	−1.119	(0.086)
Social rank	−0.573	(0.044)
Urbanization	−0.073	(0.032)
z = 52.69	df = 36	p = 0.036

The significance level of the goodness of fit test statistic indicates that the regression model does not fit the data in a satisfactory way. In order to evaluate the reasons for the lack of fit, the residuals and Cook's distances are shown in table 9.10.

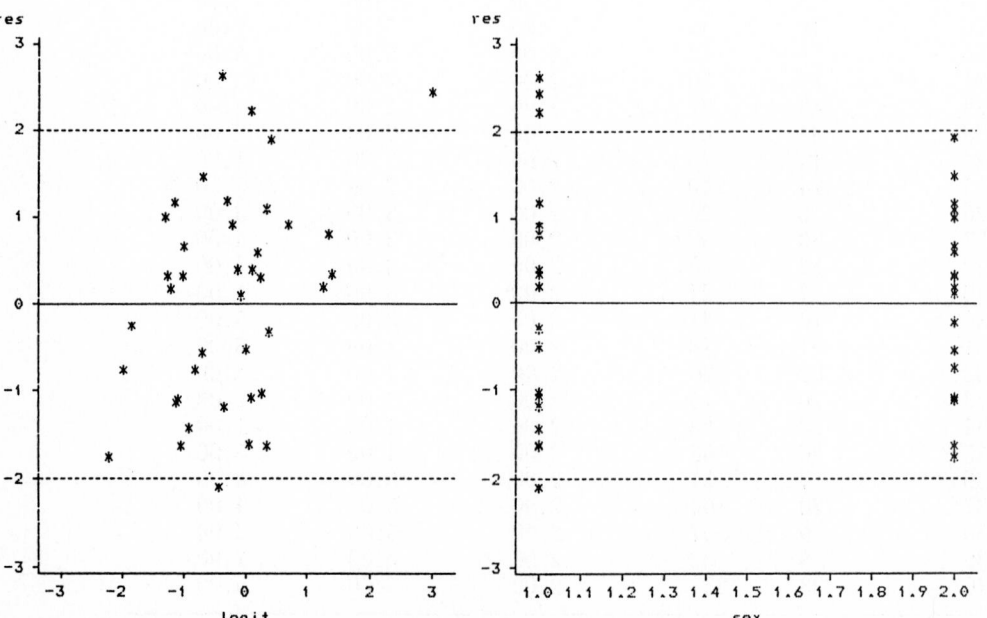

Figur 9.4. Standardized residuals plotted against the logits and Sex.

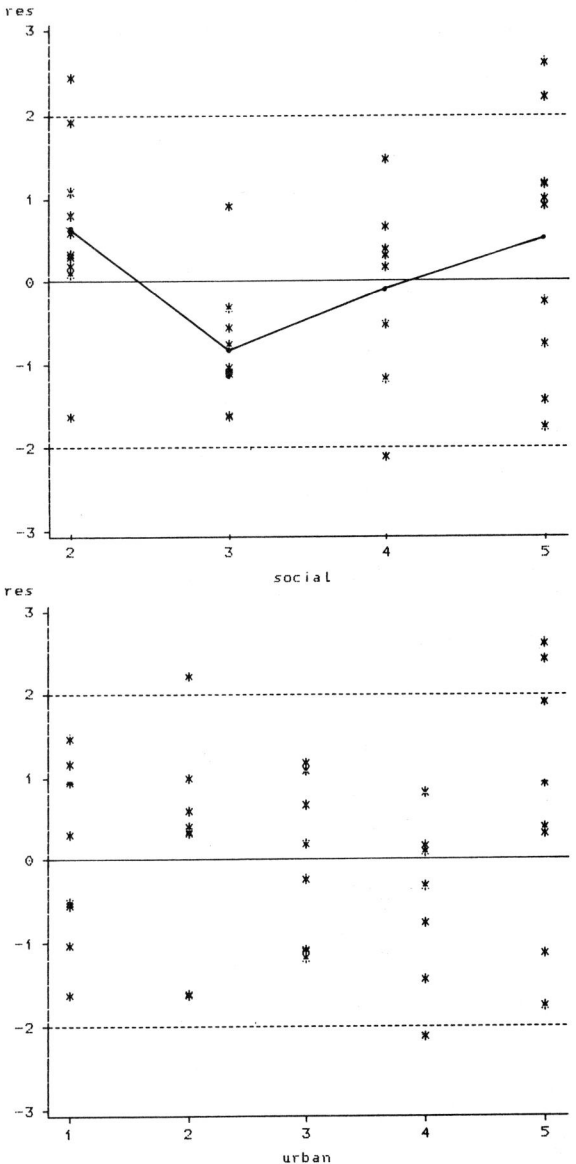

Figur 9.4. Standardized residuals plotted against Social rank and Urbanization.

Four of the residuals are larger than 2. In figur 9.4 the residuals are plotted against the logits and against the three explanatory variables. The second largest residual for case number 5 is connected with the largest logit. The Cook distance for this observation is, however, not particularly large. The residual plot for Sex shows, what is also obvious from the table, that the model fits better for women than for men.

Table 9.10. Logits L_i, standardized residuals r_i, alternative residuals r_i^* and Cook's distances C_i for the data in table 9.8.

Case	L_i	r_i	r_i^*	C_i
1	1.253	0.188	0.190	0.000
2	1.350	0.788	0.809	0.019
3	1.386	0.331	0.337	0.001
4	0.336	−1.641	−1.523	0.031
5	2.996	2.431	2.834	0.138
6	0.087	−1.090	−1.074	0.013
7	0.381	−0.333	−0.333	0.004
8	0.061	−1.625	−1.591	0.051
9	0.262	−1.021	−1.001	0.020
10	0.693	0.908	0.918	0.023
11	−0.348	−1.189	−1.193	0.027
12	−0.426	−2.125	−2.136	0.281
13	0.108	0.393	0.393	0.008
14	0.000	−0.526	−0.525	0.022
15	−0.118	0.377	0.376	0.009
16	−0.288	1.171	1.154	0.030
17	−0.908	−1.449	−1.468	0.172
18	0.091	2.218	2.174	0.141
19	−0.201	0.897	0.889	0.034
20	−0.370	2.595	2.541	0.635
21	0.336	1.088	1.090	0.032
22	−0.069	0.099	0.099	0.001
23	0.194	0.582	0.583	0.020
24	0.241	0.288	0.288	0.002
25	0.405	1.890	1.888	0.155
26	−1.099	−1.109	−1.140	0.010
27	−0.818	−0.760	−0.766	0.021
28	−1.041	−1.641	−1.685	0.064
29	−0.693	−0.557	−0.563	0.005
30	−1.131	−1.137	−1.167	0.028
31	−0.999	0.654	0.645	0.009
32	−1.208	0.174	0.174	0.002
33	−1.012	0.306	0.304	0.003
34	−0.671	1.454	1.417	0.115
35	−1.253	0.306	0.304	0.006
36	−1.838	−0.266	−0.269	0.001
37	−1.974	−0.764	−0.779	0.033
38	−1.299	0.995	0.955	0.020
39	−1.139	1.153	1.098	0.033
40	−2.216	−1.768	−1.841	0.320

It is the plot for Social rank, which reveals the most distinct model departure. The line on the graph connects the average residual value for each social rank group. Such a curved development of the averages indicate that the dependency of social rank is not linear. The plot for Urbanization does not show systematic model departures.

Social rank was scored 2,3,4,5 for the four categories, but as the residual plot shows, the logits depend in a non–linear way on this scoring. Since Social rank is categorical, we can use the dummy variabel method in section 8.2 to check the scoring. Table 9.11 shows the results from a logistic regression analysis with explanatory variables Sex and Urbanization combined with the dummy variables SocII, SocIII and SocIV defined as

$$\text{SocII} \quad = \begin{cases} 1 \text{ for Social rank } =2 \\ 0 \text{ for Social rank } =3,4 \\ -1 \text{ for Social rank } =5 \end{cases}$$

$$\text{SocIII} \quad = \begin{cases} 1 \text{ for Social rank } =3 \\ 0 \text{ for Social rank } =2,4 \\ -1 \text{ for Social rank } =5 \end{cases}$$

$$\text{SocIV} \quad = \begin{cases} 1 \text{ for Social rank } =4 \\ 0 \text{ for Social rank } =2,3 \\ -1 \text{ for Social rank } =5 \end{cases}$$

Table 9.11. A logistic regression analysis with dummy variables for Social rank.

Variable	Estimate	Standard error
Intercept	1.634	0.175
Sex	−1.144	0.087
SocII	1.079	0.096
SocIII	0.033	0.087
SocIV	−0.312	0.066
Urbanization	−0.076	0.032
z = 42.25	df = 34	p = 0.156

The estimates for Sex and Urbanization are in agreement with table 9.9 but the estimates for SocII, SocIII and SocIV show, that the proper scoring of the categories should put more distance between rank groups I–II and rank group III than between rank groups III and IV. With dummy variables SocII, SocIII and SocIV, chosen as they are, the $\hat{\beta}$'s should be compared with $\hat{\beta}=-1.079-0.033+0.312=0.800$ for rank group V. Hence the distance between the $\hat{\beta}$'s of rank groups IV and V is about the same as between rank groups III and IV. In fact with a scoring of social rank 0,3,4,5 rather than 2,3,4,5, a logistic regression model fits the data well. Returning to table 9.10, we note that the highest values of Cook's distance are for case 20 and case 40, i.e. the two cases with lowest social rank and lowest degree of urbanization. Table 9.12 summarizes different logistic regression analyses with cases or explanatory variables omitted. Both the removal of case 20 and of case 5 improves the fit of the model, but does not change the values of the β–estimates significantly. A large value of Cook's distance does not, therefore, correspond to an outlier. Case 5 is one of the observations in Social rank group I–II, which as we saw contributed most to the lack of fit. If this case is omitted the fit improves considerable.

The model without Urbanization fits the model as well as the original model. Hence Urbanization only contribute in a marginal way to explain the differences in percentage of privately employed persons. The signs of $\hat{\beta}_1$ and $\hat{\beta}_2$ are as expected. Women are more often privately employed and the higher the social rank, the higher the percentage of privately employed. $\triangle$.

Table 9.12. Logistic regression analyses with case 20, case 40 and Urbanization removed compared with the original regression analysis.

	All cases	Without case 20	Without case 5	All cases without urbanization
Intercept	3.586	3.661	3.485	3.411
Sex	−1.119	−1.071	−1.090	−1.134
Social rank	−0.573	−0.597	−0.551	−0.587
Urbanization	−0.073	−0.096	−0.083	−
z	52.69	46.16	44.79	57.87
df	36	35	35	37
Level of significance	0.036	0.098	0.124	0.016

A very critical survey of various suggestions for residuals in logistic regression analysis is due to Jennings (1986). He especially emphasize the differences of substancial character between residuals in regressions analysis based on data with normally distributed error terms and logistic regression analysis.

Johnson (1985) discussed diagnostics which are especially important when judging the influence of the various data points on predictions. Various methods for making graphs of regression diagnostics for logistic regression are surveyed by Landwehr, Pregiborn and Shoemaker (1984).

Fowles (1987) suggested to apply smoothing of the data to overcome the discrete nature of the binary responses, which is known to hamper residual plots in logistic regression.

9.3 Predictions

In the case of logistic regression the response variable is binary. The interesting quantity to predict is, therefore, the probability under the model of obtaining the response $y_\nu = 1$ given the values of the explanatory variables. Under the logistic regresssion model this probability is estimated as

$$\hat{P}_\nu = \hat{P}(z_{1\nu},...,z_{p\nu}) = \exp\left(\hat{\beta}_0 + \sum_{j=1}^{p} \hat{\beta}_j z_{j\nu}\right) / \left[1 + \exp\left(\hat{\beta}_0 + \sum_{j=1}^{p} \hat{\beta}_j z_{j\nu}\right)\right]$$

for combinations of the observed explanatory variables. Let now $z_{10},...,z_{p0}$ be any given combination of the explanatory variables, observed or not. Then the response is 1 with estimated or predicted probability

$$\hat{P}_0 = \exp\left(\hat{\beta}_0 + \sum_{j=1}^{p} \hat{\beta}_j z_{j0}\right) / \left[1 + \exp\left(\hat{\beta}_0 + \sum_{j=1}^{p} \hat{\beta}_j z_{j0}\right)\right].$$

It is a common practice also to estimate the logit

(9.32)
$$\hat{g}_0 = \ln\left(\hat{P}_0 / (1 - \hat{P}_0)\right) = \hat{\beta}_0 + \sum_{j=1}^{p} \hat{\beta}_j z_{j0} .$$

For a given set of explanatory values $z_{10},...,z_{p0}$, $\hat{g}_0$ represents a linear predictions based on the model. Approximate confidence limits for $\hat{P}_0$ or $\hat{g}_0$ can be derived from the asymptotic distribution of the ML–estimates $(\hat{\beta}_0,...,\hat{\beta}_p)$, by an application of theorem 3.7. Let $\mathrm{var}[\hat{g}_0]$ be the asymptotic variance of $\hat{g}_0$, and let

$$(9.33) \qquad g_0 = \beta_0 + \sum_{j=1}^{p} \beta_j z_{j0}.$$

Then g_0 is with approximate level of confidence $1-\alpha$ contained in the interval

$$(9.34) \qquad \hat{g}_0 \pm u_{1-\alpha/2}\sqrt{\widehat{\mathrm{var}[\hat{g}_0]}},$$

where $\widehat{\mathrm{var}[\hat{g}_0]}$ is $\mathrm{var}[\hat{g}_0]$ with the β's replaced by their estimates.

Confidence limits for $P_0 = \exp(g_0)/[1+\exp(g_0)]$ are obtained by transforming the limits in (9.34) by $P = \exp(g)/[1+\exp(g)]$.

Example 9.1 (Continued)

Table 9.13 show the confidence limits for P_i based on the observed values of temperature, moisture, fresh air and dust.

Table 9.13. Confidence limits for the probability of observing $Y_\nu = 1$ for the observed value z_1, z_2, z_4 and z_5 of explanatory variables T, M, F and D.

z_1	z_2	z_4	z_5	Lower limit	$\hat{P}_i$	Upper limit
22.0	30	8.5	0.20	0.709	0.906	0.974
21.5	25	6.1	0.08	0.542	0.705	0.829
21.5	25	4.8	0.06	0.194	0.328	0.499
18.5	25	9.2	0.07	0.421	0.626	0.793
20.0	25	8.7	0.08	0.739	0.855	0.925
20.0	25	5.2	0.12	0.162	0.301	0.488
20.5	30	13.1	0.09	0.889	0.959	0.985
21.0	30	12.5	0.06	0.824	0.929	0.973
21.5	30	8.7	0.07	0.259	0.414	0.588
21.0	30	9.3	0.09	0.362	0.513	0.661

Note that the limits are not symmetric. The confidence limits provide us with a clear picture of the predictive power of the model. With the given data the uncertainty concer–

ning the value of P_i for given values of the explanatory variables is thus between 0.025 and 0.205. $\triangle$.

Imrey, Koch, Stokes (1982) contains a number of illuminating examples of logistic regression analyses.

9.4 Polytomous response variables

The simple logistic model does not apply if the response variable has several categories. One possibility, often suggested, is to form a set of logistic transformations by taking the categories of the response variable two by two. Let Y_ν with response categories $t=1,...,T$, be a polytomous response variable corresponding to the values $z_{1\nu},...,z_{p\nu}$ of the explanatory variables. The Y_ν's are then independent with distribution

$$(9.35) \qquad P(Y_\nu = t \mid z_{1\nu},...,z_{p\nu}) = P_{\nu t}, \quad t=1,...,T,$$

satisfying $\sum\limits_{t=1}^{T} P_{\nu t}=1$. It is then possible to formulate various sets of $T-1$ logistic regression models. One possibility is

$$(9.36) \qquad \ln\left[\frac{P_{\nu t}}{1-P_{\nu t}}\right] = \beta_0^{(t)} + \sum_{j=1}^{p} \beta_j^{(t)} z_{j\nu}, \quad t=1,...,T-1.$$

Other possibilities are

$$(9.37) \qquad \ln\left[\frac{P_{\nu t}}{1-P_{\nu T}}\right] = \beta_0^{(t)} + \sum_{j=1}^{p} \beta_j^{(t)} z_{j\nu}, \quad t=1,...,T-1 \ ,$$

$$\ln\left[\frac{P_{\nu t}}{P_{\nu t+1}}\right] = \beta_0^{(t)} + \sum_{j=1}^{p} \beta_j^{(t)} z_{j\nu}, \quad t=1,...,T-1$$

or

$$\ln\left[\frac{P_{\nu t}}{\sum\limits_{1 \ge t+1} P_{\nu l}}\right] = \beta_0^{(t)} + \sum\limits_{j=1}^{p} \beta_j^{(t)} z_{j\nu}, \quad t=1,...,T-1.$$

Some of these models are only logistic regresson models in a conditional sense. The left hand side of (9.37) is thus a logistic regression model conditional upon the response being either t or T.

For each of these models it is straight forward to apply the methods in earlier sections to estimate the regression parameters. The problem is of course, that the analysis results in T−1 sets of regression coefficients connected with the different levels of the response variable.

Begg and Gray (1984) compared models (9.36) and (9.37) and demonstrated that they are identical as regards $\beta_1^{(t)},...,\beta_p^{(t)}$.

A more satisfactory solution is to introduce the levels of the response variable in the model. The model suggested by McCullogh (1980), cf. also Andersson and Blair (1982), introduces additive parameters for the levels in the following way.

(9.38)
$$\ln(\sum\limits_{l=1}^{t} P_{\nu l} / \sum\limits_{l=t+1}^{T} P_{\nu l}) = \alpha_t + \sum\limits_{j=1}^{p} \beta_j z_{j\nu}.$$

This corresponds to the logistic model

$$\ln \frac{Q_{\nu t}}{1-Q_{\nu t}} = \alpha_t + \sum\limits_{j} \beta_j z_{j\nu}$$

where $Q_{\nu t}$ is the cumulative probability

$$Q_{\nu t} = \sum\limits_{l=1}^{t} P_{\nu l}.$$

Especially if the response variable is ordinal this model seems appealing, as an ordering of the α's, e.g. $\alpha_1 < \alpha_2 < ... < \alpha_T$ entail an ordering of the response probabilities given the values of the explanatory variables. The model does not belong to the class of log–linear mo–

dels, but in practice it is relatively easy to obtain ML–estimates of the parameters.

Another possibility is to assume that a regression model describes the data for a weighted sum of the logarithms of the $P_{\nu t}$'s. This would be the case if $P_{\nu t}$ satisfies

$$(9.39) \qquad P_{\nu t} = \exp\left[(\beta_0 + \Sigma_j \beta_j z_{j\nu}) w_t\right] / \sum_{t=1}^{T} \exp[(\beta_0 + \Sigma_j \beta_j z_{j\nu}) w_t].$$

where $w_1,\ldots,w_T$ are preassigned weights. Whether an ordering of the w's here introduce an ordering of the $P_{\nu t}$'s will depend on the z's. Since $w_1,\ldots,w_T$ have known values the model with probabilities given by (9.39) is log–linear. It is possible to derive the likelihood function in case A with different sets of explanatory variables for each value of the response variable. Here we consider case B, i.e. the binomial situation (9.7), where there are n_i cases with z–values $z_{1i},\ldots,z_{pi}$ of which x_{it} have observed value $Y_\nu = t$.

For the polytomous case, the likelihood corresponding to (9.7) is

$$(9.40) \qquad L = \prod_{i=1}^{I} \left[x_{i1}^{n_i} \ldots x_{iT} \right] \prod_{t=1}^{T} P_{it}^{x_{it}},$$

where

$$(9.41) \qquad P_{it} = \exp\left[(\beta_0 + \Sigma_j \beta_j z_{ji}) w_t\right] / \sum_{t=1}^{T} \exp\left[(\beta_0 + \Sigma_j \beta_j z_{ji}) w_t\right].$$

From (9.40) the likelihood equations for the estimation of $\beta_0,\ldots,\beta_p$ are derived as

$$(9.42) \qquad \sum_{t=1}^{T} x_{\cdot t} w_t = \sum_{t=1}^{T} w_t \sum_{i=1}^{I} n_i P_{it}$$

and

$$(9.43) \qquad \sum_{t=1}^{T} w_t \sum_{i=1}^{I} x_{it} z_{ji} = \sum_{t=1}^{T} w_t \sum_{i=1}^{I} z_{ji} n_i P_{it}, \quad j=1,\ldots,p$$

since

$$E[X_{it}] = n_i P_{it}.$$

From (9.42) and (9.43) the ML–estimates for $\hat{\beta}_0$ and $\hat{\beta}_1,...,\hat{\beta}_p$ are easily computed by a suitable iterative procedure.

That the model is a regression model for weighted averages of the logarithmically transformed expected numbers follows from (9.41), which yields

$$\sum_t w_t \ln P_{it} = (\beta_0 + \sum_j \beta_j z_{ji}) \sum_t w_t^2$$

$$-\sum_t w_t \ln \sum_{s=1}^T \exp[(\beta_0 + \sum_j \beta_j z_{ji}) w_s].$$

Hence if location and scale for the weights $w_1,...,w_1$ are chosen such that

$$\sum_t w_t = 0$$

and

$$\sum_t w_t^2 = 1,$$

then

(9.44)
$$\sum_t w_t \ln P_{it} = \beta_0 + \sum_j \beta_j z_{ji}.$$

An immediate consequence of this result is that the model in case $p=1$ can be checked by plotting

$$y_i = \sum_t w_t \ln \left[\frac{x_{it}}{n_i}\right]$$

against z_{1i}, since x_{it}/n_i is an estimate for P_{it}. The points should then scatter around the line $y = \beta_0 + \beta_1 z_1$.

The goodness of fit test statistic for both model (9.38) and model (9.39) is

$$(9.45) \qquad Z = 2 \sum_{i=1}^{I} \sum_{t=1}^{T} X_{it}[\ln X_{it} - \ln(n_i \hat{P}_{it})],$$

where $\hat{P}_{it}$ for model (9.38) is

$$\hat{P}_{it} = \hat{Q}_{it} - \hat{Q}_{it-1}$$

with

$$\hat{Q}_{it} = \exp(\hat{\alpha}_t + \sum_j \hat{\beta}_j z_{ji}) / [1 + \exp(\hat{\alpha}_t + \sum_j \hat{\beta}_j z_{ji})]$$

and for (9.39)

$$\hat{P}_{it} = \exp[(\hat{\beta}_0 + \sum_j \hat{\beta}_j z_{ji}) w_t] / \sum_{t=1}^{T} \exp[(\hat{\beta}_0 + \sum_j \hat{\beta}_j z_{ji}) w_t].$$

The degrees of freedom for the approximating χ^2-distribution of (9.45) are for model (9.38)

$$df = I(T-1) - p - (T-1)$$

since the model is a product of I multinomial distributions each of dimension T and one of the α's is redundant due to the constraint $\sum_t P_{it} = 1$. For (9.39) the degrees of freedom are

$$df = I(T-1) - p - 1.$$

Amemiya (1981) and Manski (1981) discussed various models for polytomous response variables with special regard to applications in econometrics and marketing.

Andersson (1984) suggested to extend McCullogh's model to allow for parallel regression lines, i.e.

$$\ln\left(\sum_{l=1}^{t} P_{\nu l} / \sum_{l=t+1}^{T} P_{\nu l}\right) = \alpha_t + \gamma_t \sum_{j=1}^{p} \beta_j z_{j\nu}.$$

Example 9.3

We return to example 8.1. The data in table 8.2 was analysed in chapter 8 by merging the categories fulltime employed and parttime employed. Without a dichotomization the data can be analysed by either model (9.38) or by model (9.39). These analyses are summarizes in table 9.14.

Table 9.14. Parameter estimates and goodness of fit test statistics for models (9.38) and (9.39) applied to the data in table 8.2.

Parameter	Estimates			
	Urbanization scored 1,0,–1		Urbanization scored 1,0,1	
	Model(9.38)	Model(9.39)	Model (9.38)	Model (9.39)
β_0		–0.404		–0.191
α_1	0.150		0.371	
α_2	–0.986		–0.778	
β_1	0.029	–0.113	–0.471	–0.507
β_2	0.087	0.034	0.028	0.035
Goodness of fit test	19.8	26.2	11.3	18.7
Degrees of freedom	8	9	4	5
Level of significance	0.011	0.002	0.023	0.003

Table 9.14 shows that neither model with the original scoring 1,0,–1 of urbanization describe the data in an satisfactory way. One reason could be that the scoring of the three urban categories is inappropriate. The estimates of β_k^C in section 8.2 suggested that a scoring with Fyn and Copenhagen equal to +1 and Århus equal to 0 would fit the data better. With this scoring the regression analyses are summarized in the right hand half of table 9.14.

In both cases there is now a better fit than before although for model (9.39) only slightly. Both models seem to estimate the same regression model variable with β_1 approximately equal to –0.5 and β_2 close to 0.03. △.

9.5. Exercises:

9.1. The table below is typical of the way logistic regression analyses are presented in research reports. The report is from the Danish Institute for Borderregion Research located near the Danish–German border. The binary response is whether the interviewed person have changed job within a certain period or not. The headings of the table are direct translations of the Danish text in the report.

(a)   Explain the meaning of the numbers in the table.

(b)   Is there an explanation for the degrees of freedom D.F. shown in the table.

(c)   What conclusions would you draw from the table.

Variabel	Estimated coefficient (Standard deviation)		
	Model I	Model II	Model III
Constant	1.2831 (0.9010)	4.6945 (0.5783)	5.4697 (0.9560)
Sex:Men(0) Women(1)	−0.2995 (0.2874	−0.7735 (0.2622)	−1.0041 (0.3236)
Age:years	−0.0262 (0.0136)	−0.1007 (0.0127)	−0.0989 (0.0165)
Household– income:(1000 Dkr.)	0.0002 (0.0021)	−0.0036 (0.0024)	
Number of children living at home under age 18:	−0.0157 (0.1256)		
Number of hours per week partner (wife/husband) go to work:	0.0037 (0.0125)		
Length of education (month more than 10 years):	−0.0043 (0.0070)	0.0133 (0.0066)	
Likelihood ratio:	201.1	217.06	240.90
D.F.(n.k):	166	201	210
Prob.value:	0.0300	0.2079	0.706

Left hand side: Change of job = 1
                No change = 0

9.2. A psychological experiment consists of letting a person react to a sound stimulus. For each of 100 persons it is reported if they for a given sound intensity react within a given time interval. The test persons were exposed to 3 sound intensity levels in intervals of four different lenghts. The table show how many out of 100, who managed to react.

Intensity			Time interval	
	190 msek.	195 msek.	200 msek.	205 msek.
50 db	22	31	43	51
60 db	48	58	67	70
70 db	67	70	80	84

(a)   Describe the data by means of a logistic regression model.

(b)   What are the meaning of the regression coefficients?

9.3. The table below show another example of how logistic regression analysis can be presented. Apart from translation to English and minor changes in notation, nothing is changed. The variables are:

   A: Age (0 for 20–39, 1 for 40–59)

   B: Education (1 for 12 years in school, 0 for less)

   C: Vocational education completed (1 for yes, 0 for no)

   D: Marriage status (1 for yes, 0 for no)

   E: Psychiatric diagnosis (1 for yes, 0 for no)

   F: Physical handicap (1 for yes, 0 for no)

   G: Working ability restricted (1 for yes, 0 for no).

The purpose of the investigation is to predict the socioeconomic activity after the conclusion of a training program following physical inability. The response variable thus has value 1 if the training program has been a success and the person is back in a socioeconomic activity, and 0 if not.

   It follows from the text, preceeding the table, that ΔG^2 is the z–test statistic for a logistic regression model including the variables shown against the alternative that no variables

Variables	ΔG^2	df	p	0	A	B	C	D	E	F	G
None	—	—	—	0.61							
All	49.40	7	0.0005	1.38	0.42	1.64	1.55	0.75	0.34	1.01	0.4
ABCDEF	44.27	6	0.0005	1.39	0.38	1.56	1.54	0.76	0.35	0.83	
ABCDEG	49.40	6	0.0005	1.38	0.42	1.64	1.54	0.75	0.35		0.4
ABCDFG	30.66	6	0.0005	0.81	0.48	1.73	1.67	0.70		0.97	0.4
ABCEFG	48.52	6	0.0005	1.13	0.49	1.63	1.59		0.33	1.04	0.4
ABDEFG	46.42	6	0.0005	1.76	0.40	1.79		0.70	0.32	1.01	0.4
ACDEFG	46.65	6	0.0005	1.63	0.37		1.66	0.76	0.33	0.99	0.4
BCDEFG	41.81	6	0.0005	0.68		2.01	1.68	1.13	0.36	0.97	0.3
ABCEG	48.44	5	0.0005	1.14	0.49	1.63	1.59		0.33		0.4
ABEG	45.06	4	0.0005	1.42	0.48	1.79			0.32		0.4
ACEG	45.83	4	0.0005	1.35	0.43		1.71		0.32		0.4
AEG	41.10	3	0.0005	1.82	0.40				0.31		0.4

are included. The value of the z–test statistic (9.16) for a logistic regression model against a product of unconstrained binomial distributions called G^2 is reported in the text for three of the models as follows:

All:	$G^2 = 54.00,$	df=58,	$0.20 < p < 0.30$
ABCEG:	$G^2 = 54.96,$	df=60,	$0.60 < p < 0.70$
AEG:	$G^2 = 62.30,$	df=62,	$0.40 < p < 0.50$

(a)   Form a table of z–test statistics (9.16) for all the models shown, and draw your conclusion as to what variables contribute significantly.

(b)   Justify your choice in (a) by a sequence of sequential tests.

(c)   Do you need more numbers than shown in the table to justify your conclusions?

(d)   Do you think that the table shown is the most informative one can imagine.

(e)   The coefficients in the table are for the regression model

$$E[\frac{P_\nu}{1-P_\nu}] = \beta_0 \beta_1^{z_{\nu 1}} \ldots \beta_p^{z_{\nu p}}.$$

How are the β's of (9.2) related to these new β's?

9.4. Reconsider the data in exercise 6.3.

If alcohol consumption is the response variable consider the following binary constructions:

(1) Less than 1 unit = 0

1 or more units = 1

(2) Up to 2 units = 0

More than 2 units = 1

(3) Less than 1 unit = 0

More than 2 units = 1

(4) 1–2 units = 0

More than 2 units = 1.

(a) Compare logistic regressions based on the variables in (1),(2),(3) and (4), when income is scored 25,75,125 and 175, marriage status is treated as a dummy variable and urbanization is scored Copenhagen=1, Suburban Copenhagen=2, Three largest cities=3, Other cities=4 and Country=5.

(b) In model (1) study Cook's distance to see if any data points have especially large influence on the analysis.

(c) If a computer program is available compare also with McCullogh's model.

9.5. The table show for 24 persons in the city of Elsinore in 1974 the observed values of the variables:

A: Phychiatrically demanding job

B: Sex

C: Household income

D: Household wealth

E: Age

Sex is scored

Male = 1

Female = 2

Income and wealth are both measured in 1000 Dkr. Age is measured in years.

(a) Formulate a logistic regression model with variable A as response variable and B,C,D and E as explanatory variables, after a suitable dichotomization of A, e.g.

Yes and partly = 1

No = 0.

(b) Which of the explanatory variables contribute to the description of variable A.

(c) Use regression diagnostics to describe the significance of the 24 cases individually.

(d) Suggest an analysis, where A is not dichotomized or dichotomized in a different way.

Case	Psychiatrically demanding job	Sex	Income	Wealth	Age
1	yes	2	16	0	36
2	no	1	4	0	26
3	no	2	7	3	57
4	yes	1	14	3	51
5	yes	1	8	2	52
6	no	2	7	0	31
7	partly	2	12	0	28
8	partly	2	12	8	41
9	yes	1	8	5	44
10	no	2	11	0	29
11	no	2	8	0	25
12	yes	1	12	5	32
13	partly	2	10	5	27
14	no	2	7	7	42
15	yes	2	20	2	42
16	yes	1	15	0	37
17	yes	1	8	0	28
18	no	1	13	2	37
19	yes	2	14	0	27
20	no	2	12	0	41
21	no	1	8	7	56
22	no	1	5	0	40
23	no	1	6	8	43
24	partly	1	14	8	53

9.6. We are interested in how dwellings, which are plagued by noise from the street, can be characterized. To do so the following variables were reported for a random sample of 36 persons in Elsinore.

A: Noise from the street

B: Sex

C: Number of inhabitants

D: Number of rooms in the dwelling

E: Household income

F: Age of the sampled person.

These variables for all 36 cases are reported below. Variable A is scored as

Much = 2

Some = 1

None = 0

Sex is scored with

Male = 1   Female = 2

Income is reported in 1000 Dkr. and age in years old.

Case	Noise	Sex	Number of inhabitants	Number of rooms	Income	Age
1	0	2	2	3	11	59
2	0	2	4	5	0	53
3	0	2	4	5	16	36
4	2	1	3	5	4	26
5	0	2	2	2	7	57
6	0	1	3	5	14	51
7	0	1	3	3	8	52
8	0	2	2	3	5	59
9	0	2	2	2	7	31
10	1	2	2	3	9	53
11	0	1	2	3	2	25
12	0	1	2	3	6	69
13	0	2	2	3	9	63
14	2	2	10	5	13	60
15	2	1	2	2	20	69
16	0	2	2	4	13	51
17	0	2	4	4	12	28
18	0	2	2	3	12	41
19	0	1	3	5	8	44
20	2	2	4	7	11	29
21	0	2	4	3	8	25
22	0	1	4	5	12	32
23	0	2	3	6	10	27
24	0	2	2	4	3	62
25	0	2	2	4	7	42
26	0	2	4	5	20	42
27	0	1	2	5	15	37
28	0	1	3	4	8	28
29	1	1	4	5	13	37
30	0	2	4	3	14	27
31	1	2	2	4	12	41
32	1	2	3	5	10	31
33	0	1	2	3	8	56
34	0	1	4	4	5	40
35	2	1	5	4	6	43
36	2	1	2	3	14	53

(a) Analyse the data by a logistic regression model with noise as response variabel and the remaining 5 variables as explanatory, and with a suitable dichotomization of noise, e.g.

> Much and some = 1
>
> None = 0.

(b) Use Cook's distance to determine if some of the cases influence the values of the estimates especially much.

(c) Suppose variable C is the only explanatory variable in the model. Make a graph of both the logit as a function of the values of variable C and a similar graph of the probability of noise in the dwelling.

9.7. From a research report with the title "Choice criteria for the assortment decisions of a supermarket chain", the following section is quoted directly:

"The final logit model in this category includes four predictors variables:

$$X_{212} = \text{Market leader}$$
$$X_{217} = \text{Sales index of the product}$$
$$X_{221} = \text{Size of the product group}$$
$$X_{223} = \text{Change in product group size.}$$

The estimated model is given by

$$\log \frac{P_2}{1-P_2} = -1.798 - 1.427X_{212} + 0.02573X_{217} + 0.2188X_{221} + 0.4268X_{223}$$
$$(1.161) \quad (0.7154) \quad (0.01030) \quad (0.07256) \quad (0.1581)$$

Prob–values for (one–sided) testing of the coefficient are given in the following table.

$H_0: \beta_{212} = 0$	$H_0: \beta_{217}$	$H_0: \beta_{221} = 0$	$H_0: \beta_{223} = 0$
$.01 < P < .025$	$.005 < P < .01$	$P < .005$	$P < .005$

The response variable is 1: retain the product at time of change in assortment and 2: delete the product. P_2 is the probability of response 1 and $1-P_2$ the probability of response 2".

(a)   Justify that a logistic model can be applied in this situation.

(b)   Comment on the regression equation.

(c)   Comment on the test statistics.

9.8. The Danish Institute for Borderline Research investigated in 1982 the labour mobility in the borderline area between Denmark and Germany. The two tables below show the response variables "Changed job within the last 10 years" and "Been unemployed within the last 5 years" cross–classified with the response variables sex and age.

(a)   Describe each of the two data sets by a logistic regression model.

(b)   Check the fit of the models.

(c)   Check if both explanatory variables are necessary to explain the variation in the response variable.

(d)   If Age is deleted as an explanatory variable can Sex then also be deleted without changing the fit of the model.

Sex	Age	Changed job Yes	No
Men	18–19	2	2
	20–24	23	2
	25–29	27	3
	30–34	30	4
	35–39	22	8
	40–44	11	11
	45–49	11	8
	50–54	5	10
	55–59	8	9
	60–	3	8
Women	18–19	1	0
	20–24	15	2
	25–29	19	3
	30–34	12	4
	35–39	15	11
	40–44	6	4
	45–49	4	8
	50–54	5	11
	55–59	0	6
	60–	1	0

Sex	Age	Been unemployed Yes	No
Men	18–19	1	4
	20–24	15	14
	25–29	12	19
	30–34	10	25
	35–39	4	28
	40–44	6	18
	45–49	6	15
	50–54	2	15
	55–59	3	19
	60–	1	13
Women	18–19	0	2
	20–24	12	10
	25–29	8	15
	30–34	3	16
	35–39	5	23
	40–44	1	14
	45–49	2	12
	50–54	3	18
	55–59	1	6
	60–	1	1

10. Models for the Interactions

10.1. Introduction

If the statistical analysis of a contingency table is based on one of the log–linear models in chapters 5, 6 and 7, a number of natural models are easily overlooked. Many useful models can thus be expressed as structures in the log–linear interaction parameters. In this chapter a number of such models are discussed. Many of these models can be viewed as attempts to describe the non–zero interactions by a simple structure if the analysis of the data by a log–linear model has failed to give a satisfactory fit to the model. If e.g. the independence hypothesis for a two–way table has been rejected, a residual analysis will often reveal a certain structure in the two–factor interactions.

10.2. Symmetry models

The most simple alternative to the independence hypothesis in a two–way table is an assumption of symmetric expected values. Assume thus that the table is squared, i.e. I=J, and consider the hypothesis

(10.1)
$$H_S: \mu_{ij} = \mu_{ji},$$

where $\mu_{ij}=E[X_{ij}]$. The model under H_S is a log–linear model, since the log–likelihood function for the Poisson model (4.1) under (10.1) become

$$\ln L = \underset{i<j}{\Sigma\Sigma} \, (x_{ij}+x_{ji})\lambda_{ij}+\Sigma x_{ii}\lambda_{ii}-\Sigma\Sigma\ln x_{ij}!-\lambda_{..}$$

The ML–estimates for the λ_{ij}'s are thus the solutions to

$$x_{ij}+x_{ji} = E[X_{ij}] + E[X_{ji}], \quad i \neq j$$

and

$$x_{ii} = E[X_{ii}].$$

The estimated expected numbers accordingly become

(10.2)
$$\hat{\mu}_{ij} = (x_{ij}+x_{ji})/2, \quad i{\neq}j$$

and

(10.3)
$$\hat{\mu}_{ii}= x_{ii} \quad .$$

Under the symmetry hypothesis (10.1) the expected values in the diagonal cells are thus the observed values, while outside the diagonal the expected numbers are the average of the observed values in the two cells symmetric with respect to the diagonal. The table of expected values is thus made symmetric in the intuitively obvious way.

The z–test statistic for H_S is given by

(10.4)
$$Z = 2 \sum_i \sum_j X_{ij}(\ln X_{ij}-\ln\hat{\mu}_{ij}).$$

By theorem 3.11 the test statistic (10.4) is approximately χ^2–distributed. The number of degrees of freedom is equal to the number of parameters specified under the hypothesis. Under a Poisson model, there are I^2 unrestricted λ's. Under H_S all λ's above the diagonal are equal to the symmetric value under the diagonal. Hence there are $I(I{-}1)/2$ constraints on the λ's under H_S. The degrees of freedom for (10.4) are accordingly

$$df(H_S) = I(I{-}1)/2$$

according to theorem 3.9.

The symmetry hypothesis and the test statistic (10.4) was suggested by Bowker (1948).

In terms of the log–linear interactions the symmetry hypothesis is equivalent to

(10.5)
$$\begin{cases} \tau^{AB}_{ij} = \tau^{AB}_{ji} \\ \\ \tau^{A}_{i} = \tau^{B}_{i} \end{cases}$$

since

$$\mu_{ij} = \exp(\tau^{AB}_{i\ j} + \tau^{A}_{i} + \tau^{B}_{j} + \tau_{0}).$$

Complete symmetry thus means not only that the interactions are symmetric, but that also the main effects for variable A and B are equal.

If it is only assumed that the two–factor interactions are symmetric, the hypothesis become

(10.6)
$$H_{IS}: \tau^{AB}_{i\ j} = \tau^{AB}_{j\ i}$$

The hypothesis (10.6) is called **quasi–symmetry** or **interaction symmetry**. The log–likelihood function under H_{IS} is

$$\ln L = \sum_{i<j} \sum (x_{ij}+x_{ji})\tau^{AB}_{i\ j}+x_{ii}\tau^{AB}_{i\ i}+x_{i.}\tau^{A}_{i}+x_{.j}\tau^{B}_{j}+x_{..}\tau_{0}$$

$$+\{\text{terms in } \tau\}+\{\text{terms in x}\}.$$

Hence the likelihood equations become

(10.7)
$$x_{ij}+x_{ji} = E[X_{ij}]+E[X_{ji}], \quad i\neq j,\ i=1,...,I-1,\ j=1,...,I-1$$

(10.8)
$$x_{ii} = E[X_{ii}], \quad i=1,...,I-1$$

(10.9)
$$x_{i.} = E[X_{i.}], \quad i=1,...,I-1$$

(10.10)
$$x_{.j} = E[X_{.j}], \quad j=1,...,I-1$$

(10.11)
$$x_{..} = E[X_{..}].$$

Note that (10.9) and (10.10) does not follow from (10.8) and (10.7) with i=I an j=I included. Equations (10.7) to (10.11) do not have explicit solutions for the expected va–

lues, but are easily solved numerically. The goodness of fit test statistic for the model under H_{IS} is given by (10.4), but the $\hat{\mu}_{ij}$'s are now the estimated expected values under the H_{IS}. Under H_{IS} the constraint $\tau_i^A = \tau_i^B$ is lifted and since there are I–1 free parameters $\tau_1^A,...,\tau_{I-1}^A$, there are I–1 less constraints under H_{IS} than under H_S.

Hence the degrees of freedom are

$$df(H_{IS}) = I(I-1)/2 - (I-1) = (I-1)(I-2)/2.$$

The degrees of freedom are thus the number of cells in the table under the diagonal, excluding the last row.

Example 10.1.

In connection with the Danish referendum on membership of the EEC, a number of polls were taken in 1971, 1972 and 1973. The data in table 10.1 shows the observed number of answers "yes", "no" and "undecided" to the question: "Should Denmark be a member of the EEC?" for the polls taken in 1971 and 1973.

Table 10.1. Attitude towards the EEC in October 1971 and in December 1973 for a random sample of 493 Danes.

| October 1971 | December 1973 | | | Total |
	Yes	No	Undecided	
Yes	167	36	15	218
No	19	131	10	160
Undecided	45	50	20	115
Total	231	217	45	493

Source: Unpublished data from AIM, Survey Company.

Table 10.2 show the observed values of the test statistic (10.4) under the hypotheses of complete symmetry and quasi–symmetry.

Table 10.2. Goodness of fit test statistics for symmetry and quasi–symmetry and the data in table 10.1.

Hypothesis	z	df	Significance level
Symmetry	50.15	3	0.000
Quasi–symmetry	0.06	1	0.812

Under the quasi–symmetry hypothesis there is thus an almost perfect fit, while under complete symmetry, the fit is less than satisfactory.

The expected values under both symmetry hypotheses are shown in table 10.3. The totals of table 10.3 reveal that the expected distribution over the three categories is the same in 1971 as in 1973 under complete symmetry, while this is not the case under quasi–symmetry. Although the interactions are symmetric under H_{IS}, the fact that τ_i^A is not necessarily equal to τ_i^B implies that the expected number of supporters and opponents of the EEC and the undecided were not the same in 1971 and 1973.

Table 10.3. Expected numbers under the hypotheses of symmetry and quasi–symmetry for the data in table 10.1.

Symmetry:		December 1973 Yes	No	Undecided	Total
October 1971	Yes	167.00	27.50	30.00	224.50
	No	27.50	131.00	30.00	188.50
	Undecided	30.00	30.00	20.00	80.00
	Total	224.50	188.50	80.00	493.00

Quasi–symmetry:		December 1973 Yes	No	Undecided	Total
	Yes	167.00	35.56	15.44	218.00
October 1971	No	19.44	131.00	9.56	160.00
	Undecided	44.56	50.44	20.00	115.00
	Total	231.00	217.00	45.00	493.00

The estimated parameters under H_{IS} are as follows

$\hat{\tau}_{ij}^{AB}$	j=1	2	3
i=1	0.931	−0.778	−0.153
2	−0.778	0.968	−0.191
3	−0.153	−0.191	0.343

	i=1	2	3
$\hat{\tau}_i^A$	0.227	−0.216	−0.011

	j=1	2	3
$\hat{\tau}_j^B$	0.379	0.540	−0.919

The fact that the interaction parameters of a log–linear model describe the departures from independence allows for an interpretation of quasi–symmetry. Independence in a two–way table means that the expected frequencies in the rows of the table are identical. Under this assumption the probabilities of responding "yes", "no" and "undecided" in 1973 are the same whatever the opinion in 1971. Quasi–symmetry thus means that the departures from this behaviour is symmetric. The interpretation of the results of the analysis thus seems to be that the movements of opinion between October 1971 and December 1973 is symmetric apart from a trend towards fewer undecided and more opponents in December 1973. It is primarily the undecided in 1971, which have changed to a no in 1973. Denmark joined the EEC in 1972. The effect of the membership thus seems to have been a decrease in the number of undecided persons and a increase in the number of opponents of the EEC. The number of supporters of the EEC does not seem to have changed from 1971 to 1973. △.

An example of quasi–symmetry based on a conditional likelihood is discussed in McCullogh (1982).

Many models for contingency tables are connected with so–called **Markov chains**. A Markov chain is a random process, where individuals change between a limited number of states at certain points in time called **change points**. Let $p_j^{(0)}$ be the probability that an individual is in state j at time 0, $p_j^{(t)}$ the probability that an individual is in state j at change point t, and p_{ij} the (constant) probability that an individual change from state i to state j at any of the change points. Then the probabilities of being in states j=1,...,m at change point one are

(10.12)
$$p_j^{(1)} = \sum_{i=1}^{m} p_{ij} p_i^{(0)},$$

and in general

(10.13)
$$p_j^{(t)} = \sum_{i=1}^{m} p_{ij} p_i^{(t-1)}.$$

From (10.12) and (10.13) the values of $p_j^{(t)}$ can be computed recursively for any t. If t→∞, there exists under certain conditions limiting probabilities

$$(10.14) \qquad\qquad \pi_j = \lim_{t \to \infty} p_j^{(t)}.$$

The probability distribution $(\pi_1, \ldots, \pi_m)$ of being in the m states after the process has been observed over a long period is called the **equilibrium distribution**. It can be shown that the π_j's are uniquely determined by

$$(10.15) \qquad\qquad \pi_j = \sum_{i=1}^{m} p_{ij} \pi_i$$

if the limits (10.14) exist, and that the equilibrium distribution does not depend on the values of $p_1^{(0)}, \ldots, p_m^{(0)}$. One may think of the equilibrium distribution as the expected frequencies in the m states if a large number of persons start at the same time and move between the states over many change points. The data in example 10.1 can be thought of as one change point in a Markov chain. The change probabilities

$$p_{ij} = P\{\text{move to state } j | \text{ given state } i\}.$$

are called **transition probabilities**. They are connected with the expected numbers μ_{ij} in the contingency table by

$$p_{ij} = \mu_{ij} / \mu_{i.}.$$

Statistical inference in Markov chains from a contingency table point of view was first studied by Andersson and Goodman (1957).

If the matrix of transition probabilities is symmetric, i.e. if $p_{ij} = p_{ji}$, the Markov chain is called **double stochastic**. It then follows that

$$\pi_j = \sum_i p_{ij} \frac{1}{m} = \frac{1}{m} \sum_i p_{ji} = \frac{1}{m} p_{.i} = \frac{1}{m},$$

such that the equilibrium is uniform, if the Markov chain is double stochastic.

Under a log–linear parameterization, p_{ij} has the form

(10.16)
$$p_{ij} = \exp(\tau_{ij}^{AB} + \tau_j^B) / \sum_j \exp(\tau_{ij}^{AB} + \tau_j^B).$$

The condition $p_{ij} = p_{ji}$ then implies

$$\tau_{ij}^{AB} = \tau_{ji}^{AB} + \epsilon_i + \delta_j,$$

for certain values ϵ_i and δ_j. But since $\tau_{i.}^{AB} = \tau_{.i}^{AB}$, this means that $\epsilon_i + \delta_. = \epsilon_. + \delta_j = 0$. From $\tau_{..}^{AB} = 0$ then follows

(10.17)
$$\tau_{ij}^{AB} = \tau_{ji}^{AB}.$$

To a doubly stochastic Markov chain thus corresponds a contingency table, which is quasi–symmetric. From (10.16) follows that the hypothesis of quasi–symmetry entails that the Markov chain is only double stochastic if all τ_j^B's are constant, i.e.

$$\tau_j^B = 0, \quad j=1,...,I.$$

Whether it is double stochastic does not depend on the τ_i^A's.

Example 10.1. (Continued)

The observed numbers in table 10.1 correspond to a Markov chain, where the population in Danmark moved between three categories of opinion towards the EEC between Octo–

ber 1971 and December 1973. Suppose the attitude towards the EEC continue to change according to a Markov chain, which is observed for example every second year. One would then in the long run find a uniform distribution of the population over the three categories, if the Markov chain is double stochastic. Since we found that a hypothesis of quasi- - symmetry fitted the data, we must next check if $\tau_1^B = ... = \tau_3^B = 0$. The parameter estimates 0.377, 0.539 and −0.916 clearly indicate, that this is not the case. The equilibrium distribution corresponding to the transition probabilities estimated from table 10.1 is

$$(\pi_1, \pi_2, \pi_3) = (0.393, 0.534, 0.073).$$

If a Markov chain model applies to the process, the opposition to the EEC would thus gradually increase in Denmark to a level of about 53%. $\triangle$.

Symmetry models can be extended to higher dimensions in a number of ways. For a three–way table **complete symmetry** means that

(10.18)
$$\mu_{ijk} = \mu_{ikj} = \mu_{jik} = \mu_{kij} = \mu_{kji} = \mu_{jki}.$$

Quasi–symmetry may means that

(10.19)
$$\tau_{ijk}^{ABC} = \tau_{ikj}^{ABC} = \tau_{jik}^{ABC} = \tau_{kij}^{ABC} = \tau_{kji}^{ABC} = \tau_{jki}^{ABC},$$

while nothing is assumed about the two–factor interactions or about the main effects. The hypothesis may also be, however, that both the three–factor interactions and the two–factor interactions are symmetric. **Conditional symmetry** given the levels of variable C means that

$$\mu_{ijk} = \mu_{jik}, \quad k=1,...,K.$$

since the conditional mean values are $\mu_{ijk}/\mu_{..k}$.

A useful hypotheses is **marginal quasi–symmetry** between variables A and B, defined as

$$\tau^{ABC}_{i\,j\,k} = 0, \text{ for all i, j and k}$$

and

$$\tau^{AB}_{i\,j} = \tau^{AB}_{j\,i}, \text{ for all i and j,}$$

while nothing is assumed about the remaining parameters. This hypothesis is different from complete marginal symmetry between variables A and B, which requires that

$$\mu_{ij.} = \mu_{ji.}, \text{ for all i and j.}$$

Goodman (1971), (1985) discussed various symmetry and quasi–symmetry models and related them to the models of sections 10.5 and 10.6 below. Read (1978) discussed symmetry models for three–dimensional tables.

10.3. Marginal homogeneity

A useful hypothesis, which can not be expressed in terms of log–linear parameters is **marginal homogeneity**. A squared two–way table of dimension I is said to satisfy the hypothesis of marginal homogeneity if the expected values satisfy

$$(10.20) \qquad \mu_{i.} = \mu_{.i}, \quad i=1,...,I .$$

A maximalization of the likelihood under the constraints (10.20) does not lead to explicite solutions, but solutions are easily obtained by numerical methods.

Let the solutions be $\hat{\mu}_{11},...,\hat{\mu}_{II}$, then

$$(10.21) \qquad Z = 2\Sigma_i \Sigma_j X_{ij}[\ln X_{ij} - \ln\hat{\mu}_{ij}] \sim \chi^2(I-1),$$

according to the general result, theorem 3.13. The degrees of freedom are $I-1$ since one of the constraints in (10.20) is redundant such that there are $I-1$ constraints under the hypothesis.

A survey of various test for marginal homogeneity is due to White et al. (1982).

The hypothesis of marginal homogeneity is useful if one is not interested in the way variables A and B interact, but only want to check if the two variables are distributed marginally in the same way. Independence between A and B combined with marginal homogeneity implies complete symmetry.

Marginal homogeneity is of special interest in connection with Markov chains. Let as above p_{ij} be the transition probability from state i to state j, and let $p_i^{(t)}$ be the probability of an individual being in state i at change point t. Then the observed number of changes x_{ij} from state i to state j from time t to time $t+1$ satisfies

$$E[X_{ij}] = np_{ij}p_i^{(t)} = \mu_{ij}.$$

If the μ_{ij}'s satisfy the hypothesis (10.20) of marginal homogeneity, it follows that

$$p_j^{(t+1)} = \sum_{i=1}^{I} p_{ij}p_i^{(t)} = \frac{1}{n}\mu_{\cdot j} = \frac{1}{n}\mu_{j\cdot} = \sum_{l=1}^{I} p_{jl}p_j^{(t)} = p_j^{(t)}, \quad j=1,\dots,I$$

such that $p_j^{(t)}$ is independent of t and the process has reached its equilibrium. If the hypothesis of marginal homogeneity is accepted, when the contingency table is the observed changes at a given change point in a Markov chain, the data thus supports that the process has reached its equilibrium.

A test for marginal homogeneity is not identical with a marginal test of equal expected marginals, i.e. a test of homogeneity between the two marginals of the table. A basic condition for a formal homogeneity test is in this case not satisfied, since the two distributions are based on the same individuals and accordingly are not independent.

Example 10.2

In 1962 and 1965 two surveys were conducted in Denmark regarding the health of the elderly population. As part of the surveys the interviewed persons were asked to rate their health on a three level scale with categories "good", "neither good nor bad" and "bad". Table 10.4 shows the cross–classification of the responses for those 411 elderly people participating in both surveys and giving responses on the subjective health questions on both occasions.

Table 10.4. Cross–classification of responses on subjective self-rating of health for 411 elderly people in Denmark in 1962 and 1965.

Health 1962	Health 1965 Good	Neither	Bad	Total
Good	168	51	9	228
Neither	42	73	23	138
Bad	5	17	23	45
Total	215	141	55	411

Source: Data from the Danish National Institute for Social Research. Cf. Olsen and Hansen (1977).

The expected numbers under the hypothesis of marginal homogeneity are shown in table 10.5. Based on these numbers, the test statistic (10.21) has observed value

$$z = 2.92, \quad df = 2.$$

Table 10.5. Expected numbers for the data in table 10.4 under the hypothesis of marginal homogeneity.

Health 1962	Health 1965 Good	Neither	Bad	Total
Good	168.00	46.33	7.14	221.47
Neither	46.70	73.00	19.84	139.54
Bad	6.77	20.21	23.00	49.98
Total	221.47	139.54	49.98	411.00

Accordingly there does not seem to be any change in the distribution over health categories between 1962 and 1965. Note that marginal homogeneity does not imply independence between the responses. △.

In a 2x2 table, the hypotheses of symmetry and marginal homogeneity coincides.
Since

$$\mu_{1.} = \mu_{11} + \mu_{12}$$

and

$$\mu_{.1} = \mu_{11} + \mu_{21},$$

$\mu_{1.} = \mu_{.1}$ if and only if $\mu_{12} = \mu_{21}$.

Thus marginal homogeneity, i.e. $\mu_{1.}=\mu_{.1}$, implies $\mu_{12}=\mu_{21}$ or symmetry, and symmetry, i.e. $\mu_{12}=\mu_{21}$, implies marginal homogeneity or $\mu_{.1}=\mu_{1.}$. The test statistic for symmetry or for marginal homogeneity in a 2x2 table is

$$Z = 2 \sum_{i=1}^{2} \sum_{j=1}^{2} X_{ij} \left[\ln X_{ij} - \ln \frac{X_{ij}+X_{ji}}{2} \right],$$

and Z is approximately χ^2–distributed with one degree of freedom. The corresponding Pearson test statistic

$$Q = \sum_{i=1}^{2} \sum_{j=1}^{2} (X_{ij} - \frac{X_{ij}+X_{ji}}{2})^2 / (\frac{X_{ij}+X_{ji}}{2})$$

is widely known as **McNemars test statistic,** cf. Mc Nemar (1947). It is often presented as

$$q = \frac{(b-c)^2}{b+c},$$

where $b=x_{12}$ and $c=x_{21}$.

If

(10.22)
$$\begin{cases} \tau_{ij}^{AB} = 0 \\ \tau_i^A = \tau_i^B \end{cases}$$

the hypothesis of marginal homogeneity is combined with independence of the two variables. The expected numbers under (10.22) are

$$\hat{\mu}_{ij} = \frac{(x_{i.} + x_{.i})(x_{j.} + x_{.j})}{4n}$$

and the test statistic has $I^2 - (I-1) - 1 = I(I-1)$ degrees of freedom, since only $\tau_1^A, ..., \tau_{I-1}^A$ and τ_0 need to be estimated under (10.22).

10.4. Models for mobility tables

Some squared contingency tables describe the mobility pattern of a population. In a mobility table it is often assumed that the probability of observing an individual in a cell, is smaller the farther away from the diagonal the cell is located. This will be the case if the expected numbers μ_{ij} satisfy

(10.23) $$\ln \mu_{ij} = \delta |i-j| + \tau_i^A + \tau_j^B + \tau_0, \quad \delta < 0.$$

This model generates a smaller expected value in cell (i,j) the larger the value of $|i-j|$, which is a simple measure of the distance to the diagonal. The model is log–linear and the likelihood equations become

(10.24) $$\sum_{i \neq j} \sum |i-j| x_{ij} = \sum_{i \neq j} \sum |i-j| E[X_{ij}],$$

(10.25) $$x_{i.} = E[X_{i.}], \quad i=1,...,I$$

and

(10.26) $$x_{.j} = E[X_{.j}], \quad j=1,...,I.$$

which are easily solved by iterative methods.

In most mobility tables there is a clear tendency against changing group. Model (10.23) will therefore often fail to describe the data due to larger observed values than expected in the diagonal. An alternative model is to assume that (10.23) is only valid for the off–diagonal cells, i.e.

(10.27)
$$\ln\mu_{ij} = \begin{cases} \delta|\,i-j\,| + \tau_i^A + \tau_j^B + \tau_0, & i \ne j \\ \epsilon_i + \tau_i^A + \tau_i^B + \tau_0, & i = j. \end{cases}$$

This model is equivalent to (10.23) for $i \ne j$ combined with

(10.28)
$$x_{ii} = E[X_{ii}], \quad i=1,...,I,$$

since the ϵ_i's are unrestricted.

Also for (10.27) the ML–estimates for the parameters and the expected numbers are easily derived by iterative procedures. The goodness of fit test statistic is as usual given by

$$Z = 2\Sigma \; \Sigma X_{ij}(\ln X_{ij} - \ln\hat{\mu}_{ij}).$$
$$\quad\; i \;\; j$$

For model (10.23) there are $I-1$ τ^A's, $I-1$ τ^B's, one τ_0 and one δ to be estimated and the degrees of freedom for the approximating χ^2–distribution to Z is $I^2-2I=I(I-2)$. For model (10.27) there are in addition I ϵ's to be estimated and the degrees of freedom become $I(I-3)$.

Example 10.3

One of the famous data sets concerning social mobility is due to Svalastoga (1959) cf. also Haberman (1974b), chapter 6. Based on the ratings of a random sample from the Danish population, Svalastoga established five social rank groups with group I being the group of highest social ranking. This social grouping is with minor modifications still used in Denmark. As part of the investigation Svalastoga reported for all males in the sample the connection between the social rank of the interviewed and of his father. The resulting data are shown in table 10.6.

Table 10.6. The relationship between fathers and sons social rank for a random sample of sons.

Fathers social rank	Sons social rank					Total
	I	II	III	IV	V	
I	18	17	16	4	2	57
II	24	105	109	59	21	318
III	23	84	289	217	95	708
IV	8	49	175	348	198	778
V	6	8	69	201	246	530
Total	79	263	658	829	562	2391

Source: Svalastoga (1959).

By iterative procedures, the solutions to (10.24), (10.25), (10.26) and (10.28) are found to be

$$\hat{\tau}_0 = 5.184$$

$$(\hat{\tau}_1^A,...,\hat{\tau}_5^A) = (-1.280, -0.049, +0.484, +0.486, +0.360)$$

$$(\hat{\tau}_1^B,...,\hat{\tau}_5^B) = (-0.994, -0.362, 0.345, 0.552, 0.458)$$

$$\hat{\delta} = -0.822$$

$$(\hat{\epsilon}_1,...,\hat{\epsilon}_5) = (-0.020, -0.119, -0.347, -0.370, -0.498).$$

The expected numbers given these estimates are shown in table 10.7. Note that the expected numbers in the diagonal are equal to the observed numbers.

Table 10.7. Expected values under the social mobility model (10.27) for the data in table 10.6.

Sons social rank

Fathers social rank	I	II	II	IV	V	Total
I	18.00	15.19	13.55	7.33	2.93	57.00
II	27.63	105.00	105.48	57.06	22.84	318.01
III	20.71	88.62	289.00	221.15	88.52	708.00
IV	9.12	39.04	180.12	348.00	201.71	777.99
V	3.54	15.14	69.86	195.46	246.00	530.00
Total	79.00	262.99	658.01	829.00	562.00	2391.00

The test statistic based on the expected numbers in table 10.7 has observed value

$$z = 13.00, \quad df = 10.$$

With level of significance p=0.328, the model describes the data in a satisfactory way. △.

10.5. Association models

In a number of situations the relationship between two categorical variables can be modelled in terms of scores assigned to the categories of the variables. These scores may be known or unknown. With a somewhat misleading name such models are often termed models with ordered categories, cf. Agresti (1983) or Goodman (1979b). The models are now widely known as **row–column association models**, or RC–association models. The name is due to Goodman (1981a).

Consider first the case where the scores for both variable A and variable B are known. If the row scores are denoted $e_1,...,e_I$ and the column scores $d_1,...,d_J$, an association model specifies that the interactions between variables A and B have the multiplicative form

(10.29) $$\tau_{ij}^{AB} = \rho e_i d_j .$$

In order to remove the indeterminancies in the parametrization it is necessary to introduce constraints. It is common practice to use the following constraints

$$(10.30) \quad \begin{cases} \sum\limits_{i=1}^{I} e_i x_{i.}/n = \sum\limits_{j=1}^{J} d_j x_{.j}/n = 0 \\ \sum\limits_{i=1}^{I} e_i^2 x_{i.}/n = \sum\limits_{j=1}^{J} d_j^2 x_{.j}/n = 1 \end{cases}$$

The constraints (10.30) do not conform with

$$\tau_{i.}^{AB} = \tau_{.j}^{AB} = 0,$$

but with

$$\sum_{j} \tau_{ij}^{AB} x_{.j}/n = \sum_{i} \tau_{ij}^{AB} x_{i.}/n = 0 .$$

Under (10.29) the expected values satisfy

$$(10.31) \quad \ln\mu_{ij} = \rho e_i d_j + \tau_i^A + \tau_j^B + \tau_0,$$

where τ_i^A, τ_j^B and τ_0 are main effects.

Model (10.31) was termed **a uniform association model** by Goodman (1979b) in case the known scores are equidistant, i.e. the e_i's are rescaled values of $1,...,I$ and the d_j's rescaled values of $1,2,...,J$. The motivation for this name is given in section 10.7 below.

One important implication of (10.29) is that the expected value in a cell is larger than under independence if e_i and d_j have the same sign, and smaller than under independence, if e_i and d_j have different signs. In addition the larger the numerical value of e_i or d_j, the more will τ_{ij}^{AB} differ from what one gets under independence. Note, however, that the effect on the expected value μ_{ij} of a change in e_i depends on the value of d_j.

In order to illustrate the effect of the model (10.29) on the expected values, consider a hypothetical model with $\tau_i^A = \tau_j^B = 0$ for all i and j and assume that both A and B have equidistant scores, i.e. the differences $e_i - e_{i-1}$ and $\delta_j - \delta_{j-1}$ are constant. Under (10.30) and if the marginal frequencies are constant this implies that

$$(e_1,...,e_4) = (-0.671, -0.224, +0.224, +0.671)$$

and

$$(d_1,...,d_5) = (-0.632, -0.316, 0.000, +0.316 +0.632)$$

Let n=400. If $\tau_{ij}^{AB}=0$, in addition to $\tau_i^A=\tau_j^B=0$, then all expected numbers are equal to 20. If, on the other hand, $\rho=2$ the expected numbers become

$$\mu_{ij} = \exp(2e_i d_j + 2.896),$$

where the value $\tau_0=2.896$ is obtained from $\mu_{..}=400$. Note that τ_0 under (10.31) is different from τ_0 when $\rho=0$, where $\tau_0 = \ln 20 = 2.996$. The reason is that the terms $\exp(\rho e_i d_j)$ do not necessarily sum to n/(IJ). The expected values $\mu_{ij}=\exp(2e_i d_j + 2.896)$ are shown in table 10.8.

Table 10.8. Expected values under model (10.29) with $\tau_i^A=\tau_j^B=0$, $\rho=2$ and n=400. The values of e_i and d_j are shown in the marginals of the table.

	j=1	2	3	4	5	e_i
i=1	42.3	27.5	18.1	11.9	7.8	−0.671
2	24.0	20.8	18.1	15.7	13.7	−0.224
3	13.7	15.7	18.1	20.8	24.0	+0.224
4	7.8	11.9	18.1	27.7	42.3	+0.671
d_j	−0.632	−0.316	0.000	+0.316	+0.632	

Table 10.8 shows how the category scores influence the expected values. The larger the value of the product $e_i d_j$ the larger the expected number. There will accordingly be a large expected number if either both scores are large and positive or if they are both large and negative.

The log–likelihood function for (10.31) is

$$\ln L = \rho \sum_{i=1}^{I} \sum_{j=1}^{J} x_{ij} e_i d_j + \sum_i \tau_i^A x_{i.} + \sum_j \tau_j^B x_{.j} + x_{..} \tau_0$$

$$+ \{\text{terms in the parameters}\}$$

$$+ \{\text{terms in the observations}\}.$$

such that the model is log–linear. The likelihood equations become

(10.32)
$$\sum_i \sum_j x_{ij} e_i d_j = \sum_i \sum_j e_i d_j E[X_{ij}]$$

(10.33)
$$x_{i.} = E[X_{i.}], \text{ for all i,}$$

and

(10.34)
$$x_{.j} = E[X_{.j}], \text{ for all j.}$$

These equations are easily solved by iterative methods. The structure of the likelihood equations thus allow for a modification of the marginal proportional fitting procedure as suggested by Goodman (1979b). The procedure is described in section 10.6 below.

The goodness of fit of the model is tested in the usual way by the test statistic

(10.35)
$$Z = 2\sum_i \sum_j X_{ij}(\ln X_{ij} - \ln\hat{\mu}_{ij}),$$

with

$$\hat{\mu}_{ij} = \exp(\hat{\rho} e_i d_j + \hat{\tau}_i^A + \hat{\tau}_j^B + \hat{\tau}_0).$$

The number of degrees of freedom for the approximating χ^2–distribution is

$$df = (I-1)(J-1)-1,$$

since in addition to the main effects ρ needs to be estimated.

Example 10.4.

The data set in table 10.9 was first analysed by the Danish statistician Georg Rasch as an illustration of a type of statistical model closely related with the model of this section, which Rasch introduced in 1964, cf. Rasch (1966) and the discussion in Goodman (1986). The observations in table 10.9 are the number of criminal cases for young men, 15 to 19 years of age, for the years 1955 to 1958, where charges were dropped by the police before the case had lead to a verdict.

Table 10.9. Number of criminal cases for young men between age 15 and age 19 in Denmark 1955 to 1958, where charges were dropped by the police.

			Age			
Year	15	16	17	18	19	Total
1955	141	285	320	441	427	1614
1956	144	292	342	441	396	1615
1957	196	380	424	462	427	1889
1958	212	424	399	442	430	1907
Total	693	1381	1485	1786	1680	7025

Source: Rasch (1966), F.3 Tabel 1.

A first glance at the data seems to indicate that the number of dropped charges increased more rapidly with age in 1955 than in 1958. If this observation is correct, an assumption of independence between the variation over years and the variation over age does not hold. The observed value of (10.35) under independence is

$$z = 38.25, \quad df = 12.$$

The level of significance is less than 0.0005 confirming that an independence model does not describe the data. In order to get a first impression of the applicability of a multiplicative model (10.29) for the interactions, the estimates of τ_{ij}^{AB} are shown in table 10.10.

Table 10.10. Estimated interactions τ_{ij}^{AB} between year and age for the data in table 10.9.

			Age		
Year	15	16	17	18	19
1955	−0.089	−0.076	−0.041	+0.089	+0.118
1956	−0.075	−0.060	+0.018	+0.081	+0.035
1957	+0.052	+0.022	+0.051	−0.054	−0.071
1958	+0.112	+0.114	−0.028	−0.116	−0.082

The sign pattern of the estimated interactions strongly suggests a multiplicative model for the interactions. A set of equidistant row scores satisfying (10.30) have values

$$(e_1,...,e_4) = (1.568, 0.567, -0.435, -1.437).$$

The equidistant scores for the columns are

$$(d_1,...,d_5) = (1.665, 0.978, 0.291, -0.396, -1.083).$$

From these values an initial estimate of ρ is easily obtained. The definition (10.29) of ρ in terms of τ_{ij}^{AB} and the constraints (10.30) imply that

$$\sum_i \sum_j e_i d_j \tau_{ij}^{AB} x_{i.} x_{.j} / n^2 = \rho.$$

Hence an initial estimate for ρ is

$$\hat{\rho}_0 = \sum_i \sum_j e_i d_j \hat{\tau}_{ij}^{AB} x_{i.} x_{.j} / n^2.$$

From the estimates in table 10.10, we get

$$\hat{\rho}_0 = -0.063.$$

Thus an impression of the fit of model (10.31) is rendered by table 10.11 showing the values of $\hat{\rho}_0 e_i d_j$.

Table 10.11. Values of the expression $\hat{\rho}_0 e_i d_j$.

Year	Age				
	15	16	17	18	19
1955	−0.166	−0.097	−0.029	+0.039	+0.107
1956	−0.060	−0.035	−0.010	+0.014	+0.039
1957	+0.046	+0.027	+0.008	−0.011	−0.030
1958	+0.152	+0.089	+0.027	−0.036	−0.098

To a certain extent the pattern of table 10.10 is thus reproduced. The ML–estimates for ρ and the main effects τ_i^A, τ_j^B, τ_0 obtained by solving the likelihood equations (10.32), (10.33) and (10.34), are found to be

$$\hat{\rho} = -0.065,$$
$$\hat{\tau}_i^A = (-0.083, -0.082, +0.071, +0.074),$$
$$\hat{\tau}_j^B = (-0.779, -0.084, -0.008, +0.176, +0.113),$$
$$\hat{\tau}_0 = 5.927.$$

The observed value of the test statistic (10.35) is

$$z = 8.61, \quad df=11.$$

The level of significance is p=0.658, such that the model (10.31) with an equidistant scoring of the categories describes the given data adequately. The negative value of $\hat{\rho}$ shows that relative more charges were dropped for the 19 year old in 1955 than in 1958, while for 15 year old boys relatively more charges were dropped in 1958 than in 1955. △.

Consider now a mixed model with interactions

$$\tau_{ij}^{AB} = \rho \epsilon_i d_j,$$

i.e. a model with

$$(10.36) \qquad \ln\mu_{ij} = \rho\epsilon_i d_j + \tau_i^A + \tau_j^B + \tau_0,$$

where $d_1,...,d_I$ are known while the ϵ's and ρ are unknown. This model appears among other places in Haberman (1974a), Simon (1974), Plackett (1981), p.75 and Andersson (1984). It was called a **row effects association model** by Goodman (1979b) for equidistant d_j's.

For the ϵ's the same constraints as for the e's are imposed, i.e.

$$(10.37) \qquad \sum_{i=1}^{I} \epsilon_i x_{i.}/n = 0$$

and

$$(10.38) \qquad \sum_{i=1}^{I} \epsilon_i^2 x_{i.}/n = 1$$

At first glance the model is not log–linear, since the likelihood function has the form

$$\ln L = \rho\sum_i \epsilon_i \sum_j d_j x_{ij} + \sum_i \tau_i^A x_{i.} + \sum_j \tau_j^B x_{.j} + \tau_0 x_{..}.$$

The model can, however, be reparameterized through the I–1 parameters

$$\epsilon_i^* = \rho\epsilon_i, \quad i=1,...,I-1$$

satisfying $\sum_i \epsilon_i^* x_{i.}/n=0$. Since ρ is found as the square root of

$$\rho^2 = \sum_i (\epsilon_i^*)^2 x_{i.}/n.$$

and $\epsilon_i=\epsilon_i^*/\rho$, the reparameterization is one to one. Note that ρ can always be chosen larger than zero when the ϵ's are unknown. In terms of the ϵ_i^*'s the model is log–linear and the likelihood equations consists of

(10.39)
$$\sum_j d_j x_{ij} = \sum_j d_j E[X_{ij}], \quad \text{for all } i$$

together with (10.33) and (10.34).

The goodness of fit of the model is tested by the test statistic (10.35) with

$$\hat{\mu}_{ij} = \exp(\hat{\hat{\rho}} \hat{\epsilon}_i d_j + \hat{\tau}_i^A + \hat{\tau}_j^B + \hat{\tau}_0).$$

The degrees of freedom for the approximating χ^2-distribution is

$$df = I^2 - (I-1) - (J-1) - 1 - (I-1) = (I-1)(J-2),$$

since there are I–1 τ_i^A's, J–1 τ_j^B's, one τ_0 and I–1 ϵ_i^*'s to be estimated.

Row effects models when the unknown scores satisfy the order restriction $\epsilon_1 \leq \ldots \leq \epsilon_I$ were studied by Agresti, Chuang and Kezouh (1987).

The equivalent of (10.36) if the scores for variable A are known and those for variable B unknown, has

(10.40)
$$\ln\mu_{ij} = \rho e_i \delta_j + \tau_i^A + \tau_j^B + \tau_0.$$

This model will be called a **column effects association model**.

The likelihood equations for (10.40) are (10.33) and (10.34) combined with

(10.41)
$$\sum_i e_i x_{ij} = \sum_i e_i E[X_{ij}], \quad \text{for all } j$$

and the goodness of fit is tested by (10.35) with

$$\hat{\mu}_{ij} = \exp(\hat{\rho} e_i \hat{\delta}_j + \hat{\tau}_i^A + \hat{\tau}_j^B + \hat{\tau}_0).$$

The degrees of freedom for the approximating χ^2–distribution is in this case $df=(I–2)(J–1)$.

10.6. RC–association models

Consider now the situation where both sets of scores are unknown, i.e.

$$\tau_{ij}^{AB} = \rho\epsilon_i\delta_j,$$

where the mean value μ_{ij} has the form

(10.42)
$$\ln\mu_{ij} = \rho\epsilon_i\delta_j + \tau_i^A + \tau_j^B + \tau_0,$$

The ϵ's and δ's satisfy

(10.43)
$$\begin{cases} \sum_i \epsilon_i\, x_{i.}/n = \sum_j \delta_j\, x_{.j}/n = 0 \\ \sum_i \epsilon_i^2\, x_{i.}/n = \sum_j \delta_j^2\, x_{.j}/n = 1 \end{cases}$$

The model (10.42) has been studied in the literature under various forms by many authors, e.g. Rasch (1966). It was brought to general prominence in the mid– and late 1970's by Haberman (1974b), Simon (1974) and Goodman (1979b),(1981a). An excellent survey paper is due to Agresti (1983). An important reference is also the monograph by Agresti (1982). The usefulness of the model in sociological research was demonstrated by Clogg (1982b). We shall term the model (10.42) an **RC–association model** in accordance with Goodman (1981a).

The RC–association model is not log–linear in its parameters. In fact the log–likelihood function is given by

(10.44)
$$\ln L = \rho \Sigma \Sigma \epsilon_i \delta_j x_{ij} + \Sigma x_{i.} \tau_i^A + \Sigma x_{.j} \tau_j^B + x_{..} \tau_0$$

+ {terms in the parameters}

+ {terms in the observations}.

The equations necessary to solve in order to obtain the ML–estimates are, therefore, more complicated.

Theorem 10.1.

The likelihood equations for the RC–association model (10.42) have the same solutions as

(10.45)
$$\sum_{i=1}^{I} \epsilon_i x_{ij} = \sum_{i=1}^{I} \epsilon_i \mu_{ij}, \quad j=1,...,J,$$

(10.46)
$$\sum_{j=1}^{J} \delta_j x_{ij} = \sum_{j=1}^{J} \delta_j \mu_{ij}, \quad i=1,...,I,$$

(10.47)
$$x_{i.} = \mu_{i.}, \quad i=1,...,I$$

and

(10.48)
$$x_{.j} = \mu_{.j}, \quad j=1,...,J.$$

where μ_{ij} is the expected value of X_{ij} under the model.

Proof:

If the model is a Poisson model, i.e. if $X_{11},...,X_{IJ}$ are independent,

$$X_{ij} \sim Ps(\lambda_{ij}), \quad i=1,...,I, \quad j=1,...,J$$

and $\lambda_{ij} = \mu_{ij}$ satisfies (10.42), the log–likelihood function become

$$\ln L = \rho \sum_i \sum_j \epsilon_i \delta_j x_{ij} + \sum_i x_{i.} \tau_i^A + \sum_j x_{.j} \tau_j^B + x_{..} \tau_0 - \sum_i \sum_j \ln(x_{ij}!) - \sum_i \sum_j \mu_{ij}$$

Taking partial derivatives with respect to the parameters and using $\mu_{ij} = \exp(\rho \epsilon_i \delta_j + \tau_i^A + \tau_j^B + \tau_0)$ yields the likelihood equations

(10.49)
$$\frac{\partial \ln L}{\partial \rho} = \sum_i \sum_j \epsilon_i \delta_j x_{ij} - \sum_i \sum_j \epsilon_i \delta_j \mu_{ij} = 0.$$

(10.50)
$$\frac{\partial \ln L}{\partial \epsilon_i} = \rho \sum_j \delta_j x_{ij} - \rho \sum_j \delta_j \mu_{ij} = 0, \; i=1,...,I-2.$$

(10.51)
$$\frac{\partial \ln L}{\partial \delta_j} = \rho \sum_i \epsilon_i x_{ij} - \rho \sum_i \epsilon_i \mu_{ij} = 0, \; j=1,...,J-2.$$

(10.52)
$$\frac{\partial \ln L}{\partial \tau_i^A} = x_{i.} - \mu_{i.} = 0, \; i=1,...,I-1$$

(10.53)
$$\frac{\partial \ln L}{\partial \tau_j^B} = x_{.j} - \mu_{.j} = 0, \; j=1,...,J-1.$$

(10.54)
$$\frac{\partial \ln L}{\partial \tau_0} = x_{..} - \mu_{..} = 0.$$

There are only I–2 equations in (10.50) and J–2 equations in (10.51) due to the constraints (10.43). Since (10.54) is obtained by summing (10.52) over all i or (10.53) over all j, (10.47) and (10.48) have the same solutions as (10.52), (10.53) and (10.54).

Notice further that the middle term in (10.50) sum to 0 over all i due to (10.53) and that the middle term in (10.51) sum to 0 over all j due to (10.52). In addition (10.49) is the weighted sum of the middle term in (10.50) with the ϵ_i's as weights or the weighted sum of the middle term in (10.51) with the δ_j's as weights. Equations (10.49) and (10.53) thus together form two linear constraints on (10.50), which means that (10.50) is satisfied for all i if it is satisfied for i=1,...,I–2. In the same way (10.49) and (10.52) ensure that (10.51) is satisfied for all j, if it holds for j=1,...,J–2. Collecting the results, we have shown that (10.49) to (10.54) is equivalent to (10.50) to (10.53) for all i and j. ☐ .

Equations (10.45) to (10.48) can be solved by a Newton–Raphson procedure. In the examples below the estimates are obtained this way. It is, however, also possible to apply an algorithm due to Goodman (1979b), which is closely related to the iterative proportional fitting procedure for the log–linear models described in sections 3.7 and 5.3. Proportional fitting based on equation (10.47) yields the adjustments

$$(10.55) \qquad \mu_{ij}^{(n+1)} = \mu_{ij}^{(n)} (x_{i.} / \mu_{i.}^{(n)}),$$

where $\mu_{ij}^{(n)}$ is the expected number after n iterations. If only τ_i^A is changed in iteration n, this means that τ_i^A is adjusted as

$$(10.56) \qquad \tau_i^{A(n+1)} = \tau_i^{A(n)} + \ln(x_{i.} / \mu_{i.}^{(n)}),$$

Similarly τ_i^B is adjusted as

$$(10.57) \qquad \tau_j^{B(n+1)} = \tau_j^{B(n)} + \ln(x_{.j} / \mu_{.j}^{(n)}).$$

In order to derive the adjustment to ϵ_i consider the differens

$$\Delta = \sum_j \delta_j (x_{ij} - \mu_{ij}),$$

which according to (10.46) is 0, when the solutions are obtained. Neglecting the constraints between the ϵ_i's, the partial derivative of Δ with respect to ϵ_i is

$$\frac{\partial \Delta}{\partial \epsilon_i} = - \rho \sum_j \delta_j^2 \mu_{ij},$$

due to (10.42). Hence a Taylor expansion of Δ with respect to ϵ_i at the value $\epsilon_i^{(n)}$ of ϵ_i yields

$$\Delta \simeq \sum_j \delta_j (x_{ij} - \mu_{ij}^{(n)}) - (\epsilon_i - \epsilon_i^{(n)}) \sum_j \rho^{(n)} (\delta_j^{(n)})^2 \mu_{ij}^{(n)},$$

where $\ln \mu_{ij}^{(n)} = \rho^{(n)} \epsilon_i^{(n)} \delta_j^{(n)} + \tau_i^{A(n)} + \tau_j^{B(n)} + \tau_0^{(n)}$.

Hence with $\Delta = 0$ and $\epsilon_i = \epsilon_i^{(n+1)}$ we obtain the adjusted value

(10.58)
$$\epsilon_i^{(n+1)} = \epsilon_i^{(n)} + \sum_j \delta_j^{(n)} (x_{ij} - \mu_{ij}^{(n)}) / \sum_j \rho^{(n)} (\delta_j^{(n)})^2 \mu_{ij}^{(n)}$$

where $\rho^{(n)}$, $\delta_j^{(n)}$ and $\epsilon_i^{(n)}$ are the parameter values obtained after n iterations and $\mu_{ij}^{(n)}$ the expected values evaluated from these parameter. The adjustments for the δ_j's are derived in the same way as

(10.59)
$$\delta_j^{(n+1)} = \delta_j^{(n)} + \sum_i \epsilon_i^{(n)} (x_{ij} - \mu_{ij}^{(n)}) / \sum_i \rho^{(n)} (\epsilon_i^{(n)})^2 \mu_{ij}^{(n)}$$

Finally ρ is adjusted as

(10.60)
$$\rho^{(n+1)} = \rho^{(n)} + \sum_i \sum_j \epsilon_i^{(n)} \delta_j^{(n)} (x_{ij} - \mu_{ij}^{(n)}) / \sum_i \sum_j (\epsilon_i^{(n)} \delta_j^{(n)})^2 \mu_{ij}^{(n)}.$$

The algorithm introduced in Goodman (1979b) and known as **Goodman's algorithm** is defined as successive applications of (10.58) to (10.60).

The model is in the usual way checked by the test statistic

(10.61)
$$Z = 2 \sum_i \sum_j X_{ij} (\ln X_{ij} - \ln \hat{\mu}_{ij}),$$

where

$$\hat{\mu}_{ij} = \exp(\hat{\rho} \hat{\epsilon}_i \hat{\delta}_j + \hat{\tau}_i^A + \hat{\tau}_j^B + \hat{\tau}_0).$$

Since there are I–2 unconstrained ϵ_i's, J–2 unconstrained δ_j's and one ρ to be estimated in addition to the I+J–1 main effects, the number of degrees of freedom for the approximating χ^2–distribution is

$$df = IJ-(I-2)-(J-2)-1-(I+J-1) = (I-2)(J-2).$$

Example 10.5.

Table 10.12 shows a random sample of 1816 persons in Denmark in 1974, who rented their dwelling (in the sequel denoted "renters"), cross–classified according to income and wealth.

We shall analyze this data set by the RC–association model (10.42) as well as by the column effects association model (10.40), the row effects association model (10.36) and the uniform association model (10.29) with known row and column effects. Table 10.13 show the z–test statistics for these four models and for the independence model.

Table 10.12. A random sample of renters in Denmark in 1974, cross–classified according to income and wealth.

	Wealth (1000 Dkr.)					
Income (1000 DKr.)	0	1–50	50–150	150–300	300–	Totals
0–40	292	126	22	5	4	449
40–60	216	120	21	7	3	367
60–80	172	133	40	7	7	359
80–110	177	120	54	7	4	362
110–	91	87	52	24	25	279
Total	948	586	189	50	43	1816

Source: The Danish Welfare Study: Hansen (1978), table 6.H.32.

Table 10.13. Test statistics for independence and for four models applied to the data in table 10.12.

Model	z	df	Level of significance
Independence	167.99	16	0.000
Uniform association	52.07	15	0.000
Row effects association	45.52	12	0.000
Column effects association	21.13	12	0.049
RC–association	14.46	9	0.107

The known scores for the rows and columns in the middle three models of table 9.11 was chosen as interval midpoints for the income and wealth scales. Thus $(e_1,...,e_5)$ was chosen as a rescaling of $(20,50,70,95,200)$, and $(d_1...d_5)$ as a rescaling of $(0,25,100,225,450)$. The values 200 and 450 were chosen in rather arbitrary fashion.

A RC–association model thus fit the data well, which means that the association between income and wealth for renters in Denmark in 1974 is described in a satisfactory way by a model with multiplicative interactions. The column effects association model fits the data relatively well, while none of the other models fit the data. Table 10.14 shows the estimates for the row and column effects for the RC–association model together with the rescaled scores obtained from the interval mid–points for the income and wealth scales. The estimated ϵ's in the RC–association model by and large reflect the income scale, as we would expect since the column effects association model fit the data. The estimated δ_j's on the other hand do not reflect the wealth scale. △.

Extensions of the RC–association model to three–dimensional and higher dimensional contingency tables were discussed by Goodman (1986), Clogg (1982a) and Choulakians (1988).

Table 10.14. Comparison between the estimated row and column scores in the RC–association model and scores obtained by rescaling the interval mid–points of the income and wealth scales.

Income (1000 Dkr.)

	0–40	40–60	60–80	80–110	110–
$\hat{\epsilon}_i$	−1.151	−0.644	0.283	0.344	1.890
e_i	−1.012	−0.494	−0.148	0.284	2.099

Wealth (1000 Dkr.)

	0	−50	50–150	150–300	300–
$\hat{\delta}_j$	0.726	−0.194	−1.601	−2.570	−3.338
d_i	0.449	0.131	−0.822	−2.412	−5.273

10.7. Log–linear association models

In his original paper on the RC–association models, Goodman (1979b) considers a whole range of association models. The building blocks for these models are the odds–ratio's

$$\omega_{ij} = \frac{\mu_{ij}\, \mu_{i+1.j+1}}{\mu_{i+1.j}\, \mu_{i.j+1}}$$

for neighbouring cells in the contingency table.

The RC–association model is equivalently with a model, where $\ln\omega_{ij}$ is multiplicative, i.e. there exist parameters φ_i and ψ_j such that

$$\ln\omega_{ij} = \varphi_i \psi_j.$$

As we have seen these models are not log–linear. If, however, the odds–ratio itself is multiplicative, i.e. if there exist parameters α_i and β_j such that

$$\omega_{ij} = \alpha_i \beta_j,$$

then the model is log–linear. In fact, if $w_{ij} = \alpha_i \beta_j$, the logarithm to the expected value in cell (ij) has the form

(10.62)
$$\ln\mu_{ij} = \tau_0 + \tau_i^A + \tau_j^B + j\,\alpha_i + i\,\beta_j$$

for certain parameters α_i and β_j.

A model, where the expected values satisfy (10.62) was called a **RC–association model, type I** by Goodman (1979b). The log–likelihood function for this model, which we shall call a **log–linear association model**, is

$$\ln L = x_{..}\tau_0 + x_{i.}\tau_i^A + x_{.j}\tau_j^B + \sum_i \alpha_i \sum_j jx_{ij} + \sum_j \beta_j \sum_i ix_{ij}\;.$$

The likelihood equations are accordingly

(10.63)
$$x_{i.} = \mu_{i.}\,, \qquad i=1,...,I$$

(10.64)
$$x_{.j} = \mu_{.j}, \qquad j=1,...,J$$

(10.65)
$$\sum_i ix_{ij} = \sum_i i\mu_{ij}, \qquad j=1,...,J$$

and

(10.66)
$$\sum_j jx_{ij} = \sum_j j\mu_{ij}, \qquad i=1,...,I.$$

These equations can be solved by a Newton–Raphson procedure or by an algorithm similar to the Goodman algorithm (10.56) to (10.60) in section 10.6. With this algorithm, which was also suggested by Goodman (1979b), α_i is adjusted in order to satisfy as

$$\alpha_i^{(n+1)} = \alpha_i^{(n)} + \ln\left[1 + \sum_j (j-\bar{j})(x_{ij}-\mu_{ij}^{(n)})/\sum_j (j-\bar{j})^2 \mu_{ij}^{(n)}\right]$$

where $\bar{j}=(J+1)/2$ and similarly for the β's.

Under the model (10.62), the test statistic (10.61) is approximately χ^2–distributed with $(I-2)(J-2)$ degrees of freedom.

. The log–linear association model can be modified by assumming that the neighbourhood odds ratio's ω_{ij} only depend on i, only depend on j or are constant.

If ω_{ij} is independent of i, the model was termed a **column effects association** model by Goodman (1979b). The cell mean values then have the structural form

(10.67)
$$\ln\mu_{ij} = \tau_i^A + \tau_j^B + i\beta_j \, .$$

It is identical with the column effects association model in section 10.5, if the row scores are equidistant. In the same way the **row effects association** model of section 10.5 emerges if the known column scores are equidistant.

In case the neighbourhood odds ratios are constant, Goodman (1979b) termed the model a **uniform association model**. The mean values for a uniform association model have for a certain α the form

(10.68)
$$\ln\mu_{ij} = \tau_0 + \tau_i^A + \tau_j^B + ij\alpha \, ,$$

i.e. the model is an RC–association model with known and equidistant row and column scores.

10.8. Exercises

10.1. The tables below show the forecasts for production and prices for the coming three year periods in July 1956 and the reports of the actual production and prices in May 1959 for about 4000 Danish factories.

(a) Test the hypotheses of symmetry and quasi–symmetry on both tables.

(b) Compare the results for the two tables.

(c) What are the practical use of the results?

| Prices: | | Report May 1959 | | |
		Higher	No change	Lower
Forecast	Higher	209	169	6
July	No change	190	3073	184
1956	Lower	3	62	81

| Production: | | Report May 1959 | | |
		Increase	No change	Decrease
Forecast	Higher	532	394	69
July	No change	447	1727	334
1956	Lower	39	230	231

10.2. Verify that the Q–test statistic for symmetry (or marginal homogeneity) in 2x2 table can be written as

$$Q = \frac{(X_{12} - X_{21})}{X_{12} + X_{21}}.$$

Give an intuitive argument for the fact that Q only depends on the off–diagonal elements of the table.

10.3. The data in the table below correspond to those in table 10.1, but relate to a poll taken in August 1971 and the one taken in October 1971.

| | | Attitude in October 1971 | | |
		Yes	No	Undecided
Attitude	Yes	176	33	40
in August	No	21	94	32
1971	Undecided	21	33	43

(a) Test the hypotheses of symmetry and quasi–symmetry on this table.

(b) Compare with the analysis in example 10.1

(c) Test if there is marginal homogeneity in table 10.1 and in the table above.

10.4. The table shows the sample from the Danish Welfare Study cross–classified according to own educational level and fathers educational level.

		Own educational level		
		I	II	III
Fathers	I	3000	1090	218
educational	II	88	209	102
level	III	23	46	87

The educational levels are

      I: Left school with an examen

      II: Examen, but below high school

      III: High school examen.

(a)   Analyse the data by a symmetry and a quasi–symmetry model.

(b)   Interprete the parameters of the symmetry model, which fits the data, if any does.

10.5. In a methodological follow up to the Danish Welfare Study the structure of the classification in social rank groups was studied in details for all subjects in the sample with age between 40 and 59. The two tables below are based on the married women in this subsample. In the tables the womens social rank is cross–classified with their husbands social rank and theirs fathers social rank.

Husbands social rank	Womans social rank			
	I–II	III	IV	V
I–II	20	35	42	22
III	4	44	122	71
IV	6	12	49	71
V	0	6	32	146

Fathers social rank	Womans social rank			
	I–II	III	IV	V
I–II	12	17	22	3
III	8	33	85	95
IV	11	26	72	87
V	2	18	50	111

(a)   Test the hypotheses of symmetry and quasi–symmetry on both tables.

(b)   Interpret the parameters of the final model.

10.6. In 1972 100 Polish refugees in Denmark were interviewed among other things about their health. A medical examination revealed how many of 15 specific illness symptoms they suffered by, ranging from coughing and headaches to heart pain and stomach pains. The table below show the number of symptoms in three intervals 0–3, 4–6, and more than 7, cross–classified with ones own evaluation of health status.

Number of symptoms	Own evaluation of health status		
	Good	Reasonable	Bad
0–3	26	12	0
4–6	5	16	5
7–	2	10	24

(a)   For this problem describe the meaning of the hypotheses

(i)   Independence

(ii)   Symmetry

(iii)   Quasi–symmetry

(iv)   Marginal homogeneity

(b)   Test one or more of these hypotheses.

(c)   Draw your conclusions from the testing in (b) based on what you found in (a).

10.7. The table below show for the Danish Welfare Study the sample cross–classified according to marriage status and social group.

Marriage status	Social rank group			
	I–II	III	IV	V
Married	364	697	898	695
Unmarried	60	117	252	202
Widow	32	76	114	113

(a) Analyse the data by an RC–association model.

(b) Interprete the parameters.

(c) Compute the expected numbers under the RC–association model and under the independence hypothesis. Give an interpretation of the differences.

10.8. The contingency table below show the connection between urbanization and social rank in the Danish Welfare Study.

| Urbanization | Social rank group | | | |
	I–II	III	IV	V
Copenhagen	45	64	160	74
Copenhagen suburbs	99	107	174	90
Three largest cities	57	85	153	103
Cities	168	287	415	342
Countryside	83	346	361	399

(a) Show that an RC–association model does not quite fit the data.

(b) Use residuals to locate the cells that contribute most to the lack of fit.

(c) Can the fit be significantly improved by deleting a row or column?

10.9. From the Danish Welfare Study we consider the association between alcohol consumption and social rank. Alcohol consumption in the contingency table below is grouped according to number of "units" consumed pr.day. A unit is typical a beer, half a bottle of wine or 2 cl of 40% alcohol.

| Units of alcohol consumed per day | Social rank group | | | |
	I–II	III	IV	V
Under 1	98	338	484	484
1–2	235	406	588	385
More than 2	123	144	191	137

(a) Test whether an RC–association model fits the data.

(b) Suppose the average consumption in the three alcohol consumption intervals can be set to 0.5, 1.5 and 2.5. Compute the correctly scaled row scores corresponding to these values.

(c) Does an RC–association model with the fixed scores in (b) fit the data?

(d) Try to determine an average consumption value in the upper interval such that an RC–association model fits the data.

10.10. Escoufier (1982) introduced the data set below in the statistical literature. It has since proved to be one of the most frequently reanalysed data set. The table show a sample of 1660 subjects cross–classified according to mental health status and parents socio–economic status.

Parents socio–economic status

Mental health status	A	B	C	D	E	F
Well	64	57	57	72	36	21
Mild symptoms	94	94	105	141	97	71
Moderate symptoms	58	54	65	77	54	54
Impaired	46	40	60	94	78	71

10.11. Caussinus (1986) introduced the data set below showing the 2730 reported cases of cancer cross–classified according to age and type of cancer.

(a) Analyse the data by an RC–association model.

(b) Try to fit the data by a row effects association model with equidistant column scores.

Type of cancer	–49	50–59	Age 60–69	70–79	80–
A	55	175	230	381	174
B	108	138	191	334	262
C	15	18	76	194	80
D	19	41	67	117	55

10.12. The table below shows the sample in the Danish Welfare Study cross–classified according to social rank and family type.

Social rank group

Family type	I–II	III	IV	V
Single, with children	54	116	187	196
Single, without children	11	13	45	34
Married, with children	104	264	357	309
Married, without children	287	497	675	471

(a) Does an RC–association model fit the data?

(b) Describe the model departures for example by a residual diagram.

(c) Are there cells with contribute especially much to the values of the paramemter estimates.

10.13. Apply the Goodman algorithm described in section 10.6 on the data set below, with the following initial values.

$$\tau^A = (0.0, 0.0, 0.0)$$

$$\tau^B = (0.0, 0.0, 0.0, 0.0)$$

$$\epsilon = (1.0, 0.0, -1.0)$$

$$\delta = (-1.0, 0.0, 0.0, 1.0)$$

$$\rho = 1.0$$

The data are subjects in the Danish Welfare Study with incomes higher than 150.000 Dkr. cross–classified after social rank and alcohol consumption.

Social rank group

Alcohol consumption per day	I–II	III	IV	V
Less than 1 unit	58	147	174	201
1–2 units	145	197	220	155
More than 2 units	77	73	98	59

(a) Compute the expected values, the weighted sums (10.45) and (10.46) and the improved estimates in each step of the algorithm for two iterations.

(b) The ML–estimates (obtained after 6 iterations) are

$$\hat{\tau}^A = (0.095, 0.318, -0.554)$$

$$\hat{\tau}^B = (-0.461, -0.008, 0.157, -0.053)$$

$$\hat{\epsilon} = (-1.539, 0.617, 1.267)$$

$$\hat{\delta} = (1.618, 0.033, 0.083, -1.060)$$

$$\hat{\rho} = 0.203.$$

Compare these with the results in (a).

11. Correspondance Analysis

11.1. Correspondance analysis for two–way tables

A statistical technique, which is closely related to the models discussed in chapter 10, was developed in France in the 1970's. This technique known in the English speaking world as **correspondance analysis** was introduced by Benzecri (1973) as l'Analyse de Correspondance. Many authors have argued that correspondance analysis was not developed in France by Benzecri. This claim is correct in the sense that the technique is closely related to many other forms of statistical analyses, which go far back in time. Some of these connections are discussed in section 11.3 below, where also references will be given. Whatever those connections are, the name correspondance analysis and its popularity in France and neighbouring countries is certainly due to Benzecri and his students.

Correspondance analysis is a combination of a mathematical technique to explore the structure of a contingency table and a graphical technique, where the derived structure is illustrated in a diagram with points representing the categories of the variables. A basic concept in correspondance analysis is the **profile** of a category. Let x_{ij}, $i=1,...,I$, $j=1,...,J$ be the observed numbers in the cells of the contingency table. The profile of category i on variable A is then the vector

$$(f_{i1}/f_{i.},...,f_{iJ}/f_{i.}),$$

where $f_{ij}=x_{ij}/n$ and $f_{i.}$ the sum of f_{ij} over j. Correspondingly the profile of category j on variable B is the vector

$$(f_{1j}/f_{.j},...,f_{Ij}/f_{.j}).$$

Note that the concepts in correspondance analysis by tradition are formulated in terms of frequencies rather than observed numbers. If the variables A and B are independent, the estimated expected value in cell (ij) is

$$x_{i.} x_{.j}/n$$

Hence under independence between A and B, f_{ij} is equal to $f_{i.} f_{.j}$ apart from random fluctuations. The estimated common profile for the rows under independence is accordingly equal to the vector

$$(f_{.1}, \ldots f_{.J}) \, .$$

Hence we can evaluate the extent to which the profile of row category i resembles the expected profile under independence through the vector of distances

$$(11.1) \qquad h_i = \left[\frac{f_{i1}}{f_{i.}} - f_{.1}, \ldots, \frac{f_{iJ}}{f_{i.}} - f_{.J} \right]$$

between the profile of category i and the average profile of variable B. If h_i is equal to a vector of zero's apart from random fluctuations, the frequencies in row i are close to what should be expected under independence. Similarly the vector

$$(11.2) \qquad g_j = \left[\frac{f_{1j}}{f_{.j}} - f_{1.}, \ldots, \frac{f_{Ij}}{f_{.j}} - f_{I.} \right]$$

should be zero apart from random fluctuations under independence between A and B. The fundamental idea of correspondance analyses is to represent the vectors (11.1) and (11.2) by scores connected with the rows and the column categories. Consider the following decomposition

$$(11.3) \qquad \frac{f_{ij}}{f_{i.} f_{.j}} - 1 = \sum_{m=1}^{M} \lambda_m u_{im} v_{jm} \, ,$$

where M=min(I−1,J−1).

Given (11.3) the distance (11.1) between the profile of row i and the average profile of variable B can be written as

$$h_j = \sum_m \lambda_m u_{im}(f_{.1}v_{1m},\ldots,f_{.J}v_{Jm})$$

and correspondingly the distance (11.2) between the profile of column j and the average profile of variable A can be written as

$$g_j = \sum_m \lambda_m v_{jm}(f_{1.}u_{1m},\ldots,f_{I.}u_{Im}).$$

The decomposition (11.3) thus means that the profiles are closely connected with the vectors

$$u_m = (u_{1m},\ldots,u_{Im}) , \quad m=1,\ldots,M$$

of **row scores** and the vectors

$$v_m = (v_{1m},\ldots,v_{Jm}) , \quad m=1,\ldots,M$$

of **column scores**.

As we shall see, the sum in (11.3) can often be approximated by a sum of just one or two terms, in which case the representation of the profiles by scores represent a useful data reduction.

The scores u_m and v_m must satisfy the constraints.

(11.4)
$$\sum_{i=1}^I u_{im}f_{i.} = \sum_{j=1}^J v_{jm}f_{.j} = 0, \quad m=1,\ldots,M,$$

and

(11.5)
$$\sum_{i=1}^I u_{im}^2 f_{i.} = \sum_{j=1}^J v_{jm}^2 f_{.j} = 1, \quad m=1,\ldots,M.$$

In addition the relationship between the u's and v's for different values of m is determined through the **ortogonality constraints**.

$$(11.6) \qquad \sum_{i=1}^{I} u_{im} u_{i\mu} f_{i\cdot} = \sum_{j=1}^{J} v_{jm} v_{j\mu} f_{\cdot j} = 0 \ , \ \mu \neq m.$$

the constraints (11.4) and (11.5) are widely used, although other choices of constraints are possible. The possibilities were reviewed by Goldstein (1987).

It is convenient to use matrix notation to describe correspondance analysis. Let thus

$$\mathbf{F} = \{f_{ij}\}, \ i=1,...,I, \ j=1,...,J,$$

be the matrix of frequencies and

$$\mathbf{R} = \{f_{ij} - f_{i\cdot} f_{\cdot j}\}, \ i=1,...,I, \ j=1,...,J.$$

the matrix of residuals.

The marginals $f_{i\cdot}$ and $f_{\cdot j}$ are collected in the diagonal matrices

$$\mathbf{C}_I = \begin{bmatrix} f_{1\cdot} & & 0 \\ & \ddots & \\ 0 & & f_{I\cdot} \end{bmatrix}$$

and

$$\mathbf{C}_J = \begin{bmatrix} f_{\cdot 1} & & 0 \\ & \ddots & \\ 0 & & f_{\cdot J} \end{bmatrix}.$$

Let further $\mathbf{\Lambda}$ be a diagonal matrix of dimension m with $\lambda_1,...,\lambda_m$ in the diagonal and $\mathbf{U}$ and $\mathbf{V}$ be matrices of dimension I.M and J.M, respectively, which have the u_{im}'s and v_{jm}'s as elements. Then the decomposition (11.3) can be expressed as

$$(11.7) \qquad \mathbf{C}_I^{-1} \mathbf{F} \mathbf{C}_J^{-1} - 1_I 1_J' = \mathbf{U\Lambda V'},$$

where $1_J'=(1,...,1)$ and $1_I'=(1,...,1)$, or as

$$(11.8) \qquad \mathbf{C}_I^{-1} \mathbf{R} \mathbf{C}_J^{-1} = \mathbf{U\Lambda V'}.$$

The constraints (11.5) and (11.6) can be collected in the orthogonality conditions

(11.9)
$$U'C_I U = I,$$

and

(11.10)
$$V'C_J V = I,$$

where I is the identity matrix.

Equation (11.8) shows that the decomposition (11.3) is obtained thorugh an **eigenvalue/eigenvector decomposition** or **single value decomposition** of the **scaled residual matrix** $C_I^{-1} R C_J^{-1}$. This means that the λ's and the scores in (11.3) under proper normalizations are uniquely determinated. The exact formulation of this result is given in theorem 10.1.

Theorem 10.1.

There is a unique solution to (11.8), under the constraints (11.9) and (11.10). The elements of Λ are the square roots of the eigenvalues of the matrix

(11.11)
$$D = C_I^{-1} R C_J^{-1} R'$$

or equivalently the square roots of the eigenvalues of

(11.12)
$$E = C_J^{-1} R' C_I^{-1} R.$$

The columns of U are the eigenvectors of D and the columns of V the eigenvectors of E.

Proof

Equation (11.8) yields

$$C_I^{-1} R C_J^{-1} C_J V = U \Lambda V' C_J V = U \Lambda,$$

or

$$C_I^{-1} R V = U \Lambda$$

due to (11.10). By a symmetric calculation

$$C_J^{-1}R'U = V\Lambda.$$

When V is eliminated from these equations, we get due to (11.9)

$$C_I^{-1}RC_J^{-1}R'U = U\Lambda^2,$$

such that Λ^2 contains the eigenvalues of D and the columns of U are the corresponding eigenvectors. If U is eliminated rather than V, we get

$$C_J^{-1}R'C_I^{-1}RV = V\Lambda^2,$$

showing that E has the same eigenvalues as D. Since

$$\sum_{i=1}^{I}\left[\frac{f_{ij}}{f_{i.}f_{.j}}-1\right]f_{i.} = \sum_{j=1}^{J}\left[\frac{f_{ij}}{f_{i.}f_{.j}}-1\right]f_{.j} = 0$$

D has at most rank I–1 and E at most rank J–1. Hence M≤min(I–1,J–1). △.

Equations

(11.13) $$C_I^{-1}RC_J^{-1}R'U = U\Lambda^2.$$

and

(11.14) $$C_J^{-1}R'C_I^{-1}RV = V\Lambda^2.$$

represent the **correspondance analysis solutions** based on the residual matrix R.

It is an important property in correspondance analysis that the matrix F with elements f_{ij} has the same correspondance analysis solution as R.

Since

$$R = F - C_I 1_I 1_J' C_J,$$

$$C_I^{-1} R C_J^{-1} R'U = C_I^{-1}(F - C_I 1_I 1_J' C_J) C_J^{-1}(F' - C_J 1_J 1_I' C_I)U = C_I^{-1} F C_J^{-1} F'U - C_I^{-1} F 1_J 1_I' C_I U$$

$$- 1_I 1_J' F'U + 1_I 1_J' C_J 1_J 1_I' C_I U.$$

Condition (11.4) can also be written as

$$1_J 1_I' C_I U = 0$$

and $1_J 1_I' F' = 1_J 1_I' C_I$ implies that

$$1_I 1_J' F'U = 0.$$

Hence

$$C_I^{-1} R C_J^{-1} R'U = C_I^{-1} F C_J^{-1} F'U.$$

The correspondance analysis solution can thus be expressed either as (11.13) or as

(11.15)
$$C_I^{-1} F C_J^{-1} F'U = U\Lambda^2.$$

To equation (11.14) corresponds in the same way

(11.16)
$$C_J^{-1} F' C_I^{-1} FV = V\Lambda^2.$$

As mentioned we are interested in approximations to the left hand side of (11.13) with a moderate value M_0 of M, usually $M_0 = 1$ or 2. The next theorem shows that the eigenvector/eigenvalue decomposition is useful also in this context.

Theorem 11.2

The weighted sum of squares of deviations

$$(11.17) \qquad Q(M_0) = \sum_{i=1}^{I} \sum_{j=1}^{J} f_{i\cdot} f_{\cdot j} \left[\frac{f_{ij}}{f_{i\cdot} f_{\cdot j}} - 1 - \sum_{m=1}^{M_0} \lambda_m u_{im} v_{jm} \right]^2$$

is minimized if $\lambda_1^2, \ldots, \lambda_{M_0}^2$, are the M_0 largest eigenvalues of E or D, $(u_{1m}, \ldots, u_{Im})$ is the eigenvector of D corresponding to λ_m^2 and $(v_{1m}, \ldots, v_{Jm})$ the eigenvector of E corresponding to λ_m^2.

Proof

The derivative of $Q(M_0)$ with respect to u_{im} is

$$\partial Q(M_0)/\partial u_{im} = -2\lambda_m \sum_{j=1}^{J} \left[\frac{f_{ij}}{f_{i\cdot} f_{\cdot j}} - 1 \right] f_{i\cdot} f_{\cdot j} v_{jm}$$

$$+ 2\lambda_m \sum_{j=1}^{J} f_{i\cdot} f_{\cdot j} v_{jm} \sum_{\mu=1}^{M_0} \lambda_\mu u_{i\mu} v_{j\mu}$$

which due to (11.5) and (11.6) reduces to

$$\sum_{j=1}^{J} v_{jm} \left[f_{ij} - f_{i\cdot} f_{\cdot j} \right] \frac{1}{f_{i\cdot}} = \lambda_m u_{im}.$$

or in matrix notation

$$(11.18) \qquad C_I^{-1} R \begin{bmatrix} v_{1m} \\ \vdots \\ v_{Jm} \end{bmatrix} = \lambda_m \begin{bmatrix} u_{1m} \\ \vdots \\ u_{Im} \end{bmatrix}.$$

In the same way $\partial Q(M_0)/\partial v_{jm}$ leads to

$$(11.19) \qquad C_J^{-1} R' \begin{bmatrix} u_{1m} \\ \vdots \\ u_{Im} \end{bmatrix} = \lambda_m \begin{bmatrix} v_{1m} \\ \vdots \\ v_{Jm} \end{bmatrix}.$$

If (11.18) is multiplied by λ_m and the left hand side of (11.19) is inserted in (11.18), we get

$$C_I^{-1} R C_J^{-1} R' \begin{bmatrix} u_{1m} \\ \vdots \\ u_{Im} \end{bmatrix} = \lambda_m^2 \begin{bmatrix} u_{1m} \\ \vdots \\ u_{Im} \end{bmatrix}.$$

It follows that a necessary condition for $Q(M_0)$ to attain its minimum is that λ_m^2 is an eigenvalue of D and $(u_{1m},...,u_{Im})$ the corresponding eigenvector. It is somewhat more involved to prove that the minimum is attained, when $\lambda_1^2,...,\lambda_{M_0}^2$ are the M_0 largest eigenvalues. $\square$.

If $M_0=1$ or $M_0=2$ provide us with a reasonable good approximation to (11.3), i.e. if

$$(11.20) \qquad \frac{f_{ij}}{f_{i.}f_{.j}} - 1 \simeq \sum_{m=1}^{M_0} \lambda_m u_{im} v_{jm},$$

it is common practice in correspondance analysis to display the result of the analysis graphically in a **correspondance analysis diagram**. For $M_0=2$ the French tradition is to plot the points $(u_{11}^*,u_{12}^*),...,u_{I1}^*,u_{I2}^*)$, where

$$u_{im}^* = \lambda_m u_{im}$$

in a two–dimensional diagram and correspondingly, $(v_{11}^*,v_{12}^*),...,(v_{J1}^*,v_{J2}^*)$ with

$$v_{im}^* = \lambda_m v_{jm}$$

in a two–dimensional diagram. The v^*'s can be plotted in the same diagram as the u^*'s, or in a separate plot. If $M_0=1$ provide a satisfactory approximation in (11.20), the values of $u_{11}^*,...,u_{I1}^*$ and $v_{11}^*,...,v_{J1}^*$ are plotted on a line. Since the most interesting case is $M_0=2$, we concentrate on this case in the following. Cases where M_0 is higher than 2 are difficult to visualize in graphical displays.

Since for $M_0=2$ each point in a correspondance analysis diagram relate to the category of a variable, we can write an abbreviation for the label of the category next to the point in the diagram. As an illustration, fig.11.1 shows an imaginary example with three categories for each variable.

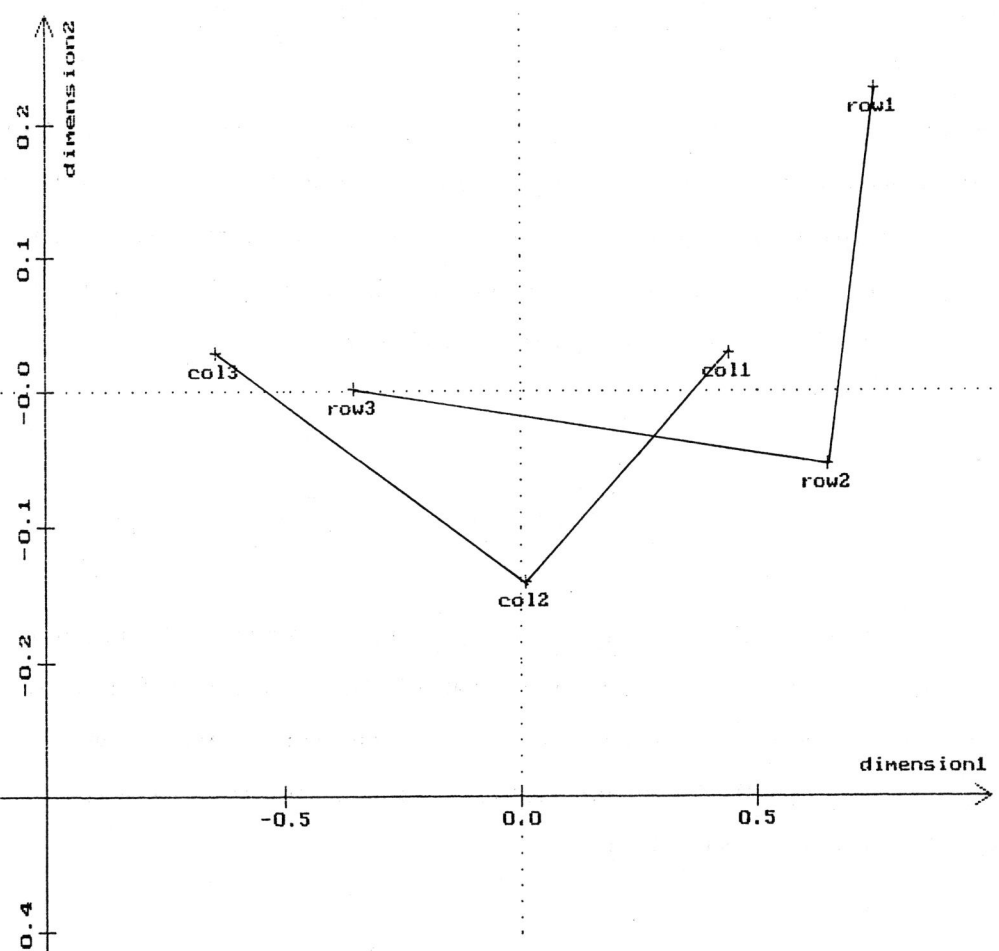

Fig. 11.1. An example of a correspondance analysis diagram.

The interpretation of a correspondance analysis diagram is based on three properties. First consider the distances (11.1) from the profile of category i, to the average profile for variable B. From (11.5) and (11.6) follows under the approximation (11.20) for $M_0=2$

(11.21)
$$\sum_{j=1}^{J}\left[\frac{f_{ij}}{f_{i\cdot}}-f_{\cdot j}\right]^2\frac{1}{f_{\cdot j}}=\sum_{j=1}^{J}\left[\frac{f_{ij}}{f_{i\cdot}f_{\cdot j}}-1\right]^2 f_{\cdot j}\simeq$$

$$\sum_{m=1}^{2}\sum_{\mu=1}^{2}\lambda_m\lambda_\mu u_{im}u_{i\mu}\sum_{j=1}^{J}f_{\cdot j}v_{jm}v_{j\mu}=\sum_{m=1}^{2}\lambda_m^2 u_{im}^2=\sum_{m=1}^{2}(u_{im}^*)^2.$$

Hence the larger the distance from the point (u_{i1}^*,u_{i2}^*) to origo, the larger is the weighted discrepancy between the profile of category i and the average profile of variable B. It follows that points far away from origo indicates a clear deviation from what we would expect under independence, while a point near origo indicates that the frequencies in row i of the contingency table fits the independence hypothesis well.

Secondly, since

(11.22)
$$\sum_{j=1}^{J}\left[\frac{f_{ij}}{f_{i\cdot}}-\frac{f_{sj}}{f_{s\cdot}}\right]^2\frac{1}{f_{\cdot j}}\simeq\sum_{m=1}^{2}\lambda_m^2(u_{im}-u_{sm})^2=\sum_{m=1}^{2}(u_{im}^*-u_{sm}^*)^2$$

the distance between the points (u_{i1}^*,u_{i2}^*) and (u_{s1}^*,u_{s2}^*) is a measure of the discrepancy between the profiles of categories i and s on variable A. This means that the further apart on the correspondance analysis diagram the points representing categories i and s for variable A, the larger the dissimilarities between the corresponding row frequencies.

Thirdly it follows from (11.20) that

$$\frac{f_{ij}}{f_{i\cdot}f_{\cdot j}}-1\simeq\sum_{m=1}^{2}\lambda_m u_{im}v_{jm}=\sum_{m=1}^{2}u_{im}^*v_{jm}^*/\lambda_m$$

such that the difference between $f_{ij}/(f_{i\cdot}f_{\cdot j})$ and the number 1 to be expected under independence is proportional to the cosine to the angle between the points (u_{i1}^*,u_{i2}^*) and

(v_{j1}^*, v_{j2}^*). Hence the smaller the angle between the lines connecting the points representing category i on variable A and category j on variable B with origo, the larger the difference between the observed frequency and what should be expected under independence , i.e. the case of largest similarity between categories i and j. If cosine to the angle is 0, the observed frequency is equal to what we should expect under independence. This means that if the points corresponding to row category i and column category j are in "orthogonal" positions, the cell representing the joint occurrence of the two categories does not contribute to any dependency between the variables. One has to be very careful with this interpretation because under complete independence all λ's are zero and the points are scattered at random in the diagram.

The extreme case, where the points corresponding to categories i and j are on opposite sides of origo corresponds to $f_{ij}=0$, i.e. the observed frequency in the cell is zero, or a situation with the largest dissimilarity between categories i and j.

These rules for interpretation of a correspondance analysis graph applied to fig. 11.1 yield, that category 2 for variable A and category 1 for variable B are similar, reflecting the fact that cell (2,1) contains a much higher frequency than should be expected under independence. The same is true for category 3 on both variables. Cell (1,2) contains very few observations. The diagram reveals on the other hand that row 1 and column 3 of the table have profiles which differ most from the picture to be expected under independence.

The appropriateness of representing the categories of variables A and B by points in the same diagram is an issue which has received much discussion. In the examples below we have chosen to represent the categories of A and B in the same diagram in accordance with the main French tradition. This is partly justified by the rules of interpretation described above. A recent discussion of the diagrams used in correspondance analysis and possible alternatives is due to Greenacre and Hastie (1987).

Another issue which has received much attention is an appropriate choice of the dimension M_0. The dominating tradition in correspondance analysis is to use $M_0=2$, where the diagram is two–dimensional. Whether the choice of dimension is appropriate is judged by the values of the ratios

$$(11.23) \qquad r_m^2 = \frac{\lambda_1^2 + \ldots + \lambda_m^2}{\lambda_1^2 + \ldots + \lambda_M^2},$$

for $m=1,2,\ldots,M$ where $\lambda_1 > \lambda_2 > \ldots > \lambda_M$. The ratio r_m^2 is interpreted as the amount of the variation accounted for by the first m dimensions. The choice of dimension $M_0 = m$ is considered satisfactory if r_m^2 is close to 1 and largely unchanged if m is further increased.

The choice of M_0 and the appropriateness of the model can also be based on statistical inference principles.

Suppose a parametric multinomial model can be assumed for the data of the contingency table. Then under the regularity and identifiability conditions, discussed in section 3.5,

$$(11.24) \qquad Z = 2 \sum_{i=1}^{I} \sum_{j=1}^{J} X_{ij}(\ln X_{ij} - \ln \hat{\mu}_{ij})$$

follows an approximative χ^2-distribution if $\hat{\mu}_{ij}$ is the mean value of X_{ij} with the parameters of the model replaced by their ML–estimates. The degrees of freedom for the χ^2-distribution depend on the number of estimated parameters in the model. For a correspondance analysis a statistical model can be infered from (11.3). Since (11.3) implies

$$nx_{ij} = x_{i.} x_{.j}(1 + \sum_{m=1}^{M} \lambda_m u_{im} v_{jm}),$$

we get the approximative relationship

$$(11.25) \qquad x_{ij} \simeq \frac{x_{i.} x_{.j}}{n}(1 + \sum_{m=1}^{M_0} \lambda_m u_{im} v_{jm})$$

for a correspondance analysis solution with $M_0 < M$. If (11.25) is regarded as an estimation equation for the unknown parameters $\lambda_m, u_{im}, v_{jm}$, $i=1,\ldots,I$, $j=1,\ldots,J$, $m=1,\ldots,M_0$, theorem

11.2 shows that the values obtained through a correspondance analysis can be regarded as weighted least squares estimates of the parameters. It also follows that the right hand sides in (11.25) are approximations to the expected numbers. Hence

$$(11.26) \qquad \hat{\mu}_{ij} = \frac{x_{i.}\, x_{.j}}{n}\left(1+ \sum_{m=1}^{M_0} \lambda_m u_{im} v_{jm}\right),$$

are the estimated expected numbers. The appropriate number of degrees of freedom are obtained by counting the estimated parameters in (11.26) as shown in table 11.1. In table 11.1 the terms $2M_0$, which are subtracted in the row score count and the column score count, account for the constraints (11.4) and (11.5). The terms $M_0(M_0-1)/2$ account for the constraints (11.6). The degrees of freedom for (11.24) is the according to the general result equal to

$$(11.27) \qquad df = IJ - 1 - (M_0+1)(I+J) + M_0^2 + 2M_0 + 2 = (I-M_0-1)(J-M_0-1).$$

Table 11.1. Parameter count for the correspondance analysis model.

Parameters	Notation	Number of parameters
Row marginals	$x_{i.}/n$	$I-1$
Column marginals	$x_{.j}/n$	$J-1$
Eigen values	λ_m	M_0
Row scores	u_{im}	$IM_0 - 2M_0 - \dfrac{M_0(M_0-1)}{2}$
Column scores	v_{jm}	$JM_0 - 2M_0 - \dfrac{M_0(M_0-1)}{2}$
Total		$(I+J)(M_0+1) - 2M_0 - M_0^2 - 2$

Gilula and Haberman (1986) has shown that the χ^2-approximation to (11.24) is valid under the hypothesis $H_0: \lambda_m = 0$ for $m > M_0$ if $\lambda_1 > \lambda_2 > ... \lambda_{M_0} > 0$. The correspondance analysis description of the data with M_0 terms in the approximation (11.20), can thus be accepted if the level of significance of (11.24) is small as compared with a χ^2 distribution with $(I-M_0-1)(J-M_0-1)$ degrees of freedom. Note that the validity of the χ^2-approximation may be invalid if $\lambda_m = 0$ for some $m \leq M_0$ or if $\lambda_m = \lambda_l$ for $m \leq M_0$, $l \leq M_0$ and $l \neq m$.

The Pearson test statistic

$$(11.28) \qquad Q = \sum_{i=1}^{I} \sum_{j=1}^{J} (X_{ij} - \frac{X_{i.} X_{.j}}{n})^2 / (\frac{X_{i.} X_{.j}}{n})$$

is a measure of how well the independence model fits the data. The observed value q of Q satisfies

$$q = \sum_{i=1}^{I} \sum_{j=1}^{J} (\frac{X_{ij}}{X_{i.}} - \frac{X_{.j}}{n})^2 (\frac{X_{i.} n}{X_{.j}}) = n \sum_{i=1}^{I} f_{i.} \sum_{j=1}^{J} (\frac{f_{ij}}{f_{i.}} - f_{.j})^2 \frac{1}{f_{.j}},$$

and the symmetric expression in i and j.

Hence q can be written

$$(11.29) \qquad q = n \sum_{j=1}^{J} f_{.j} \sum_{i=1}^{I} (\frac{f_{ij}}{f_{.j}} - f_{i.})^2 \frac{1}{f_{i.}},$$

or as

$$q = n \sum_{i=1}^{I} f_{i.} \sum_{j=1}^{J} (\frac{f_{ij}}{f_{i.}} - f_{.j})^2 \frac{1}{f_{.j}}.$$

Pearsons test statistic can thus be interpreted as a weighted sum of distances between the column profiles and the estimated common column profile under independence or as a weighted sum of the distances between the row profiles and the estimated common row profile under independence. The more the profiles thus differ from the expected profiles under independence, the larger the value of the Pearson test statistic.

Also the additive components

$$q_i = \sum_{j=1}^{J} (x_{ij} - \frac{x_{i.} x_{.j}}{n})^2 / (\frac{x_{i.} x_{.j}}{n}) = \sum_{j=1}^{J} nf_{.j} f_{i.} (\frac{f_{ij}}{f_{i.} f_{.j}} - 1)^2$$

in Pearsons test statistic have an interpretation in terms of the parameters of a correspondance analysis model. From (11.3) follows that

$$q_i = n \sum_{m=1}^{M} \sum_{\mu=1}^{M} \lambda_m \lambda_\mu f_{i.} \sum_{j=1}^{J} u_{im} u_{i\mu} v_{jm} v_{j\mu} f_{.j}$$

or according to (11.5) and (11.6)

$$(11.30) \qquad q_i = nf_{i.} \sum_{m=1}^{M} \lambda_m^2 u_{im}^2 = nf_{i.} \sum_{m=1}^{M} (u_{im}^*)^2.$$

Apart from the factor $nf_{i.}$, the partial sum q_i is thus the distance from origo to the point $(u_{i1}^*,...,u_{iM}^*)$.

Under the correspondence analysis model the Pearson test statistic is

$$(11.31) \qquad Q = \sum_{i=1}^{I} \sum_{j=1}^{J} (X_{ij} - \hat{\mu}_{ij})^2 / \hat{\mu}_{ij}$$

where $\hat{\mu}_{ij}$ is the expected number (11.26) in cell (ij) under the model. The test statistic Q has, as (11.24) an approximative χ^2–distribution with $(I-M_0-1)(J-M_0-1)$ degrees of freedom if the conditions of Gilula and Haberman (1986) are satisfied. The tradition in correspondence analysis is to use the quantity

$$(11.32) \qquad Q_0 = \sum_{i=1}^{I} \sum_{j=1}^{J} (X_{ij} - \hat{\mu}_{ij})^2 / (\frac{X_{i.} X_{.j}}{n})$$

where $\hat{\mu}_{ij}$ in the denominator of (11.31) are replaced by the expected values under independence, to measure how well the model fits the data. The popularity of (11.32) as a measure of goodness of fit follows from the following expression for the observed value of Q_0, derived from (11.3), (11.26) and (11.32)

$$
(11.33) \qquad q_0 = n \sum_{i=1}^{I} \sum_{j=1}^{J} \left(\frac{f_{ij}}{f_{i.} f_{.j}} - 1 - \sum_{m=1}^{M_0} \lambda_m u_{im} v_{jm}\right)^2 f_{i.} f_{.j}
$$

$$
= n \sum_{i=1}^{I} \sum_{j=1}^{J} f_{i.} f_{.j} \left(\sum_{m=M_0+1}^{M} \lambda_m u_{im} v_{jm}\right)^2 = n \sum_{m=M_0+1}^{M} \lambda_m^2 ,
$$

where the last reduction is due to (11.5) and (11.6).

It follows that the observed value of Q_0 is large if there are large eigenvalues connected with dimensions, which are not included in the estimated expected values. The observed value of Q_0 is on the other hand small, if the eigenvalues of the not included dimensions are small. From (11.33) follows that the value of (11.23) for $m=M_0$ is

$$
r_{M_0}^2 = \frac{q - q_0}{q} = 1 - \frac{q_0}{q}.
$$

If q_0 can be regarded as a measure of how much variation the data exhibit as compared to a correspondance analysis model of dimension M_0 and q is a measure of variation as compared to independence, then $r_{M_0}^2$ is the percentage of the variation in the data, which is explained by the correspondance analysis model. The closer $r_{M_0}^2$ is to 1, the more satisfactory is the fit by a correspondance analysis model as compared to the fit obtainable under independence. There are no general rules for how close $r_{M_0}^2$ should be to 1 before one can claim that a correspondance analysis model gives a satisfactory description of the data.

The contribution of dimension m to the partial sum q_i is according to (11.30) $n \lambda_m^2 f_{i.} u_{im}^2$. If this value is compared to other categories, one gets the **measure of contribution**

$$D^A_{im} = n\lambda^2_{m\,i.} f_{i.} u^2_{im} / \sum_{i=1}^{I} n\lambda^2_{m\,i.} f_{i.} u^2_{im}$$

of row category i to dimension m. Due to (11.5), D^A_{im} reduces to

(11.34) $$D^A_{im} = f_{i.} u^2_{im}.$$

The contribution of column category j to dimension m is in the same way defined as

(11.35) $$D^B_{jm} = f_{.j} v^2_{jm}.$$

A second measure of contribution is

(11.36) $$d^A_{im} = n\lambda^2_{m\,i.} f_{i.} u^2_{im} / \sum_{m=1}^{M} n\lambda^2_{m\,i.} f_{i.} u^2_{im} = \lambda^2_m u^2_{im} / \sum_m \lambda^2_m u^2_{im}$$

which measure the contribution to row category i from dimension m. The contribution to column category j from dimension m is

(11.37) $$d^B_{jm} = \lambda^2_m v^2_{jm} / \sum_m \lambda^2_m v^2_{jm}.$$

Example 11.1.

Table 11.2 shows a random sample of the Danish population in 1976 between age 20 and age 64 cross–classified according to income and occupation. The headings in table 11.2 are the official translations to English of the categories in the Danish official employment statistic.

Table 11.2. A random sample of 4013 Danes in 1976 between age 20 and age 64 cross–classified according to income and occupation.

Occupation:

Monthly Income: – Dkr –	Self employed	Salaried employees	Workers	Pensioners	Unem– ployed	Students
0–3192	108	226	308	406	61	174
3193–4800	82	242	359	37	64	9
4801–5900	52	257	320	9	2	0
5901–7488	44	362	230	9	2	0
7489–	147	423	75	4	1	0

Source: Data from the Danish Welfare Study: Hansen (1978), vol.II, tables 6.H.1 and 6.H.9.

The elements $f_{ij}/(f_{i.}f_{.j})-1$ of the scaled residual matrix $C_I^{-1}RC_J^{-1}$ are shown in table 11.3.

Table 11.3. The residual matrix $C_I^{-1}RC_J^{-1}$ for the data in table 11.2.

Monthly Income – D.Kr.–	Self employed	Salaried employees	Workers	Pensioners	Unem– ployed	Students
0–3192	–0.22	–0.53	–0.25	1.73	0.47	1.97
3193–4800	–0.04	–0.19	0.41	–0.60	1.49	–0.75
4801–5900	–0.25	0.07	0.55	–0.88	–0.90	–1.00
5901–7488	–0.37	0.49	0.10	–0.88	–0.90	–1.00
7489–	1.10	0.73	–0.64	–0.95	–0.95	–1.00

The eigenvalues of the residual matrix and the values of r_m^2 for m=1,2,3 and 4 are shown in table 11.4.

Table 11.4. Eigenvalues and the amount of variation accounted for (r_m^2) for the data in table 11.2

		Dimension			
		m=1	2	3	4
Eigenvalue	λ_m	0.571	0.301	0.152	0.063
Variation accounted for	r_m^2	0.735	0.939	0.991	1.000

The values of r^2_m in table 11.4 strongly suggest that a correspondance analysis model with $M_0=2$ fits the data well. It does not seem that $M_0=1$ is sufficient to explain the variation in the data. Since some of the cellls have very small expected numbers, the χ^2–approximation is not valid for the Pearson test statistic. The scores u_{im} and v_{jm} for dimensions 1 and 2 are summarized in table 11.5.

Table 11.5. Row scores (u_{im}) and column scores (v_{jm}) for a correspondance analy-sis solution with $M_0=2$ and the data in table 11.2.

	Dimension	
Row	m=1	2
i=1	1.420	0.283
2	−0.265	−1.182
3	−0.677	−0.955
4	−0.825	−0.023
5	−0.991	1.847
Column	m=1	2
j=1	−0.346	1.186
2	−0.737	0.673
3	−0.188	−1.312
4	2.069	0.498
5	0.885	−1.497
6	2.342	0.701
λ^2_m	0.326	0.091

Fig. 11.2 shows the correspondance analysis diagram for the model with $M_0=2$. Note that it is $u^*_{im}=\lambda_m u_{im}$ and $v^*_{jm}=\lambda_m v_{jm}$ which are plotted.

Fig. 11.2 shows that income and occupation follows very closely. High incomes are connected with the two highest occupation groups, while students and pensioners as ex-pected mostly are in the lower income brackets. Most surprising is the distance between workers and salaried employees and the fact that workers and unemployed are rather close incomewise.

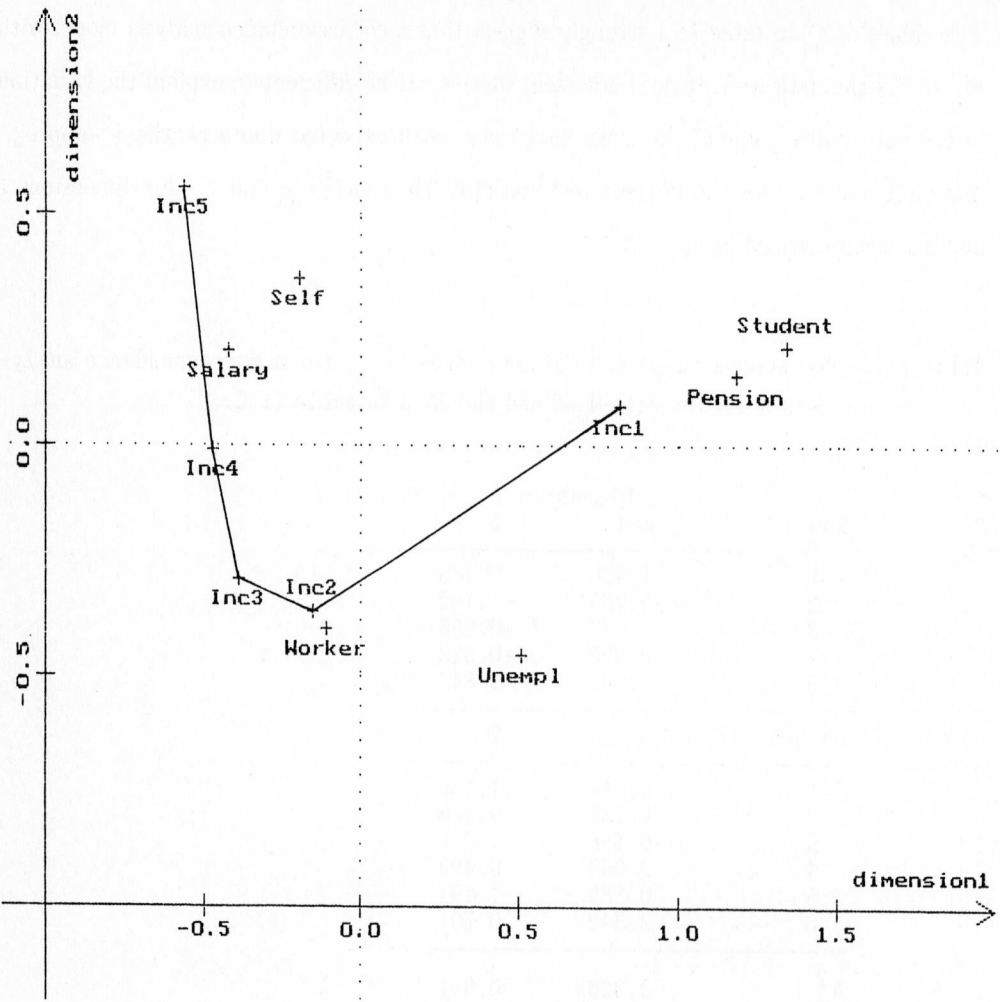

Fig. 11.2. Correspondance analysis diagram for the data in table 11.6 with $M_0=2$.

To evaluate the contributions of the categories of A and B to the dimensions, the measures of contribution D^A_{im}, D^B_{jm}, d^A_{im} and d^B_{jm} are for m=1 and 2 shown in table 11.6.

The numbers in table 11.6 show that among the income brackets, it is the lowest, which contributes the most to dimension one and the highest, which contributes the most to dimension two. Among the occupation groups pensioners contribute the most to dimension one, while workers contribute the most to dimension two. Note that the measures of contribution D^A_{im} and D^B_{jm} reflects both the position of the category on the diagram and the marginal frequency. Thus the coordinate of unemployed on the second axes in figure 11.2 has a larger value than the coordinate of workers, but workers contribute more due to the larger percentage of workers in the sample. A similar argument applies to pensioners and students on the first axes. The values of d^A_{im} show that dimension one gives the dominating contribution to the position of the lowest income brackets on fig.11.2. For the highest income bracket, on the other hand, dimension one and two provide equal contribution. △.

Table 11.6. The measures of contribution for dimensions 1 and 2 for the data in table 11.2.

	The contribution, D^A_{im}, from category i to dimension m		The contribution, d^A_{im}, to category i from dimension m	
	m=1	2	m=1	2
i=1	0.644	0.026	0.988	0.011
2	0.014	0.276	0.111	0.614
3	0.073	0.145	0.556	0.307
4	0.110	0.000	0.825	0.000
5	0.159	0.553	0.499	0.481

	The contribution, D^B_{jm}, from category j to dimension m		The contribution, d^B_{jm}, to category j from dimension m	
	m=1	2	m=1	2
j=1	0.013	0.152	0.161	0.526
2	0.204	0.170	0.795	0.184
3	0.011	0.554	0.068	0.913
4	0.496	0.029	0.981	0.016
5	0.025	0.073	0.277	0.221
6	0.250	0.022	0.971	0.024

Example 11.2

As another example consider the data in table 11.7 on the connection between frequency

of attending meetings and social rank in Denmark.

The values of the eigenvalues λ_m and the variance accounted for r_m^2 for each dimension m are shown in table 11.8.

Table 11.7. A sample of 1779 persons 40–59 years old in Denmark in 1976 cross-classifed according to social rank and according to frequency of attending meetings outside working hours.

Attend meetings outside working hours

Social group:	One or more times a week	One or more times a month	Approx.once every second month	A few times a year	Never	Total
I	17	27	13	24	25	106
II	25	57	17	49	55	203
III	38	91	41	217	213	600
IV	22	33	21	133	222	431
V	9	21	17	87	305	439
Total	111	229	109	510	820	1779

Source: Data from the Danish Welfare Study: Hansen (1984), bilagstabel 14.

Table 11.8. Eigenvalues λ_m and variance accounted for r_m^2 for each dimension m=1,2,3 and 4 and the data in table 11.7.

Dimension

	m=1	2	3	4
λ_m	0.354	0.139	0.048	0.021
r_m^2	0.850	0.981	0.997	1.000

Table 11.8 strongly suggests that a model with $M_0=2$ fits the data well while a model with $M_0=1$ would not suffice. The Pearson test statistic (11.31) for $M_0=1$ has observed value

$$q = 42.12, \text{df} = 9,$$

with approximate level of significance less than 0.0005. For $M_0=2$ the Pearson test statistic (11.31) has observed value

$$q = 4.14, \quad df = 4,$$

with approximate level of significance p=0.388. Hence a description with $M_0=2$ is satisfactory, while a description of $M_0=1$ is not. The scores u_{im} and v_{jm} for dimensions m=1 and 2 are shown in table 11.9.

Table 11.9. Row scores (u_{im}) and column scores (v_{jm}) for a correspondance analysis solution with $M_0=2$ and the data in table 11.7.

	Dimension	
u_{im}	m=1	2
i=1	−1.761	−1.640
2	−1.526	−1.333
3	−0.464	1.049
4	0.459	0.500
5	1.314	−0.912
v_{jm}	m=1	2
j=1	−1.622	−1.198
2	−1.652	−0.860
3	−0.929	−0.393
4	−0.234	1.548
5	0.950	−0.508
λ_m^2	0.125	0.019

The correspondance analysis diagram is shown in fig.11.3.

From fig.11.3 it can be concluded that the categories for both variables can be scaled. The scaling is represented in the diagram by the connecting lines. These scalings can not be represented in one dimension, i.e. by the projections of the points onto the first axes. After such a projection it would for example appear that categories 3 and 4 of the meetings variable are equally close to social group III, while in fact category 4 is close to social group III and category 3 is as close to social group II as to social group III. Three groupings seem to emerge: Persons in social groups I and II predominantly attend meetings at least once a month. Most persons in social group III only attend meetings a few times a year, while most people in social group V never attends meetings outside working hours. Persons in social group IV seems to be divided between those who never attend meetings and those who attend a meeting a few times a year. Finally to attend a meeting approximately every second month does not seem to be typical of any of the social groups. In a comparison of categories within a variable, the profiles of social groups I and II seems to be similar, and the profiles of "one or more times a week" and "one or more times a month" are similar. △.

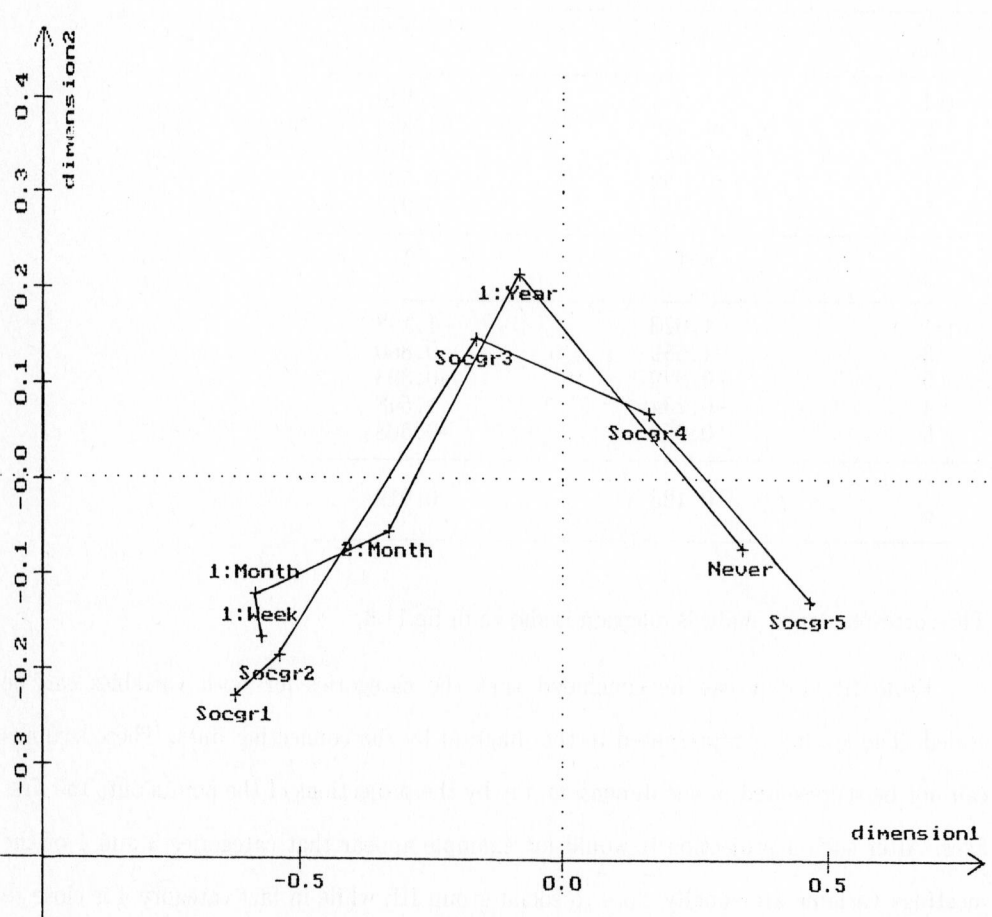

Figur 11.3. Correspondance analysis diagram for the data in table 11.7.

There are a number of textbooks in English dealing with correspondance analysis, e.g. Greenacre (1984), Lebart, Morineau and Warwick (1984), Nishisato (1980), Gifi (1981) and van der Heijden (1987). Important references in French are Escoufier (1982), and Escofier (1979). Correspondance analysis of incomplete tables was studied by de Leeuw and van der Heijden (1988).

11.2. Correspondance analysis for multiway tables

Correspondance analysis for a two–way table was defined as the eigenvalue/eigenvector decomposition (11.8) of the residual matrix $\mathbf{R}$.

Hence the problem for multi–way tables is to find an appropriate two–way matrix to perform the decomposition on. One approach is to consider the so–called **Burt–matrix**. For three variables the Burt–matrix $\mathbf{B}$ is defined as

$$(11.38) \qquad \mathbf{B} = \begin{bmatrix} \mathbf{C}_I & \mathbf{F}_{AB} & \mathbf{F}_{AC} \\ \mathbf{F}'_{AB} & \mathbf{C}_J & \mathbf{F}_{BC} \\ \mathbf{F}'_{AC} & \mathbf{F}'_{BC} & \mathbf{C}_K \end{bmatrix},$$

where $\mathbf{F}_{AB} = \{f_{ij.}\}$, $\mathbf{F}_{AC} = \{f_{i.k}\}$ and $\mathbf{F}_{BC} = \{f_{.jk}\}$. For the Burt matrix, the equivalent of $\mathbf{C}_I$ is

$$(11.39) \qquad \mathbf{C}_B = 3 \begin{bmatrix} \mathbf{C}_I & \mathbf{0} & \mathbf{0} \\ \mathbf{0} & \mathbf{C}_J & \mathbf{0} \\ \mathbf{0} & \mathbf{0} & \mathbf{C}_k \end{bmatrix}$$

since the marginals of $\mathbf{B}$ are $3x_{i..}$, $3x_{.j.}$ and $3x_{..k}$. The eigenvector/eigenvalue decomposition of the Burt matrix is, therefore,

$$(11.40) \qquad \mathbf{C}_B^{-1} \mathbf{B} \mathbf{C}_B^{-1} = \mathbf{U\Lambda U'},$$

where $\mathbf{\Lambda}$ is an M–dimensional diagonal matrix of eigenvalues and $\mathbf{U}$ a $(I+J+K)xM$ dimensional matrix with the eigenvectors as columns. The number M is the rank of $\mathbf{B}$. The eigenvectors are normed by

(11.41) $$U'C_B U = I.$$

From (11.40) and (11.41) follow that the eigenvectors and eigenvalues satisfy the matrix equation

(11.42) $$C_B^{-1} BU = U\Lambda,$$

and hence

(11.43) $$C_B^{-1} B C_B^{-1} BU = U\Lambda^2,$$

which corresponds to (11.13) and (11.14).

The eigenvalue/eigenvector decomposition (11.40) of the Burt–matrix produces a simultaneous set of scores for the categories of all three variables. The correspondance analysis diagram will thus represent the categories of all three variables in the same diagram. Since the Burt–matrix is constructed from the marginal frequency tables F_{AB}, F_{AC} and F_{BC}, the category scores produced by a correspondance analysis based on the Burt–matrix should be related to the category scores obtained from three separate correspondance analyses based on the marginal tables. It can in fact be proved that if correspondance analyses of the marginal tables produce the same eigenvalues and eigenvectors, they will be equal to the ones obtained from the Burt–matrix. To be precise if u_{im}^{AB}, v_{jm}^{AB} are the eigenvectors obtained from F_{AB}, u_{im}^{AC}, v_{km}^{AC} those obtained from F_{AC} and u_{jm}^{BC}, v_{km}^{BC} those obtained from F_{BC} and

$$u_{im}^{AB} = u_{im}^{AC} \text{ for all } i \text{ and } m,$$
$$v_{jm}^{AB} = u_{jm}^{BC} \text{ for all } j \text{ and } m,$$
$$v_{km}^{AC} = v_{km}^{BC} \text{ for all } k \text{ and } m,$$

then the eigenvectors obtained from the Burt–matrix are

$$(u_{1m}^{AB}, \ldots, u_{Im}^{AB}, v_{1m}^{AB}, \ldots, v_{Jm}^{AB}, v_{1m}^{BC}, \ldots, v_{Km}^{BC})$$

for m=1,...,min(I–1,J–1,K–1).

If the eigenvectors obtained from the analyses of the marginal tables are not identical, the eigenvectors from an analysis of the Burt–matrix tend to be average values of the eigenvectors from the marginal tables.

To illustrate these results consider the Burt–matrix

(11.44)
$$B = \begin{bmatrix} C_I & F \\ F' & C_J \end{bmatrix}$$

for just two variables. The diagonal matrix of marginals is here

(11.45)
$$C_B = 2 \begin{bmatrix} C_I & O \\ O & C_J \end{bmatrix}$$

The matrix of eigenvalues Λ and the matrix of eigenvectors U are derived from (11.42), which in case of two variables due to (11.44) and (11.45) takes the form

$$\frac{1}{2} \begin{bmatrix} C_I^{-1} & 0 \\ 0 & C_J^{-1} \end{bmatrix} \cdot \begin{bmatrix} C_I & F \\ F' & C_J \end{bmatrix} \cdot \begin{bmatrix} U_1 \\ U_2 \end{bmatrix} = \begin{bmatrix} U_1 \Lambda \\ U_2 \Lambda \end{bmatrix},$$

where $U'=(U_1',U_2')$ and U_1 and U_2 are matrices of dimension I.M and J.M respectively.

Hence (11.42) is equivalent with the combined equations

(11.46)
$$\begin{cases} C_I^{-1}FU_2 = U_1(2\Lambda-I) \\ \\ C_J^{-1}F'U_1 = U_2(2\Lambda-I) \end{cases}$$

Solved with respect to U_1, (11.46) yields

$$(11.47) \qquad \mathbf{C}_I^{-1}\mathbf{F}\mathbf{C}_J^{-1}\mathbf{F}'\mathbf{U}_1 = \mathbf{U}_1(2\mathbf{\Lambda}-\mathbf{I})^2.$$

A comparison with (11.15) then shows that $\mathbf{U}_1$ is equal to the matrix $\mathbf{U}$ of eigenvectors obtained by a decomposition of $\mathbf{F}$. The eigenvalues $\lambda_1^2,...,\lambda_M^2$ obtained from a decomposition of $\mathbf{F}$ and the eigenvalues, say, $\delta_1^2,...,\delta_N^2$ obtained from a decomposition of the Burt–matrix are connected by

$$2\delta_m - 1 = \lambda_m$$

or

$$(11.48) \qquad \delta_m^2 = (\frac{\lambda_m + 1}{2})^2.$$

It turns out that if $\lambda_1^2 > ... > \lambda_M^2$, then $\delta_1^2,...,\delta_M^2$ given by (11.48) are the M largest eigenvalues of the Burt–matrix. There may , however, be eigenvalues, $\delta_{M+1}^2,...,$ different from zero. These are, as can be seen from (11.48), less than 0.25. The eigenvalues obtained from the Burt–matrix are, accordingly, not as immediately appealing as indicators of the dimensionality of the correspondance analysis model as those obtained from $\mathbf{F}$.

Greenacre (1988) suggested to perform the correspondance analysis on the Burt–matrix with the "diagonal"–matrices $\mathbf{C}_I, \mathbf{C}_J, \mathbf{C}_K$, etc. omitted.

Correspondance analysis for a multi–way table can also be defined by relating the correspondance analysis to a log–linear model as described by van der Heijden and de Leeuw (1985), Goodman (1986) and van der Heijden, Falguerolles and de Leeuw (1987).

Suppose for example that variables A and B in a three–way table are conditionally independent given the third variable C. In the terminology of chapter 5, this is the hypothesis

$$(11.49) \qquad H_0: A \otimes B \,|\, C$$

Under H_0 the expected numbers are

$$(11.50) \qquad \hat{\mu}_{ijk} = x_{i.k} x_{.jk} / x_{..k}.$$

These expected numbers are of course just the expected numbers in the K conditional two–way tables formed by variables A and B given the levels k=1,...,K of variable C. It follows that we can study the structure of the deviations of the observed frequencies from the expected under H_0 by performing a correspondance analysis on each of the K conditional two–way tables. This type of analysis effectively utilise the frequencies of the three–way table while an analysis of the Burt–matrix only utilise the frequencies in the marginal two–way tables. Although a correspondance analysis of the K conditional two–way subtables given C thus utilise the three–dimensional character of the table, it can not account for all the variability in the numbers. In fact it only accounts for that variability, which is not accounted for by the model A⊗B|C. In terms of a log–linear model, the three–factor interactions and the two–factor interactions between A and B are zero under (11.49). Hence a correspondance analysis of the conditional two–way tables given variable C is a method of studying the structure of the two–factor interactions between A and B separatedly, if we assume that the three–factor interactions are all zero.

Consider next the hypothesis

(11.51) $$H_1: A⊗B,C$$

Under H_1 the expected numbers are

(11.52) $$\hat{\mu}_{ijk} = x_{i..} x_{.jk}/n.$$

These expected numbers can, however, also be interpreted as independence in a two–way table, where variable A is cross–classified with a new variable BC obtained by combining the categories of variables B and C. One can thus study the variability in the contingency table, not accounted for by the hypothesis (11.51) by merging the categories of variables B an C into a new variable BC with JK categories.

Example 11.3

The data in table 11.10 is also from the Danish Welfare Study, 1976. The table shows the sample cross–classified according to the variables

A: Income

B: Wealth

C: Ownership of dwelling.

Table 11.10. A random sample from the Danish Welfare Study in 1976 cross–classified according to income, wealth and ownership of dwelling.

| A: Income | B: Wealth | C: Ownership | |
| | | Own dwelling | Rent dwelling |
– 1000 Dkr.–	–1000 Dkr.–		
0–40	0	69	291
	0–50	70	126
	50–150	98	22
	150–300	75	5
	300–	35	4
40–60	0	68	215
	0–50	76	120
	50–150	113	21
	150–300	87	7
	300–	56	3
60–80	0	98	171
	0–50	64	133
	50–150	153	40
	150–300	120	7
	300–	54	7
80–110	0	110	176
	0–50	100	120
	50–150	155	54
	150–300	115	7
	300–	73	4
110–	0	103	90
	0–50	64	87
	50–150	122	52
	150–300	131	24
	300–	151	25

Source: Data from the Danish Welfare Study: Hansen (1978), table 6.H.32.

In order to compare the different ways of analyzing a three–way table by correspondance analysis, six different correspondance analysis diagram are shown below in figures 11.4 to 11.9.

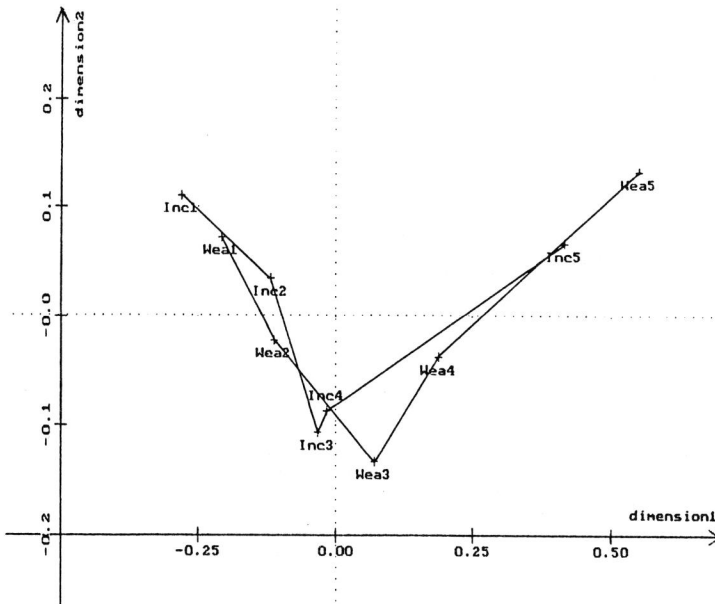

Figur 11.4. Correspondance analysis of the marginal table of variables A and B.

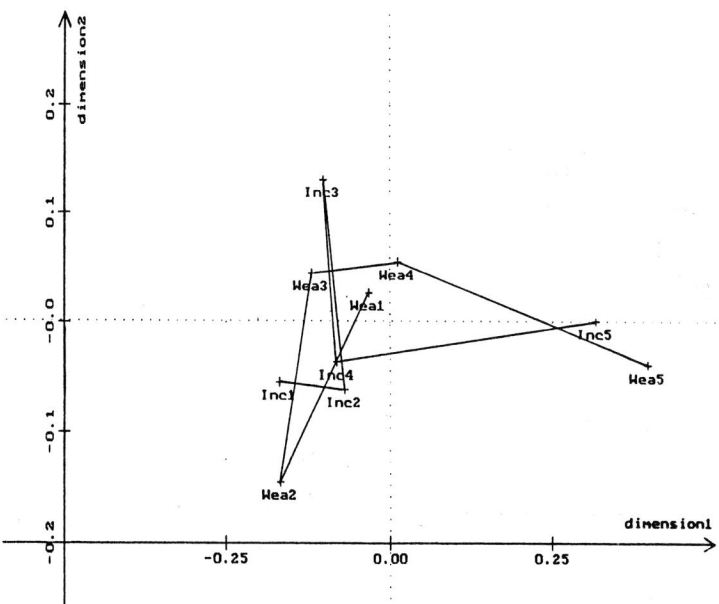

Fig.11.5. Correspondance analysis of the conditional table of variable A against variable B given variable C at level 1, i.e. income against wealth for owners only.

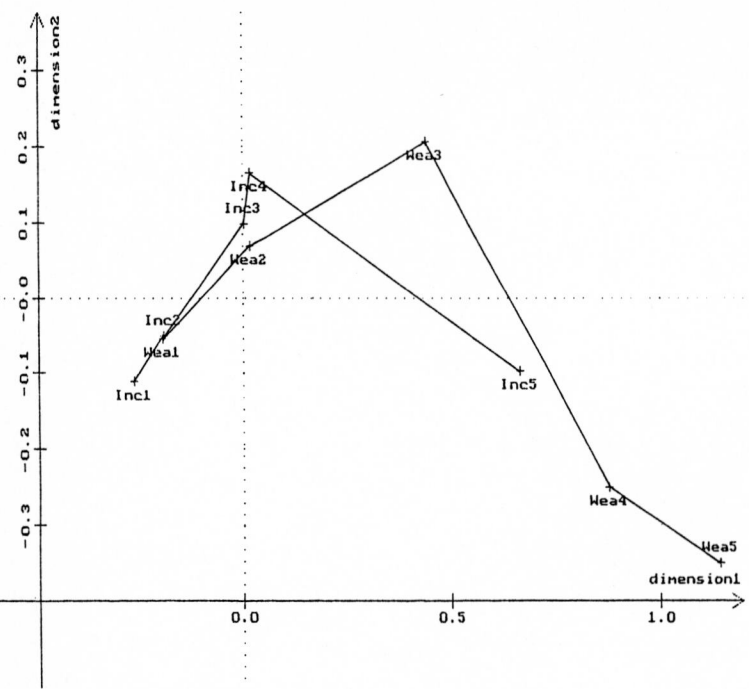

Fig.11.6. Correspondance analysis of the conditional table of variable A against variable B given variable C at level 2, i.e. income against wealth for renters only.

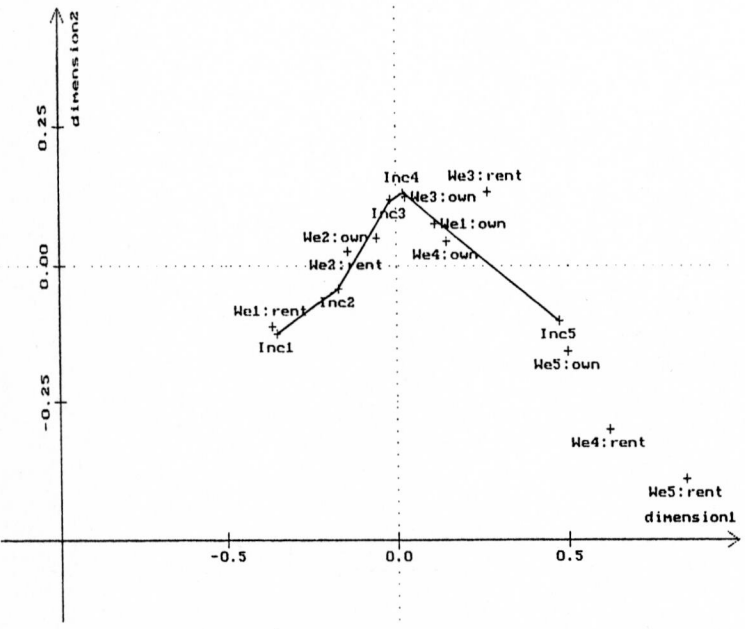

Fig.11.7. Correspondance analysis of variable A against variable B and C combined, i.e. income against wealth and ownership combined.

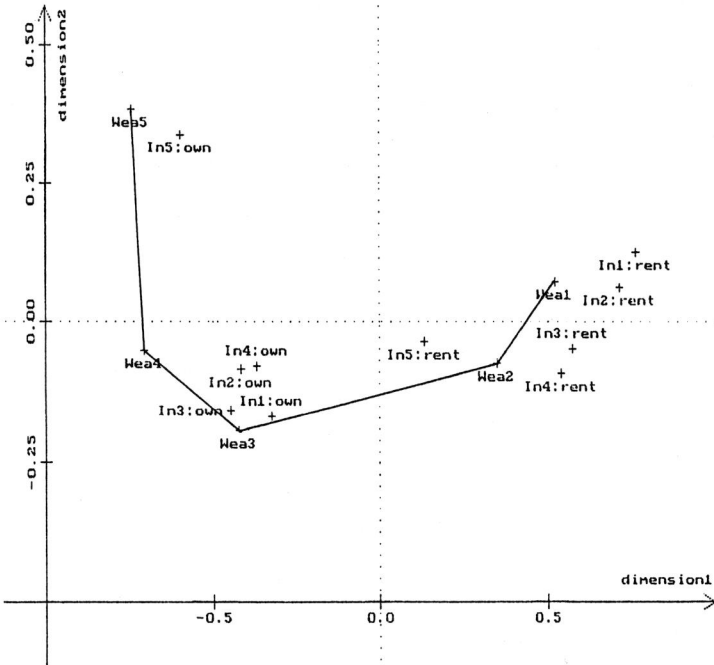

Fig.11.8. Correspondance analysis of variable B against variables A and C combined, i.e. wealth against income and ownership combined.

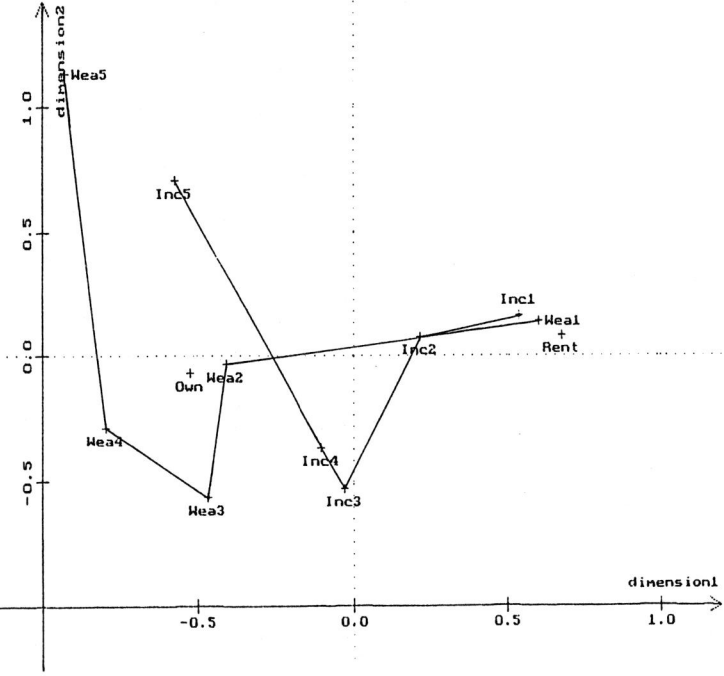

Fig.11.9. Correspondance analysis of the Burt–matrix.

As should be expected the various diagrams show different patterns.

Firstly fig. 11.4 shows that there is a strong connection between income and wealth: high incomes correspond to high wealth and low incomes to low wealth. Fig. 11.5 and 11.6 show, however, that the relationship between income and wealth is more complicated. Thus if only owners are included in the analysis, the relationship disappear except for the highest income and wealth brackets. One reason is that elderly people with declining incomes still are wealthy due to the savings implicit in owning their house or apartment. The diagram for renters alone, fig. 11.6, looks more like the over–all picture in fig. 11.4. The difference between owners and renters become more clear in fig. 11.8, which shows that there are real differences between the distribution of wealth between owners and renters. The renters are thus concentrated in the two lowest wealth brackets and the owners in the three highest. Note also in fig. 11.8 as in fig. 11.5 and 11.6, that income is scaled in relation to wealth for renters while, this is not the case for owners. The diagram fig. 11.9 for the Burt–matrix shows that there is a basic difference between owners and renters, but otherwise fig. 11.9 only shows that both wealth and income are scalable.

This example illustrates how the Burt–matrix can not reveal the characteristic features for understanding the data structure, which emerge from correspondance analysis decompositions, which utilise the multi–dimensionality of the data, here especially manifest in figure 11.5 and 11.8.

That the scores obtained from decomposing the Burt–matrix are average values of what one gets from correspondance analyses of the marginal two–way tables can be illustrated by table 11.11. The table show the scores for the income categories obtained in three ways:

(a)   from a correspondance analysis of a two–way table of income against wealth

(b)   from a correspondance analysis of a two–way table of income against ownership

(c)   from a correspondance analysis of the Burt–matrix.

The scores for the wealth categories are in the same way obtained from two–ways tables against income and ownership and from the Burt–matrix.

Table 11.11. Scores for income and wealth obtained from correspondance analyses of marginal two–way tables and from the Burt–matrix.

Income	0–40	40–60	60–80	80–110	110–
Against wealth	0.616	0.268	0.082	−0.004	−0.895
Against ownership	0.821	0.278	−0.075	−0.250	−0.675
The Burt–matrix	0.743	0.300	−0.038	−0.137	−0.780

Wealth	0	0–50	50–150	150–300	300–
Against income	0.453	0.253	−0.184	−0.395	−1.176
Against ownership	0.492	0.350	−0.410	−0.695	−0.657
The Burt–matrix	0.485	0.331	−0.381	−0.651	−0.755

As can be seen from table 11.11 the scores obtained from an analysis of the Burt–matrix are with one exception intermediate values between the scores obtained from analyzing the marginal two–way tables. △.

11.3. Comparison of models

In this section we compare the correspondance analysis model with the association models discussed in chapter 10. A number of authors have made such comparisons and commented on differences and similarities. Important references are Goodman (1985), (1986) and van der Heijden and de Leeuw (1985). Cf. also Choulakian (1988) and Tennenhaus and Young (1985).

Suppose for a two–way contingency table one wants to determine scores for the row categories and scores for the column categories such that there is maximum correlation between these scores. If the row scores are denoted $u_1, ..., u_I$ and the column scores $v_1, ..., v_J$, the covariance between the scoring of variables A and B is

$$\text{cov}(A,B) = \sum_{i=1}^{I} \sum_{j=1}^{J} (u_i - \bar{u})(v_j - \bar{v}) p_{ij},$$

where p_{ij} is the probability of observing a unit in cell (i,j),

$$\bar{u} = \sum_{i=1}^{I} u_i p_i.$$

and

$$\bar{v} = \sum_{j=1}^{J} v_j p_{.j} .$$

If it is further assumed that the scores are scaled to satisfy

(11.53)
$$\sum_j v_j p_{.j} = \sum_i u_i p_{i.} = 0$$

and

(11.54)
$$\sum_j v_j^2 p_{.j} = \sum_i u_i^2 p_{i.} = 1 ,$$

then $E[A] = \bar{u} = 0$, $E[B] = \bar{v} = 0$, $var[A] = \sum (u_i - \bar{u})^2 p_{i.} = 1$ and $var[B] = \sum (v_j - \bar{v})^2 p_{.j} = 1$. The correlation coefficient ρ_{AB} between A and B then becomes

$$\rho_{AB} = \sum_i \sum_j u_i v_j p_{ij} .$$

If the constraints (11.53) and (11.54) are replaced by the data dependent constraints

(11.55)
$$\sum_j v_j f_{.j} = \sum_i u_i f_{i.} = 0$$

and

(11.56)
$$\sum_j v_j^2 f_{.j} = \sum_i u_i^2 f_{i.} = 1,$$

where $f_{ij} = x_{ij}/n$, then

(11.57)
$$r_{AB} = \sum_i \sum_j u_i v_j f_{ij}$$

is the empirical correlation coefficient between A and B. The quantity r_{AB} is called the

canonical correlation coefficient between A and B if the scores are chosen such that r_{AB} is maximized. The scores that maximize (11.57) are called the **canonical scores**. The next theorem shows that correspondance analysis with $M_0=1$ is equivalent to determining the canonical correlation.

Theorem 11.3

Let λ^2 be the largest eigenvalue of the matrix D defined by (11.11). Then λ is the canonical correlation coefficient and the eigenvector $(u_1,...,u_I)$ corresponding to λ has as elements the canonical scores for variable A. Similarly if E is the matrix (11.12), the eigenvector $(v_1,...,v_J)$ corresponding to λ, has as elements the canonical scores for variable B.

Theorem 11.3 throws new light on some of the elements of correspondance analysis. Not only is the largest eigenvalue λ_1 the canonical correlation, but the row scores $u_{11},...,u_{I1}$ and the column $v_{11},...,v_{J1}$ for dimension m=1 represents a scoring of the variables, which makes them correlate as much as possible.

The result in theorem 11.3 can be extended. Thus if $u_{11},...,u_{I1}$ and $v_{11},...,v_{J1}$ are the canonical scores and $u_{12},...,u_{I2}$ and $v_{12},...,v_{J2}$ two other set of scores, which satisfy the constraints (11.5) and (11.6), then these new scores maximize (11.57) subject to (11.5) and (11.6) and the constrained maximum value of r_{AB} is the second largest eigenvalue λ_2 of E or D. Note that the maximization of (11.57) is **conditional** upon the constraint (11.6). The conditions in (11.6) are often termed **orthogonality constraints**. Thus if (11.6) is satisfied, i.e. if

$$\sum_i u_{i1} u_{i2} f_{i.} = 0$$

then in an I–dimensional Euclidean space the lines connecting the points $x_1=(\sqrt{f_{1.}}\,u_{11},...,\sqrt{f_{I.}}\,u_{I1})$ and $x_2=(\sqrt{f_{1.}}\,u_{12},...,\sqrt{f_{I.}}\,u_{I1})$ to origo are arthogonal.

The scores obtained from the second dimension in the correspondance analysis are thus scores that make the variables correlate as much as possible, given that the new set of scores are orthogonal to the canonical scores obtained from the first dimension.

The category scores which generate the canonical correlation coefficient and thus the parameters of the correspondance analysis model for $M_0=1$ are closely related to the parameters of the RC–association model discussed in chapter 10. According to Goodman (1981a), the scores $u_{11},...,u_{I1}$, and $v_{11},...,v_{J1}$, which defines the canonical correlation coefficient are under certain conditions approximately equal to the estimated parameters $\hat{\epsilon}_1,...,\hat{\epsilon}_I$ and $\hat{\delta}_1,...,\hat{\delta}_J$ of the RC–association model.

This is illustrated by table 11.12, where the estimated parameters of examples 10.4, 10.5, 11.1 and 11.2 are compared. The comparisons in table 11.12 clearly supports the result in Goodman (1981a) that a canonical correlation analysis and an analysis by the model (10.42) leads to essential the same results. Only for the data in example 11.1 is there a rather poor match between the estimates. In this case the RC–association model fits the data very badly, however.

Caussinus (1986) has shown that these results are only valid if the canonical correlation is clearly larger than the second largest eigenvalue of $\mathbf{E}$ or $\mathbf{D}$. If $\lambda_1 \simeq \lambda_2$ one may get quite different results. The intuitive reason is that if $\lambda_1 \simeq \lambda_2$ it can be a matter of chance which eigenvector is chosen to represent the first axes and which to represent the second axes.

Gilula (1986) used the scores $u_{1m},...,u_{Im}$ and $v_{1m},...,v_{Jm}$ from a canonical correlation analysis to determine if the categories of a two–way table can be grouped. The rule is that two rows i and l can be grouped if $u_{im}=u_{lm}$ for all m.

Table 11.12. A comparison of parameters estimated under an RC–association model with the scores from a correspondance analysis with $M_0=1$.

Example 10.4:	i = 1	2	3	4		
$\hat{\epsilon}_i$:	1.390	0.902	−0.644	−1.367		
u_{i1}:	1.253	0.828	−0.554	−1.213		

	j=1	2	3	4	5	
$\hat{\delta}_j$:	−1.334	−1.210	−0.349	0.840	0.741	
v_{j1}:	−1.433	−1.297	−0.333	1.001	0.888	

Example 10.5:	i = 1	2	3	4	5	
$\hat{\epsilon}_i$:	1.151	0.644	−0.283	−0.344	−1.890	
u_{i1}:	0.880	0.636	−0.007	−0.053	−2.174	

	j=1	2	3	4	5	
$\hat{\delta}_j$:	0.726	−0.194	−1.601	−2.570	−3.338	
v_{j1}:	0.644	−0.052	−1.453	−2.877	−3.761	

Example 11.1:	i = 1	2	3	4	5	
$\hat{\epsilon}_i$:	1.084	0.462	0.018	−0.319	−2.174	
u_{i1}:	1.420	−0.265	−0.677	−0.825	−0.991	

	j=1	2	3	4	5	6
$\hat{\delta}_j$:	0.426	0.442	0.056	−1.694	−0.700	−2.927
v_{j1}:	0.346	0.737	0.188	−2.069	−0.885	−2.342

Example 11.2:	i = 1	2	3	4	5	
$\hat{\epsilon}_i$:	−1.633	−1.429	−0.513	0.355	1.407	
u_{i1}:	−1.761	−1.526	−0.464	0.459	1.314	

	j=1	2	3	4	5	
$\hat{\delta}_j$:	−1.628	−1.657	−0.874	−0.245	0.951	
v_{j1}:	−1.622	−1.652	−0.929	−0.234	0.950	

11.4. Exercises

11.1. We return to the data in exercise 10.7.

(a) Analyse the data by correspondance analysis.

(b) In how many dimensions is it appropriate to draw the correspondance analysis diagram.

(c) Draw the diagram and interprete it.

(d) Compare with the results of the analysis in exercise 10.7.

11.2. Reanalyse the data in exercises 10.10 and 10.11 by correspondance analysis and compare the results with the results obtained earlier.

11.3. We return to the data in exercise 10.9.

(a) Perform a correspondance analysis on the data.

(b) Discuss whether the correspondance analysis diagram should be one– or two–dimensional.

(c) Compare the results of the correspondance analysis with those obtained in exercise 10.9.

11.4. Discuss the relationship between social class and family type based on a corrrespondance analysis of the data on exercise 10.4.

11.5. Denmark is often divided in 10 regions, which differ in many respects. In the table below the sample of the Danish Welfare Study is cross–classified according to social class and geographical region.

(a) Perform a correspondance analysis of the data.

(b) Compute the standardized residuals for the independence hypothesis and compare the pattern in these with the results of the correspondance analysis.

(c) Describe the connection between a subjects social rank and the region he or she lives in.

(d) Comment on any peculiarities in the structure of the estimated parameters.

| | Social rank group | | | |
Region	I–II	III	IV	V
Copenhagen	58	77	169	75
Suburban Copenhagen	87	93	165	89
Zeeland	82	147	194	142
Lower Islands	19	52	60	60
Bornholm	4	3	9	7
Fuen	35	74	120	103
South Jutland	15	45	67	58
East Jutland	74	158	197	206
West Jutland	33	97	118	111
North Jutland	49	144	165	159

11.5. The table below is once again from the Danish Welfare Study. It shows social rank cross–classified with age for both men and women. Since the table is three–dimensional the task is to analyse the table by various forms of correspondance analysis and compare the results.

| | | Social rank group | | | |
Sex	Age:	I–II	III	IV	V
Women:	20–29	21	84	202	94
	30–39	48	88	164	112
	40–49	24	48	131	111
	50–59	11	39	82	110
	60–69	4	22	33	39
Men:	20–29	52	80	199	132
	30–39	129	177	198	130
	40–49	87	126	107	105
	50–59	51	137	95	123
	60–69	29	89	53	54

(a) Analyse the two conditional tables of social rank cross–classified with age for men and for women.

(b) Analyse the table where sex and age are merged into one variable.

(c) Analyse the table where sex and social rank are merged into one variable.

(d) Analyse the Burt matrix.

(e) Compare the results.

11.6. The table below show for the sample in the Danish Welfare Study, social rank cross–classified with income for each of the four family types in exercise 10.11.

Family type	Income –1000 kr.–	Social rank group			
		I–II	III	IV	V
Single, with children	0–50	0	6	2	5
	50–100	1	7	22	20
	100–150	1	4	16	13
	150–	52	99	147	158
Single, without children	0–50	0	1	3	1
	50–100	0	0	3	3
	100–150	2	0	4	3
	150–	9	12	35	27
Married, with children	0–50	0	21	23	26
	50–100	10	63	108	130
	100–150	20	56	98	56
	150–	74	124	126	97
Married, without children	0–50	6	62	46	25
	50–100	51	144	259	222
	100–150	85	108	185	88
	150–	145	183	185	136

(a)   Analyse by correspondance analysis the four two–dimensional tables of income against social rank separately for each family type.

(b)   Discuss if there are other tables a correspondance analysis can meaningful be based on.

(c)   Comment on the diagrams derived from the various analyses.

11.7. The table below show the wishes as regards the length of the regular sports feature "Sports Saturday" broadcast Saturday on Danish TV. The wishes are cross–classified with how organised the sports activities are for those in the sample, who are sportsactive.

Degree of organization in sports	Wishes as regards length of "Sports Saturday"		
	2 hours or less	2½ to 3½ hours	4 hours or more
Unorganized	47	42	9
Organized at exercise level	77	38	38
Organized at competition level	29	70	70

(a) Perform a correspondance analysis on these data.

(b) Interprete the correspondance diagram.

11.8. In the Danish Welfare Study all persons in the sample was asked if their work was physically demanding and if it was phychiatrically strenuous. In the table below these two variables are cross–classified for men and women separately.

	Work physically demanding	Work psychiatrically strenuous		
		Yes	Yes, sometimes	No
Men	Yes	113	163	370
	Yes, sometimes	45	106	280
	No	229	343	568
Women	Yes	100	109	202
	Yes, sometimes	33	89	179
	No	100	179	524

(a) Perform correspondance analysis on the table of physically demanding against psychiatrically strenuous work for the total sample and describe the association between the variables.

(b) Perform correspondance analysis on the table in (a), but now for men and women separately.

(c) Analyse the Burt–matrix and compare the results with those in (a) and (b).

12. Latent Structure Analysis

12.1. Latent structure models

The structure of a log–linear model can be described on an association diagram by the lines connecting the points. Especially for higher order contingency tables the structure on an assocition diagram can be very complicated, implicating a complicated interpretation of the model. Adding to the interpretation problems for a multi–dimensional contingency table is the fact, that the decision to exclude or include a given interaction in the model can be based on conflicting significance levels depending on the order in which the statistical tests are carried out. These decisions are thus based on the intuition and experience of the data analyst rather than on objective criteria. Hence a good deal of arbitrariness is often involved, when a model is selected to describe the data. We recall for example from several of the examples in the previous chapters that the log–linear model often gave an adequate description of the data judged by a direct test of the model against the saturated model, while among the sequence of successive tests leading to the model, there were cases of significant levels.

In addition to these problems it is important to emphasize that log–linear models only describe structures that can be formulated in terms of interactions. Chapter 10 dealt with models which introduced symmetric or multiplicative structures in the interactions. Some of these models were not log–linear. In this chapter, we shall consider models, which explain the dependencies in an association diagram through in a common dependency on a so–called latent variable.

Consider fig.12.1 showing in (a) an association diagram for a four–way contingency table with a relatively complicated internal dependency structure, the only independencies being

$$D \otimes B \,|\, C$$

and

$$A \otimes D \,|\, C$$

On fig. 12.1(b), the dependency structure is described by the common dependency of all four variables on a fifth variable θ, which in contrast to variables A,B,C and D is unobservable. Such a variable is called a **latent variable**. Given the value of θ, the observable variables A,B, C and D are independent, i.e.

$$A \otimes B \otimes C \otimes D \,|\, \theta.$$

This interpretation is valid whether θ can be observed or not, but if θ is unobservable the problem arises how to estimate θ and how to check whether the structure (b) on fig.12.1 describes the data. Statistical models, where the necessary simplicity of the model is obtained through the assumed presence of a latent variable, are called **latent structure models**. A statistical analysis, which is based on a latent structure model is called a **latent structure analysis**.

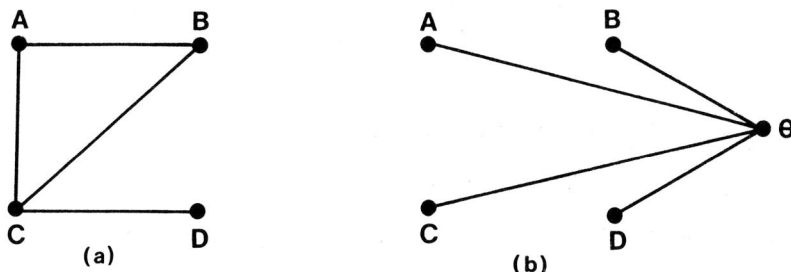

Fig. 12.1. The association diagram for a four–way contingency table with a latent structure as compared with a complicated association diagram.

The statistical analysis is different for a latent structure model with a continuous latent variable and for a latent structure model with a discrete latent variable. If the latent variable is assumed to be discrete, the model is called a **latent class model**, while the model is called a **continuous latent structure model** if the latent variable is continuous.

12.2. Latent class models

For the majority of cases met in practice, the latent variable is an individual parameter in the sense that for each individual in the sample there is associated a value of the latent variable. The variation of the latent variable thus reflects differences between the individuals. When it is assumed that the latent variable is a discrete variable this mean that the sample can be divided into **latent classes**. All members of a latent class have the same value of the latent variable. It will be assumed that there is a finite number M of latent classes.

Consider for the sake of illustration a four–dimensional contingency table obtained through the cross–classification of a simple random sample of size n according to four categorical variables A,B,C and D. The basic assumption in latent class analysis is conditional independence of variables A,B,C and D given the latent variable as illustrated on fig.12.1(b). Let p_{ijklm} be the probability that a randomly selected individual have variables A,B,C and D at levels i,j,k, and l and at the same time belong to latent class m. Then conditional independence given the latent variable means that

(12.1)
$$p_{ijklm} = \pi^A_{im} \, \pi^B_{jm} \, \pi^C_{km} \, \pi^D_{lm} \varphi_m,$$

where π^A_{im}, π^B_{jm}, π^C_{km}, and π^D_{lm} are conditional probabilities of the respective levels of the observable variables given latent class m. For example

π^A_{im} = P (observing variable A at level i given that the individual belongs to latent
class m),

while

φ_m = P (a random selected individual belongs to latent class m).

The marginal probability p_{ijkl} of being in cell (ijkl) is obtained from (12.1) by summing over the latent classes, i.e.

(12.2)
$$p_{ijkl} = \sum_{m=1}^{M} \pi^A_{im} \pi^B_{jm} \pi^C_{km} \pi^D_{lm} \varphi_m$$

In a latent class model the mean values in the cells are thus given by

(12.3)
$$\mu_{ijkl} = np_{ijkl},$$

where the cell probabilities have the form (12.3)

The parameters of a latent class models are the conditional probabilities π^A_{im}, π^B_{jm},... of observing the four variables of the contingency table given latent class m and the marginal probabilities φ_m of an individual belonging to the m latent classes.

The latent class model was introduced by Lazarsfeld (1950), cf. Henry and Lazarsfeld (1968). Early papers on estimation and identification problems are due to Andersson (1954) and McHugh (1956). The connection to modern contingency table theory and the introduction of the widely used EM–algorithm is due to Goodman (1974).

An alternative and equivalent parameterization is in terms of the two–factor interactions between variables A, B, C, D and the latent variable θ and the main effects of all five involved variables if the extended model (12.1) is log–linear. Since $A \otimes B \otimes C \otimes D \mid \theta$,

(12.4)
$$\ln\mu_{ijklm} = \tau^{A\theta}_{im} + ... + \tau^{D\theta}_{lm} + \tau^A_i + ... + \tau^D_l + \tau^\theta_m + \tau_0.$$

from which the marginal cell probability p_{ijkl} is derived as

(12.5)
$$p_{ijkl} = \frac{1}{n} \sum_{m=1}^{M} \mu_{ijklm}.$$

The parameters of (12.2) are derived from (12.4) and (12.5) as

$$\varphi_m = \mu_{....m}/n,$$

since $\pi^A_{.m} = \pi^B_{.m} = \pi^C_{.m} = \pi^D_{.m} = 1$, and

$$\pi^A_{im} = \frac{\mu_{i....m}}{n} \frac{1}{\varphi_m},$$

$$...$$

$$\pi^D_{lm} = \frac{\mu_{...lm}}{n} \frac{1}{\varphi_m}.$$

For the interpretation of the statistical results of a latent class analysis, the parameterization in (12.2) is, however, the most convenient one.

It is a necessary condition for a proper estimation of the parameters and for testing the goodness of fit of the model, that the model is identifiable in the sense that the number of parameters is smaller than the number of cells. For a four–dimensional table with 2 levels for each variable, the number of free parameters in a multinomial model is 15, while the number of parameters in (12.2) for M=2, 3 and 4 are

$$2 \cdot 4 + 1 = 9 \text{ for } M=2$$

$$3 \cdot 4 + 2 = 14 \text{ for } M=3$$

$$4 \cdot 4 + 3 = 19 \text{ for } M=4$$

Hence a latent class model is only identifiable for M=2 and 3. If there are 3 levels for each of the four variables the number of free multinomial parameters is 80, while the number of parameters in (12.2) for M=2 to 9 are

$$
\begin{aligned}
M &= 2: & 2 \cdot 2 \cdot 4 + 1 &= 17 \\
M &= 3: & 2 \cdot 3 \cdot 4 + 2 &= 26 \\
M &= 4: & 2 \cdot 4 \cdot 4 + 3 &= 35 \\
M &= 5: & 2 \cdot 5 \cdot 4 + 4 &= 44 \\
M &= 6: & 2 \cdot 6 \cdot 4 + 5 &= 53 \\
M &= 7: & 2 \cdot 7 \cdot 4 + 6 &= 62 \\
M &= 8: & 2 \cdot 8 \cdot 4 + 7 &= 71 \\
M &= 9: & 2 \cdot 9 \cdot 4 + 8 &= 80
\end{aligned}
$$

such that for a 3x3x3x3 contingency table a latent class model with up to 8 classes is identifiable.

In many situations the relatively large number of parameters in a latent class model is prohibitive for an effective testning of the model. Often, however, the problem at hand suggests a number of constraints linking parameters together. Goodman (1974), Dayton and Macready (1980) and Formann (1985), (1988) contains several examples hereof.

12.3. Continuous latent structure models

The simple expression (12.2) for the cell probabilities of an observed contingency table with a latent class structure is only valid when the latent variable is discrete. In the more general case, where θ is a continuous variable, the probabilities $\varphi_1,...\varphi_m$ of the latent classes are replaced by a **latent population density** $\varphi(\theta)$, which describes the variation of θ over the given population. The interpretation of a latent population density is straight forward. For any given interval (a,b) on the **latent scale,**

(12.6)   P(a randomly selected individual from the population has $\theta\epsilon(a,b)$)

$$= \int_a^b \varphi(\theta)d\theta.$$

The function $\varphi(\theta)$ thus describes the distribution of individuals on the latent scale. Fig. 12.2 shows a situation, where the latent population density is a normal density.

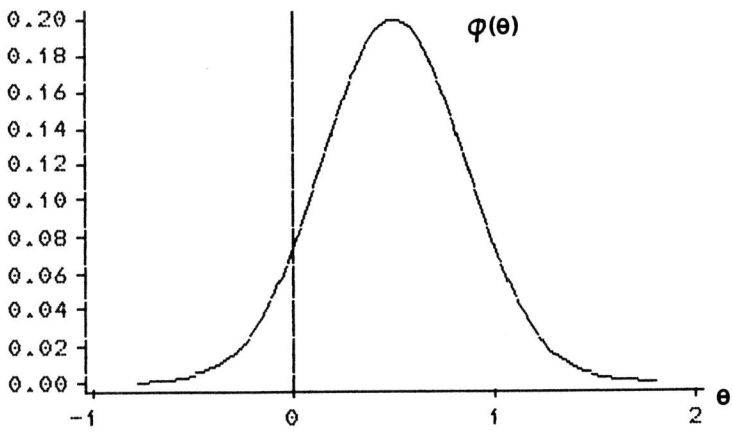

Fig. 12.2. A normal latent density.

If the latent variable has a normal distribution, the distribution of θ is adequately described by the **latent mean value** μ and the **latent variance** σ^2.

Let $p_{ijkl}(\theta)$ be the probability of observing an individual in cell (ijkl) given that the value of the latent variable is θ.

The basic assumption of latent structure analysis is conditional independence given the level of θ. Hence for a four–way table

$$(12.7) \qquad p_{ijkl}(\theta) = \pi_i^A(\theta)\pi_j^B(\theta)\pi_k^C(\theta)\pi_l^D(\theta),$$

where

$$
(12.8) \qquad
\begin{cases}
\pi_i^A(\theta) = P(\text{variable A is at level } i, \text{ given that the} \\
\qquad\qquad \text{latent variable has value } \theta). \\
\dots \\
\pi_1^D(\theta) = P(\text{variable D is at level } 1, \text{ given that the} \\
\qquad\qquad \text{latent variable has value } \theta).
\end{cases}
$$

The marginal cell probabilities p_{ijkl} are then obtained as

$$(12.9) \qquad p_{ijkl} = \int \pi_i^A(\theta) \cdot \ldots \cdot \pi_1^D(\theta)\varphi(\theta)\mathrm{d}\theta.$$

The sum in (12.2) is thus replaced by an integral with the latent population density as integration factor.

In constrast to (12.2), where all parameters are identifiable and can be estimated if M is sufficiently small, the parameters of (12.9) are functions of θ with a continuous variation. Hence in order for (12.9) to be a parametric model, the functional form of $\varphi(\theta)$ must be known apart from a finite number of parameters.

Model construction in latent structure analysis is thus tantamount to modelling the conditional probabilities (12.8) and to specify the functional form of the density $\varphi(\theta)$.

Consider first the binary case, where I=J=K=L=2, i.e. where the contingency table is a 2x2x2x2–table. For variable A there are then two conditional probabilities $\pi_1^A(\theta)$ and $\pi_2^A(\theta)$ to be determined. Since

$$\pi_2^A(\theta) = 1 - \pi_1^A(\theta)$$

only $\pi_1^A(\theta)$ needs to be specified, however.

The arguments behind the choice of functional form for $\pi_1^A(\theta)$ is illustrated by the following example.

Example 12.1:

In order to measure the consumer complain behaviour in a given population, a sample of 600 individuals were exposed to 6 situations, where a purchased item did not live up to the expectations of the consumer. Each individual in the sample was asked whether he or she (a) intended to complain to the shop, where the item had been purchased, (b) never was going to visit the shop again or (c) did not intend to take any measures. Table 12.1 shows the observed distribution over all possible response patterns for the binary response, "complain", equal to response (a) and "no complain", equal to response (b) or (c), for four of these consumer situations.

A latent structure model is reasonable in this case, since it may be theorized that the individuals can be ordered along a complain scale with people, who never would dream of complaining, at the lower end, and people, who complain if there is the slightest thing wrong, at the upper end. Such a theory would entail that the probabilities $\pi_i^A(\theta),...,\pi_1^D(\theta)$, where A,B,C, and D are the variables representing the 4 consumer situations, are increasing functions of θ, with values close to 0 for low θ-values and values approaching 1 for high θ-values.

All 4 situations may not, on the other hand, have the same tendency to provoke a complaint. Hence we may also theorize that

$$\pi_1^A(\theta) < \pi_1^B(\theta) < ... < \pi_1^D(\theta),$$

(or some other order of the variables) for all values of θ. Properly transformed, it may even be the case that the differences between the values of $\pi_1^A(\theta)$ to $\pi_1^D(\theta)$ can be described by a set of parameters $\epsilon_A,...,\epsilon_D$ such that for a suitable function g

$$(12.10) \qquad g(\pi_1^S(\theta)) - g(\pi_1^T(\theta)) = \epsilon_S - \epsilon_T, \quad S, T=A,B,C,D.$$

Table 12.1. The observed distribution over all possible response patterns for four binary variables concerning consumer complain behaviour.

Response pattern Variable A B C D	Observed numbers
1 1 1 1	193
1 1 1 2	227
1 1 2 1	13
1 1 2 2	58
1 2 1 1	21
1 2 1 2	40
1 2 2 1	4
1 2 2 2	20
2 1 1 1	5
2 1 1 2	5
2 1 2 1	1
2 1 2 2	3
2 2 1 1	0
2 2 1 2	2
2 2 2 1	0
2 2 2 2	8

Source: Poulsen (1981).

Should this be the case, the ϵ's can be interpretated as the complain level of the variable independently of the level of θ. The response probabilities should thus satisfy:

(i) The response probabilities $\pi_1^A(\theta)$, $\pi_1^B(\theta), \pi_1^C(\theta)$ and $\pi_1^D(\theta)$ are increasing functions with limits 0 for $\theta \to -\infty$ and 1 for $\theta \to +\infty$.

(ii) Equation (12.10) is satisfied for a suitable function g and suitable parameters $\epsilon_A, \ldots, \epsilon_D$. $\triangle$.

In example 12.1 it was argued that the response probabilities should satisfy (i) and (ii). There are, however, many such functions. Among the most commonly used are

(12.11) $$\pi_1^S(\theta) = \exp(\epsilon_S+\theta)/(1+\exp(\epsilon_S+\theta)) \; , \; S=A,B,C,D$$

and

(12.12) $$\pi_1^S(\theta) = \phi(\epsilon_S+\theta), \; S=A,B,C,D.$$

where ϕ is the cumulative distribution function for the normal standard distribution. In (12.11) the function g is the logistic function $g(x)=\ln(\frac{x}{1-x})$. In (12.12) g is the probitfunction $g(x)=\Phi^{-1}(x)$. Model (12.11) is called the **Rasch model** (Rasch (1960)), and (12.12) the **Lord–Lawley model** (Lord (1952), Lawley (1943)). Model (12.11) is also referred to as the one–parameter logistic model, or for short the 1PL–model.

As in example 12.1 the purpose of formulating a latent structure model is often to construct an index from the observed discrete variables, which can be used as an instrument to measure the underlying latent variable. Statistical models, where there for each individual exists a simple sufficient statistic for the latent variable are accordingly basic to latent structure analysis. One class of models, which satisfy this requirement for the binary case is the Rasch–model (12.11).

For a Rasch model the probability of response (ijkl) for an individual with latent variable θ is

$$p_{ijkl}(\theta) = \exp(\epsilon_A z(i)+...+\epsilon_D z(l)+t\theta)/[(1+e^{\epsilon_A+\theta})...(1+e^{\epsilon_D+\theta})],$$

where $z(1)=1$, $z(2)=0$ and

$$t =z(i)+...+z(l).$$

Hence the probability distribution of the response form an exponential family and the score $t=z(i)+...+z(l)$, or the number of 1–responses, is a sufficient statistic for θ.

The generalization to the polytomous case of the Rasch model is

(12.13) $$\begin{cases} \pi_i^A(\theta) = \exp(\epsilon_{Ai}+w_i\theta)/\sum_i\exp(\epsilon_{Ai}+w_i\theta) \\ ... \\ \pi_1^D(\theta) = \exp(\epsilon_{Dl}+w_1\theta)/\sum_l\exp(\epsilon_{Dl}+w_1\theta), \end{cases}$$

where $\epsilon_{Ai},...,\epsilon_{Dl}$ are item parameters and the w's known weights. If the parameters satisfy (12.13) the cell probabilities given the value θ of the latent variable, become

$$(12.14) \qquad p_{ijkl}(\theta) = \exp(\epsilon_{Ai}+...+\epsilon_{Dl}) \cdot \exp(\theta(w_i+...+w_l))/$$

$$[\underset{i}{\Sigma}\exp(\epsilon_{Ai}+w_i\theta)...\underset{l}{\Sigma}\exp(\epsilon_{Dl}+w_l\theta)].$$

Also in the polytomous case the distribution (12.14) of the response for an individual with latent variable θ forms an exponential family. The sufficient statistic for θ is the score

$$t = w_i+...+w_l.$$

Since the weights $w_1,...,w_l$ are known quantities, the score t is a function of the response (ijkl) and hence an observable quantity. We return to models, where the weights are parameters to be estimated in section 12.9.

From (12.14) the cell probabilities of the observed table are derived as

$$(12.15) \qquad p_{ijkl} = \exp(\epsilon_{Ai}+...+\epsilon_{Dl}) \cdot \int e^{\theta t}[\underset{i}{\Sigma}\exp(\epsilon_{Ai}+w_i\theta)...\underset{l}{\Sigma}\exp(\epsilon_{Dl}+w_l\theta)]^{-1} \varphi(\theta)d\theta,$$

where $\varphi(\theta)$ is the latent population density. If the latent density belongs to a parametric family, then (12.15) represents a parametric model for the observed contingency table. Let for convenience

$$H(\theta) = \underset{i}{\Sigma}\exp(\epsilon_{Ai}+w_i\theta)...\underset{l}{\Sigma}\exp(\epsilon_{Dl}+w_l\theta)$$

then (12.15) can be written

$$(12.16) \qquad p_{ijkl}(\epsilon_{A1},...,\epsilon_{DL},\alpha,\beta) = \exp(\epsilon_{Ai}+...+\epsilon_{Dl})\int e^{\theta t}H^{-1}(\theta)\varphi(\theta|\alpha,\beta)d\theta,$$

where $\varphi(\theta)=\varphi(\theta|\alpha,\beta)$ depends on the two unknown parameters α and β. We shall term a model with response probability (12.16) a **score model**. The parameterization in (12.13) is arbitrary up to linear transformations of the ϵ's and θ. The indeterminancies implicit in the sums $\epsilon_{Ai}+w_i\theta,...,\epsilon_{Dl}+w_l\theta$ are usually resolved by letting

$$\sum_i \epsilon_{Ai}+...+\sum_l \epsilon_{Dl} = 0,$$

and

$$\epsilon_{AI} =...= \epsilon_{DL} = 0.$$

Although simple in form, the model (12.16) is complicated as a basis for statistical analyses, primarily because it contains two sets of parameters, which are different in nature, but also because any statistical procedure rely on numerical evaluations of the integral in (12.16).

The literature contains a number of suggestions for latent structure models that simplifies the structure of the response probabilities $\pi_i^A(\theta),...,\pi_1^D(\theta)$. If the responses are binary, i.e. if $I=J=K=L=2$, the most widely used models are the Rasch model (12.11) or the Lord–Lawley model (12.12). Both these models can be extended by allowing the response probabilities to depend on a third parameter δ_S called the **item discriminating power** for item S. The resulting model

$$(12.17) \qquad \pi_1^S(\theta) = \exp\{(\epsilon_S+\theta)\delta_S\}/[1+\exp\{(\epsilon_S+\theta)\delta_S\}]$$

corresponding to the Rasch model is called the **Birnbaum model** (Birnbaum (1968)), or the **two–parameter logistic (2PL) model**.

The Lawley–Lord model extended with item discriminating powers is

$$(12.18) \qquad \pi_1^S(\theta) = \Phi((\epsilon_S+\theta)\delta_S)$$

Lord (1957) suggested also to include a **guessing parameter** ω. If an individual can not solve a problem in an aptitude test, he or she tries to guess the answer in which case the answer is correct with probability 0.5 in the binary case. If guessing is present and the Rasch model holds, when no guessing takes place, then

$$(12.19) \qquad \pi_1^S(\theta) = 0.5\omega + (1-\omega)\exp(\epsilon_S+\theta)/[1+\exp(\epsilon_S+\theta)].$$

Andrich (1978b), (1982) suggested to consider **rating models**, where the item para–ameters are one–dimensional and the choice between response categories are reflected in rating parameters $\tau_1,...,\tau_I$ for the I response categories. The response probability for item S in Andrich's model is

$$(12.20) \qquad \pi_i^S(\theta) = \exp(i(\theta+\epsilon_S)-\sum_{j=1}^{i}\tau_j)\frac{1}{C},$$

where

$$C = \sum_{i=1}^{I}\exp(i(\theta+\epsilon_S)-\sum_{j=1}^{i}\tau_j).$$

Masters (1982) suggested a **partial credit model**, where credit parameters $\delta_{S1},...,\delta_{SI}$ are assigned to each of the categories, such that the response probability become

$$(12.21) \qquad \pi_i^S(\theta) = \exp(i\theta+\sum_{j=1}^{i}\delta_{Si})/C,$$

where

$$C = \sum_{i=1}^{I}\exp(i\theta+\sum_{j=1}^{i}\delta_{Si}).$$

Both the rating scale model (12.20) and the partial credits model (12.21) are special cases of (12.13), with the weights $w_1,...,w_I$ chosen as $w_i=i$. In (12.20) the ϵ's are, however, assumed to have the structure

$$\epsilon_{Si} = i\epsilon_S - \sum_{j=1}^{i} \tau_j \, ,$$

where the τ's are independent of the item S. In (12.21) the credit parameters are connected to the ϵ's via the differences

$$\epsilon_{Si} - \epsilon_{Si-1} = \delta_{Si}.$$

Fischer (1977), (1983) suggested to let the item parameters be linear functions of explanatory parameters $\beta_{i1},...,\beta_{ip}$ such that

$$\epsilon_{Si} = a_{i0} + \sum_j a_{Sj}\beta_{ij}.$$

Samejima (1969) suggested to let the cumulative response probabilities

$$\omega_i^S(\theta) = \sum_{j=i}^{I} \pi_i^S(\theta).$$

satisfy the logistic form

$$\omega_i^S(\theta) = \exp(\epsilon_{Si} + \theta\delta_S)/[1 + \exp(\epsilon_{Si} + \theta\delta_S)].$$

In this so–called **graded response model**, differences between, rather than the response probabilities themselves, are logistic transformations of additive expressions, similar to the Birnbaum model.

Useful surveys of different models for the response probabilities are due to Thissen and Steinberg (1986) and Masters and Wright (1984).

In order to fully specify the probabilities (12.9), we must choose a suitable latent density $\varphi(\theta)$. Bock (1972), Andersen and Madsen (1977) and Sanathanan and Blumenthal (1978) all chose a normal density. Bartholomew (1980) discussed varies choices including the inverse Cauchy distribution and listed criterions for how to choose $\varphi(\theta)$. He argues that the logistic density,

$$\varphi(\theta) = \frac{1}{\sigma} \exp(\frac{\theta-\mu}{\sigma})/[1+\exp(\frac{\theta-\mu}{\sigma})]^2.$$

should be preferred.

Muthen (1978), (1979), Christofferson (1975) and Christofferson and Muthen (1981) discussed models, where θ is a multivariate latent variable, $\varphi(\theta)$ the multivariate normal density and the response probabilities $\pi_i^S(\theta)$ are generalized probit functions.

We now return to the score model (12.16). In order to simplify matters, consider the binary case I=J=K=L=2 and assume that $\varphi(\theta)=\varphi(\theta|\mu,\sigma^2)$ is a normal density with mean value μ and variance σ^2. In the binary case one can without loss of generality put $w_1=1$ and $w_2=0$. The probabilities (12.13) then become

$$\pi_1^A(\theta) = e^{\epsilon_A+\theta}/(1+e^{\epsilon_A+\theta})$$

$$\pi_2^A(\theta) = 1-\pi_1^A(\theta) = 1/(1+e^{\epsilon_A+\theta})$$

...

$$\pi_1^D(\theta) = e^{\epsilon_D+\theta}/(1+e^{\epsilon_D+\theta})$$

$$\pi_2^D(\theta) = 1/(1+e^{\epsilon_D+\theta}),$$

where $\epsilon_A=\epsilon_{A1}$ and $\epsilon_D=\epsilon_{D1}$. The cell probabilities $p_{ijkl}(\theta)$ follow from (12.14) as

(12.22) $$p_{ijkl}(\theta) = \exp(\epsilon_A z(i)+...+\epsilon_D z(l))e^{\theta t}H^{-1}(\theta),$$

where

$$z(i) = \begin{cases} 1 \text{ for } i=1 \\ 0 \text{ for } i=2 \end{cases}$$

and

(12.23) $$H(\theta) = \left[1+e^{\epsilon_A+\theta}\right]...\left[1+e^{\epsilon_D+\theta}\right],$$

Since $w_1=1$ and $w_2=0$, the score t can be written

$$(12.24) \qquad\qquad t = z(i)+...+z(1),$$

i.e. the score is simply the number of 1–responses. The cell probability p_{ijkl} is in the binary case given by

$$(12.25) \qquad p_{ijkl} = \exp(\epsilon_A z(i)+...+\epsilon_D z(1)) \int e^{\theta t} H^{-1}(\theta)\varphi(\theta|\mu,\sigma^2)d\theta.$$

It follows that the likelihood function is

$$(12.26) \qquad L = \Pi\ \Pi\ \Pi\ \Pi\ p_{ijkl}^{x_{ijkl}} = \exp(\epsilon_A \Sigma x_{i...} z(i)+...+\epsilon_D \Sigma x_{...1} z(1)).$$
$$\cdot \prod_{t=0}^{4} \left[\int e^{\theta t} H^{-1}(\theta)\varphi(\theta|\mu,\sigma^2)d\theta \right]^{n_t},$$

where n_t is the number of individuals in the sample with score t.

The likelihood function (12.26) depends, since $z(2)=0$, on two sets of statistics.

(i) The sizes n_t of the score groups.

(ii) The item totals $x_{1...},...,x_{...1}$.

As seen from the likelihood function these statistics are jointly sufficient for the parameters $\epsilon_A,...,\epsilon_D$, μ and σ^2. Hence the data can for the score model in the binary case be summarizes in the score group totals and the variable totals.

In the general case the likelihood function for the score model is

$$(12.27) \qquad L = \exp(\Sigma \epsilon_{Ai} x_{i...} + \Sigma \epsilon_{Dl} x_{...1}) \cdot \prod_t \left[\int e^{\theta t}\, H^{-1}(\theta)\varphi(\theta|\mu,\sigma^2)d\theta \right]^{n_t}$$

where

$$H(\theta) = \sum_i \exp(\epsilon_{Ai}+w_i\theta)...\sum_l \exp(\epsilon_{Dl}+w_l\theta),$$

and n_t is the size of the group of individuals for which the score is $w_i+...+w_l=t$.

In the general case the sufficient statistics are accordingly the variable totals $x_{i...}, i=1,...,I,...,x_{...l}, l=1,...,L$ and the score group totals.

Example 12.1. (Continued.

For the data in table 12.1, the sufficient statistics are

$$x_{1...} = 576$$
$$x_{.1..} = 505$$
$$x_{..1.} = 493$$
$$x_{...1} = 237$$

and

$$n_0 = 8$$
$$n_1 = 25$$
$$n_2 = 108$$
$$n_3 = 266$$
$$n_4 = 193$$

From these numbers inference about the parameters can be drawn. △.

In the binary case with $I=J=K=L=2$, the observed numbers can be viewed as a 2x2x2x2–dimensional contingency table. An alternative way to view the table is, however, to regard the indices i,j,k and l of cell (i,j,k,l) as the response to the binary variables A,B,C and D. For each individual, we have thus observed a **response vector** (i,j,k,l). Table 12.1 is with this interpretation a list of the number of individuals for each of the

$2^4=16$ possible response vectors for a 4–dimensional table formed by four binary variables. For an individual with position θ on the latent scale the probability $p_{ijkl}(\theta)$ of the reponse (i,j,k,l) is thus given by (12.22).

The probabilities $\pi_1^A(\theta),...,\pi_1^D(\theta)$ of response 1 as functions of θ are called **item characteristic curves**. The name has its origin in psychometrics, were variables are called items.

In example 12.1 it was argued that the item characteristic curve should be an increasing function of θ. This is the case for both the Rasch–model (12.11) where the item characteristic curve is logistic and the Lord–Lawley model (12.12), where the item characteristic curve is the cumulative standard normal distribution function, often called the **probit–function**. An example of an item characteristic curve for the Rasch model is shown in fig.12.3.

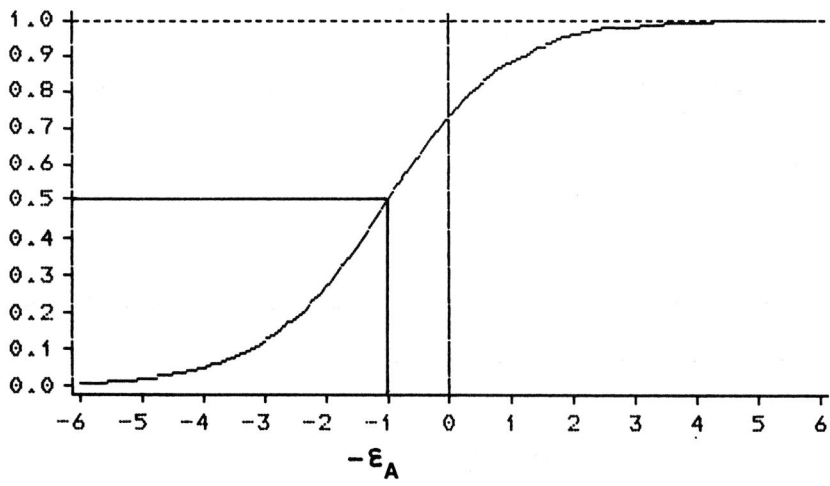

Fig. 12.3. A logistic item characteristic curve

The parameter ϵ_A is connected with variable A and is called an **item parameter**. It describes the tendency of the variable to provoke the response 1 rather than the response 2, since the larger the value of ϵ_A the further the item characteristic curve is located to the left on fig. 12.3, and the larger is the probability of the response 1 for all values of θ.

12.4. The EM–algorithm

Latent structure models are typical examples of models, which can be extended by adding variables, such that the resulting model is less complicated and hence more tractable statistically. The original model is then a case of **incomplete data** within the framework of the extended model. Consider thus fig. 12.1 (b), where the model is a simple independence model if it is possible to observe θ as a fifth variable. The observed distribution over the four observable variables then represents an incomplete dataset in relation to the model, which include the latent variable.

If the extended model is log–linear, estimates of the parameters can be derived through the so–called EM–algorithm, whether the original model is log–linear or not. The EM–algorithm was suggested in its most general form by Dempster, Laird and Rubin (1977). The basis for the algorithm, as used in this chapter, is the following theorem, first discussed by Sundberg (1974).

Theorem 12.1.

Let the random variables X and Y jointly follow a log–linear model with point probability

$$(12.28) \qquad f(x,y \,|\, \tau) = e^{\tau t(x,y) - K(\tau)} h(x,y),$$

such that τ is the canonical parameter and $t(x,y)$ the sufficient statistic. The ML–estimate for τ, if only X is observed, is then the solution to the equation

$$(12.29) \qquad E[T \,|\, X] = E[T].$$

Proof: Since only x is observed the likelihood function is

$$(12.30) \qquad L(\tau) = f(x) = e^{-K(\tau)} \sum_{y} e^{\tau t} h(x,y).$$

Hence the likelihood equation become

(12.31) $\qquad \partial \ln L(\tau)/\partial \tau = -K'(\tau) + \Sigma_y t e^{\tau t} h(x,y)/\Sigma_y e^{\tau t} h(x,y) = 0.$

But since

$$f(y|x) = e^{\tau t} h(x,y)/\Sigma_y e^{\tau t} h(x,y),$$

equation (12.31) can be written as

$$E[T|X] = K'(\tau).$$

Equation (12.29) then follows from theorem 3.1, since T is sufficient for τ in the complete model (12.28). ☐ .

For vector valued canonical parameters the likelihood equations are

(12.32) $\qquad E[T_m|X] = E[T_m], \quad m=1,...,M,$

when $T_1,...,T_M$ are the sufficient statistics for the canonical parameters $\tau_1,...,\tau_M$ and X the observable random variable.

The EM–algorithm works if the likelihood equations are easy to solve in the extended model, i.e. if for known numbers $t_1^*,...,t_M^*$ the equations

$$t_m^* = E[T_m] \qquad , m=1,...,M$$

are easy to solve in terms of the canonical parameters.

The EM–algorithm contains two steps, an E–step and a M–step:.

The E–step: For estimates $\hat{\tau}_1^{(n)},...,\hat{\tau}_M^{(n)}$ of the canonical parameter after n iterations, intermediate values

$$t_m^{(n)} = E[T_m | X], \quad m=1,...,M,$$

are obtained by computing the M conditional expectations of the T_m's given X.

The M–step: Given the intermediate values $t_1^{(n)},...,t_m^{(n)}$ obtained in the E–step, the likelihood function in the extended model is then maximized by solving the m equations

(12.33) $$E[T_m] = t_m^{(n)}, \quad m=1,...,M,$$

with respect to $\tau_1,...,\tau_M$.

The solutions $\hat{\tau}_1^{(n+1)},...,\hat{\tau}_M^{(n+1)}$ to (12.33) are then used as entries for the E–step of iteration n+1, etc.

The solutions to the likelihood equations are reached when the estimates $\hat{\tau}_1^{(n)},...,\tau_M^{(n)}$ do not change. The rate of convergence is usually very slow and often critically dependent on the values chosen for the τ's in the initial E–step. Conditions for the convergence of the EM–algorithm are discussed in Dempster, Laird and Rubin (1977) and in Wu (1983).

12.5. Estimation in the latent class model

For the latent class model, the ML–estimates can either be obtained through the EM–algorithm or by an application of the Newton–Raphson procedure.

The extended model needed for the EM–algorithm is the five–dimensional contingency table formed by the four observed categorical variables A,B,C,D and the discrete latent variable θ. The observed numbers in the extended model are

x_{ijklm} = number of individuals from latent class m observed in cell (i,j,k,l).

The cell probabilities are given by

$$\pi_{ijklm} = \pi^A_{im}\pi^B_{jm}\pi^C_{km}\pi^D_{lm}\varphi_m.$$

Hence the log–likelihood function become

$$\ln L = \sum_i\sum_j\sum_k\sum_l\sum_m x_{ijklm}\ln(\pi_{ijklm})$$

or

(12.34) $$\ln L = \sum_i x_{i...m}\ln\pi^A_{im} + ... + \sum_l x_{...lm}\ln\pi^D_{lm} + \sum_m x_{....m}\ln\varphi_m.$$

The model is thus log–linear in the extended model and the EM–algorithm applies. The canonical parameters are $\ln\pi^A_{im},...,\ln\pi^D_{lm}$ and $\ln\varphi_m$ and the corresponding sufficient statistics are $x_{i...m},...,x_{...lm},x_{....m}$.

It follows then from theorem 12.1, that the likelihood equations are

(12.35) $$\begin{cases} E[X_{i...m}|x_{1111},...,x_{IJKL}] = E[X_{i...m}] \\ ... \\ E[X_{...lm}|x_{1111},...,x_{IJKL}] = E[X_{...lm}] \end{cases}$$

and

(12.36) $$E[X_{....m}|x_{1111},...,x_{IJKL}] = E[X_{....m}]$$

The left hand sides of (12.35) and (12.36) are obtained by observing that $X_{ijkl1},...,X_{ijklM}$ given $X_{ijkl.}=x_{ijkl}$ is distributed as a multinomial distribution with and probability parameters

(12.37) $$\pi^*_{ijklm} = \pi_{ijklm}/\pi_{ijkl.} = \pi^A_{im}...\pi^D_{lm}\varphi_m / \sum_{m=1}^M \pi^A_{im}...\pi^D_{lm}\varphi_m,$$

Hence with $\mu^*_{ijklm}=x_{ijkl}\pi^*_{ijklm}$ the likelihood equations (12.35) and (12.36) become

$$(12.38) \qquad \begin{cases} \overset{*}{\mu}_{i\ldots m}/n = \pi^{A}_{i\,m}\varphi_{m} \\ \ldots \\ \overset{*}{\mu}_{\ldots lm}/n = \pi^{D}_{l\,m}\varphi_{m} \end{cases}$$

and

$$(12.39) \qquad \overset{*}{\mu}_{\ldots.m}/n = \varphi_{m}.$$

The EM–algorithm starts by selecting a set of initial values for $\pi^{A}_{i\,m},\ldots,\pi^{D}_{l\,m}$ and φ_{m}. These values are inserted on the right hand side of (12.37) and $\overset{*}{\pi}_{ijklm}$ is computed for all i,j,k, l and m. By summation intermediate values $t^{(1)}_{i\ldots m},\ldots,t^{(1)}_{\ldots lm}$ and $t^{(1)}_{\ldots.m}$ are then obtained from $\overset{*}{\mu}_{ijklm}{=}x_{ijkl}\overset{*}{\pi}_{ijklm}$ as

$$t^{(1)}_{i\ldots m} = \overset{*}{\mu}_{i\ldots m}/n$$

$$\ldots$$

$$t^{(1)}_{\ldots lm} = \overset{*}{\mu}_{\ldots lm}/n$$

and

$$t^{(1)}_{\ldots.m} = \overset{*}{\mu}_{\ldots.m}/n$$

in the E–step. In the M–step new values for $\pi^{A}_{i\,m},\ldots,\pi^{D}_{l\,m}$ and φ_{m} are computed as

$$\pi^{A}_{i\,m} = t^{(1)}_{i\ldots m}/t^{(1)}_{\ldots.m}$$

$$\ldots$$

$$\pi^{D}_{l\,m} = t^{(1)}_{\ldots lm}/t^{(1)}_{\ldots.m}$$

according to (12.38) and

$$\varphi_{m} = t^{(1)}_{\ldots.m},$$

according to (12.39).

Convergence is often very slow, but to compensate the computations in each iteration are extremely simple. It is thus a definite advantage to have good initial values, but the EM–algorithm usually converges even when the initial values are far from the solutions.

The EM–algorithm for the latent class model was proposed by Goodman (1974), reprinted in Goodman (1978).

The Newton–Raphson method also applies. With good initial values convergence is much more rapid than for the EM–algorithm. The Newton–Raphson procedure has the advantage, that it as a byproduct produces an estimate of the information matrix and hence estimates of the standard errors for the ML–estimates. How to determine the information matrix, when using the EM–algorithm, was shown by Louis (1982). The application of the Newton–Raphson procedure to solve the likelihood equations was studied by Formann (1985).

Example 12.1. (Continued)

For the data in table 12.1 a latent class model with 3 latent classes have 14 parameters, such that the model is barely identifiable with $2^4-1=15$ free multinomial parameters. With M=2 latent classes, there are, however, only 9 parameters, leaving $15-9=6$ degrees of freedom to check the model. Table 12.2 shows the estimates of the parameters of a 2–class latent class model.

Table 12.2. ML–estimates of the parameters for a two–class latent class model applied to the data in table 12.1. Standard errors are shown in parantheses.

	m=1	2
$\hat{\pi}_{1m}^A$	0.982	0.863
	(0.009)	(0.043)
$\hat{\pi}_{1m}^B$	0.893	0.619
	(0.020)	(0.070)
$\hat{\pi}_{1m}^C$	0.940	0.312
	(0.027)	(0.154)
$\hat{\pi}_{1m}^D$	0.471	0.065
	(0.033)	(0.065)
$\hat{\varphi}_m$	0.812	0.188
	(0.062)	(0.062)

With the same initial values derived from a rather crude approximation, the EM–algorithm required 250 iterations, while the Newton–Raphson procedure reached the solutions in 46 iterations. The amount of time required on a medium size personal computer was about 30 seconds in both cases.

A latent class model thus divided the individuals into two classes of sizes roughly 80% and 20%. In class 1 almost all individuals complain except in situation D, where less than half the individuals are expected to complain. Individuals in latent class 2 act more differentiated in the 4 situations. In situation A 86% are expected to complain, while in situation D only 6% can be expected to complain. The parameter estimates in table 12.2 suggest that the individuals can be ordered according to their tendency to complain, but also that the four situations corresponding to the four variables A,B,C and D can be ordered according to their tendency to provoke complaints.

We pursue this in the next section. $\triangle$.

Example 12.2.

In 1984 the Danish Institute for Working Environment Research collected data on work hazards. All persons in a sample of some 6000 employees were interviewed. Among the questions posed was a battery concerning exposure to various work hazards. From among these we consider five questions, namely:

Have you within the last year been exposed to:

A: exhausting physical work.

B: noise in the room or from machines.

C: smoke.

D: annoying chemical substances, not labelled as dangerous.

E: organic dissolutors.

The answers to these five questions from a subsample of 1200 workers are shown in table 12.3 as observed numbers for each of the $2^5=32$ possible response patterns.

The purpose of the study was among other things to form an index of degree of expo–

sure to work hazards in order to compare the degree of exposure for different types of employment. A latent structure model, therefore, seems appropriate. It may for example be attempted to fit a latent class model with M=3 latent classes to the data. The solutions to (12.35) and (12.36) extended to five variables with two levels each are shown in table 12.4.

Table 12.3. Observed response patterns for 1200 workers on 5 questions concerning work hazards.

Response pattern Variable A B C D E	Observed numbers
1 1 1 1 1	7
1 1 1 1 2	2
1 1 1 2 1	3
1 1 1 2 2	18
1 1 2 1 1	6
1 1 2 1 2	7
1 1 2 2 1	5
1 1 2 2 2	73
1 2 1 1 1	0
1 2 1 1 2	2
1 2 1 2 1	1
1 2 1 2 2	7
1 2 2 1 1	1
1 2 2 1 2	6
1 2 2 2 1	2
1 2 2 2 2	118
2 1 1 1 1	0
2 1 1 1 2	2
2 1 1 2 1	0
2 1 1 2 2	25
2 1 2 1 1	5
2 1 2 1 2	8
2 1 2 2 1	6
2 1 2 2 2	136
2 2 1 1 1	0
2 2 1 1 2	0
2 2 1 2 1	1
2 2 1 2 2	18
2 2 2 1 1	5
2 2 2 1 2	14
2 2 2 2 1	8
2 2 2 2 2	714

Source: Unpublished data from the Danish Institute for Working Environment Research.

Table 12.4. Estimated parameters for a 3–class latent class model fitted to the data in table 12.3, with standard errors in parantheses.

	m=1	2	3
$\hat{\pi}_{1m}^{A}$	0.123	0.410	0.619
	(0.021)	(0.090)	(0.099)
$\hat{\pi}_{1m}^{B}$	0.058	0.749	0.727
	(0.062)	(0.141)	(0.078)
$\hat{\pi}_{1m}^{C}$	0.014	0.202	0.290
	(0.011)	(0.062)	(0.075)
$\hat{\pi}_{1m}^{D}$	0.018	0.033	0.653
	(0.006)	(0.054)	(0.220)
$\hat{\pi}_{1m}^{E}$	0.009	0.022	0.583
	(0.004)	(0.044)	(0.218)
$\hat{\varphi}_{m}$	0.717	0.232	0.051
	(0.092)	(0.093)	(0.030)

The population is thus divided into three latent classes. The largest one, covering roughly two thirds of the population, is rarely exposed to work hazards. A second group, covering about a quarter of the population, is more exposed to work hazards but mostly the mild types A, B and C. Finally there is a small group, about 5% of the population, which has a relatively high degree of exposure to all types work hazards.

One advantage of a latent class model is thus the easy interpretability of the parameters. A disadvantage is the relatively large number of parameters. In table 10.3 there are 17 free parameters to describe a contingency table with 32 cells of which many have small expected numbers. Since there is obviously some sort of structure in table 10.4, it should be possible to describe the data with considerable less parameters. △.

With the EM–algorithm it is easy to estimate the parameters of models with constraints on the probability parameters. Suppose for example that we want to estimate the parameters under the hypotheses

$$\pi^A_{1m} = \pi^B_{2m}$$

$$\pi^A_{2m} = \pi^B_{1m}$$

$$\pi^C_{1m} = \pi^D_{2m}$$

$$\pi^C_{2m} = \pi^D_{1m}$$

It then follows directly from the log–likelihood function (12.34) that the sufficient statistic for π^A_{1m} is $x_{1...m}+x_{.2..m}$, for π^A_{2m} is $x_{2...m}+x_{.1..m}$, etc. According to equations (12.35) and (12.38) the EM–algorithm then applies, but with some of the equations merged. Estimation in a latent class with linear parameter constraints is accordingly carried out by applying the linear constraints to the equations of the EM–algorithm.

Example 12.3:

The so–called Coleman–data (cf. Goodman (1978), p.285.) is an example of a data set, where a latent class model with constraints is appropriate. A sample of 3398 schoolboys were asked about their membership of (yes or no) and their attitude towards (favourable or not) the leading crowd in their school class. The questions were asked at two different points in time.

Table 12.5 show the number of schoolboys for each of the 16 possible response patterns with 1 for "yes" and 2 for "no" on the membership question and 1 for "favourable" and 2 for "unfavourable" on the attitude question.

The analysis of the data, shown below is essentially due to Goodman (1974). With 15 free parameters in a saturated model for the contingency table, the maximum number of latent classes, for which the parameters of an unconstrained model can be identified, is $M=3$. For $M=3$ there are, however, $3\cdot 4=12$ π's and 2 φ's, so the model is almost saturated. For $M=2$ the estimates obtained by the EM–algorithm are

$$\hat{\pi}^A_{11} = 0.769 \qquad\qquad \hat{\pi}^A_{12} = 0.101$$

$$\hat{\pi}^B_{11} = 0.645 \qquad\qquad \hat{\pi}^B_{12} = 0.467$$

$$\hat{\pi}^C_{11} = 0.889 \qquad\qquad \hat{\pi}^C_{12} = 0.090$$

$$\hat{\pi}^D_{11} = 0.674 \qquad\qquad \hat{\pi}^D_{12} = 0.499$$

$$\hat{\varphi}_1 = 0.401 \qquad\qquad \hat{\varphi}_2 = 0.599$$

As will be shown in the next section this model does not fit the data very well. Instead Goodman (1974) suggested that the questions were influenced by two latent variables, θ_1 which affects the response on the membership questions and θ_2 which affects the response to the attitude questions. One can then form four latent classes by assuming that each of the latent variables has two levels. The latent classes are described in table 12.6.

The two questions asked at two different points in time make up 4 binary variables:

A: The membership at first interview

B: The attitude at first interview

C: The membership at second interview

D: The attitude at second interview.

Table 12.5. The observed responses for a sample of 3398 schoolboys on two binary questions concerning membership and attitude towards "the leading crowd", asked at two successive points in time.

First interview Membership	First interview Attitude	Second interview Membership	Second interview Attitude	Observed number
1	1	1	1	458
1	1	1	2	140
1	1	2	1	110
1	1	2	2	49
1	2	1	1	171
1	2	1	2	182
1	2	2	1	56
1	2	2	2	87
2	1	1	1	184
2	1	1	2	75
2	1	2	1	531
2	1	2	2	281
2	2	1	1	85
2	2	1	2	97
2	2	2	1	338
2	2	2	2	554

Table 12.6. Four latent classes formed by two latent variables: θ_1 affecting the membership questions and θ_2 affecting the attitude questions.

	Class number	θ_2 at level 1	θ_2 at level 2
θ_1 at level	1	1	2
	2	3	4

Since θ_1 affects variables A and C and has the same level for latent classes 1 and 2 and for latent classes 3 and 4, we must have

(12.40)
$$\begin{cases} \pi^A_{11} = \pi^A_{12} \\ \pi^C_{11} = \pi^C_{12} \\ \pi^A_{13} = \pi^A_{14} \\ \pi^C_{13} = \pi^C_{14} \end{cases}$$

Similarly, since latent variable θ_2 affects variables B and C and has the same level for latent classes 1 and 3 and for latent classes 2 and 4

(12.41)
$$\begin{cases} \pi^B_{11} = \pi^B_{13} \\ \pi^D_{11} = \pi^D_{13} \\ \pi^B_{12} = \pi^B_{14} \\ \pi^D_{12} = \pi^D_{14} \end{cases}$$

Under the constraints (12.40) and (12.41), the corresponding equations of the EM– algorithm are merged. In (12.38) the first and third equations for m=1 and 2 are thus merged into

$$\begin{cases} (\overset{*}{\mu}_{1\ldots1} + \overset{*}{\pi}_{1\ldots2})/n = \pi^A_{11}\varphi_1 + \pi^A_{12}\varphi_2 \\ (\overset{*}{\mu}_{\ldots1.1} + \overset{*}{\mu}_{\ldots1.2})/n = \pi^C_{11}\varphi_1 + \pi^C_{12}\varphi_2 \end{cases},$$

The ML–estimates obtained this way are shown in table 12.7, columns 2 to 5.

Table 12.7. ML–estimates of a 2–class latent class fitted to the data in table 12.3 under to sets of constraints.

	Constraints (12.40) and (12.41)				Constraints (12.40), (12.41) and (12.42)			
	m=1	2	3	4	m=1	2	3	4
$\hat{\pi}^A_{1m}$	0.754	0.754	0.111	0.111	0.827	0.827	0.096	0.096
$\hat{\pi}^B_{1m}$	0.806	0.267	0.806	0.267	0.821	0.287	0.821	0.287
$\hat{\pi}^C_{1m}$	0.910	0.910	0.075	0.075	0.827	0.827	0.096	0.096
$\hat{\pi}^D_{1m}$	0.832	0.302	0.832	0.302	0.821	0.287	0.821	0.287
$\hat{\varphi}_m$	0.272	0.128	0.232	0.368	0.271	0.130	0.228	0.371

A latent class model with the constraints (12.40) and (12.41) fits the data well. The interesting problem for the data in table 12.5 is whether the schoolboys have changed their feelings of membership and their attitude towards the leading crowd. If no changes have taken place the conditional probabilities for variable A should be the same as those for variable C, and the same for variables B and D. The following identities should thus hold in addition to (12.40) and (12.41)

(12.42)
$$
\begin{cases}
\pi^A_{11} = \pi^C_{11}, \ \pi^B_{11} = \pi^D_{11} \\[2mm]
\pi^A_{12} = \pi^C_{12}, \ \pi^B_{12} = \pi^D_{12} \\[2mm]
\pi^A_{13} = \pi^C_{13}, \ \pi^B_{13} = \pi^D_{13} \\[2mm]
\pi^A_{14} = \pi^C_{14}, \ \pi^B_{14} = \pi^D_{14}
\end{cases}
$$

Under these constraints, the parameters estimated by the EM–algorithm are also shown in table 12.7.

As we shall see later a model with these estimated parameters does not fit the data very well and it can be concluded, that a change in attitude has in fact taken place. △.

Applications of latent class model to mobility data was studied by Clogg (1981). Latent class analysis when the data is from several distinct groups was discussed by Clogg and Goodman (1984), 1986) and by Formann (1985).

12.6. Estimation in the continuous latent structure model

For the continuous latent structure model (12.16), the EM–algorithm does not apply unless the extended model obtained by adding the latent parameter θ to the observed response is log–linear. The extended model is a mixed discrete/continuous model with probability density

(12.43)
$$P_{ijkl\theta} = \exp(\epsilon_{Ai} + ... + \epsilon_{Dl}) e^{\theta t} H^{-1}(\theta) \varphi(\theta | \alpha, \beta),$$

defined as a function of θ that satisfies

$$P(\text{variables A to D at levels i to l and } \theta_1 < \theta < \theta_2) = \int_{\theta_1}^{\theta_2} P_{ijkl\theta} \, d\theta.$$

$$= \exp(\epsilon_{Ai} + ... + \epsilon_{Dl}) \int_{\theta_1}^{\theta_2} e^{\theta t} H^{-1}(\theta) \varphi(\theta | \alpha, \beta) d\theta.$$

The likelihood corresponding to (12.43) is log–linear if

$$\ln\varphi(\theta | \alpha, \beta) - \ln H(\theta)$$

where

$$H(\theta) = \sum_i \exp(\epsilon_{Ai} + w_i \theta) ... \sum_l \exp(\epsilon_{Dl} + w_l \theta)$$

is linear in $\epsilon_{Ai}, ..., \epsilon_{Dl}$, α and β. A density $\varphi(\theta | \alpha, \beta)$ which satisfies this requirement with-out restrictions on the ϵ's does not exist. The likelihood equations are, however, easily derived if $\varphi(\theta)$ is a normal density. From (12.27) follows that the log–likelihood function is

$$\ln L = \sum_i \epsilon_{Ai} x_{i...} + ... + \sum_l \epsilon_{Dl} x_{...l} + \sum_t n_t \ln \int e^{\theta t} H^{-1}(\theta) \varphi(\theta | \mu, \sigma^2) d\theta$$

Hence the likelihood equations pertaining to the ϵ's are of the typical form

(12.44)
$$\frac{\partial \ln L}{\partial \epsilon_{Ai}} = x_{i...} - \sum_t n_t \int \left[\exp(\epsilon_{Ai} + w_i \theta) / \sum_i \exp(\epsilon_{Ai} + w_i \theta) \right] e^{\theta t} H^{-1}(\theta) \varphi(\theta | \mu, \sigma^2) d\theta /$$

$$\int e^{\theta t} H^{-1}(\theta) \varphi(\theta | \mu, \sigma^2) d\theta = 0,$$

while the derivatives with respect to μ and σ^2 become

(12.45) $\qquad \dfrac{\partial \ln L}{\partial \mu} = \sum_t n_t \int e^{\theta t} H^{-1}(\theta) \varphi'_\mu(\theta|\mu,\sigma^2) d\theta / \int e^{\theta t} H^{-1}(\theta) \varphi(\theta|\mu,\sigma^2) d\theta = 0$

and

(12.46) $\qquad \dfrac{\partial \ln L}{\partial \sigma^2} = \sum_t n_t \int e^{\theta t} H^{-1}(\theta) \varphi'_{\sigma^2}(\theta|\mu,\sigma^2) d\theta / \int e^{\theta t} H^{-1}(\theta) \varphi(\theta|\mu,\sigma^2) d\theta = 0.$

These likelihood equations are not so complicated to solve as they may seem to be at first glance. We note first that the density $\varphi(\theta|t)$ of θ given the score t is equal to

(12.47) $\qquad \varphi(\theta|t) = e^{\theta t} H^{-1}(\theta) \varphi(\theta|\mu,\sigma^2) / \int e^{\theta t} H^{-1}(\theta) \varphi(\theta|\mu,\sigma^2) d\theta,$

Hence (12.44) can be written as

$$\frac{\partial \ln L}{\partial \epsilon_{Ai}} = x_{i...} - \sum_t n_t E\left[\exp(\epsilon_{Ai} + w_i \theta) / \sum_i \exp(\epsilon_{Ai} + w_i \theta) | t \right] = 0 \ ,$$

i.e. a weighted sum of conditional expectations. It follows moreover from (12.14) that

$$E\left[X_{i...} | \theta \right] = \sum_{jkl} p_{ijkl}(\theta) = \exp(\epsilon_{Ai} + w_i \theta) / \sum_i \exp(\epsilon_{Ai} + w_i \theta).$$

The likelihood equations pertaining to the ϵ's thus have the form

(12.48) $\qquad \begin{cases} x_{i...} = \sum_t n_t E[E[X_{i...} | \theta] | t] \\[2mm] x_{...1} = \sum_t n_t E[E[X_{...1} | \theta] | t] \end{cases}$

Although more complicated than the equations (12.35) for the EM–algorithm they suggest that the likelihood equations can be solved by successively adjusting the item totals $x_{i...}, \ldots, x_{...1}$, for given values of μ and σ^2. The likelihood equations obtained by setting (12.45) and (12.46) equal to zero can be solved by means of the EM–algorithm.

For fixed values of $\epsilon_{Ai},...,\epsilon_{Dl}$, the extended model (12.43) is namely log–linear if φ is a normal density. The sufficient statistics are $\Sigma\theta$ and $\Sigma\theta^2$, where the summations are over all individuals in the sample. Since $t=w_i+...+w_1$ is sufficient for θ for each individual, the conditional mean values of $\Sigma\theta$ and $\Sigma\theta^2$ given the observations are computed as

$$E\left[\Sigma\theta\,|\,\text{the observations}\right] = \Sigma_t n_t E[\theta\,|\,t]$$

and

$$E\left[\Sigma\theta^2\,|\,\text{the observations}\right] = \Sigma_t n_t E\left[\theta^2\,|\,t\right].$$

The density of θ given t was derived above as (12.47). Hence the likelihood equations according to the EM–algorithm are

(12.49) $\quad \Sigma_t n_t \int \theta e^{\theta t} H^{-1}(\theta)\varphi(\theta\,|\,\mu,\sigma^2)d\theta / \int e^{\theta t} H^{-1}(\theta)\varphi(\theta\,|\,\mu,\sigma^2)d\theta = nE\left[\theta\right] = n\mu.$

and

(12.50) $\quad \Sigma_t n_t \int \theta^2 e^{\theta t} H^{-1}(\theta)\varphi(\theta\,|\,\mu,\sigma^2)d\theta / \int e^{\theta t} H^{-1}(\theta)\varphi(\theta\,|\,\mu,\sigma^2)d\theta$

$$= nE\left[\theta^2\right] = n(\mu^2 + \sigma^2).$$

In the binary case I=J=K=L=2, $w_1=1$ and $w_2=0$. Hence

$$E[X_{1...}\,|\,\theta] = e^{\epsilon_A+\theta} / \left[1+e^{\epsilon_A+\theta}\right]$$

$$...$$

$$E[X_{...1}\,|\,\theta] = e^{\epsilon_D+\theta} / \left[1+e^{\epsilon_D+\theta}\right]$$

The likelihood equations (12.48), therefore, become

$$(12.51) \quad \begin{cases} x_{1\ldots} = \Sigma n_t \int \dfrac{e^{\epsilon_A + \theta}}{1 + e^{\epsilon_A + \theta}} e^{\theta t} H^{-1}(\theta) \varphi(\theta) d\theta \Big/ \int e^{\theta t} H^{-1}(\theta) \varphi(\theta) d\theta \\ \ldots \\ x_{\ldots 1} = \Sigma n_t \int \dfrac{e^{\epsilon_D + \theta}}{1 + e^{\epsilon_D + \theta}} e^{\theta t} H^{-1}(\theta) \varphi(\theta) d\theta \Big/ \int e^{\theta t} H^{-1}(\theta) \varphi(\theta) d\theta \end{cases}$$

Application of the EM–algorithm to solve the likelihood equations was suggested by Sanathanan and Blumenthal (1978), cf. also Rigdon and Tsutakawa (1983).

Example 12.1 (Continued):

For the data in table 12.1, we found the values

$$x_{1\ldots} = 576, \qquad x_{.1..} = 505$$

$$x_{..1.} = 493 \qquad x_{...1} = 237$$

$$n_0 = 8, \qquad n_1 = 25$$

$$n_2 = 108, \qquad n_3 = 266$$

$$n_4 = 193.$$

Based on these numbers the solutions to (12.49) to (12.51) are

$$\hat{\epsilon}_A = 2.012 \qquad \hat{\epsilon}_B = 0.281$$

$$\hat{\epsilon}_C = 0.107, \qquad \hat{\epsilon}_D = -2.400$$

$$\hat{\mu} = 1.843, \qquad \hat{\sigma}^2 = 1.567.$$

The estimated ϵ's confirm the results of the latent class analysis, that consumer situation A can be expected draw most complains and situation D the smallest number of complains.

The value of $\hat{\mu}$ means that the θ's have center of gravity well above 0. There is thus a general tendency to complain. The estimated value of the latent variance of σ^2 is evidence, however, of a rather large variation around the mean. Note that an individual with $\theta = 0$ will complain with probability $1/2$ on a neutral item with $\epsilon = 0$. The estimated distribution has 7% of its mass below 0. Hence the population contain only 7% consumers, who complain less often than they do not complain. △.

Example 12.2. (Continued):

For the data in table 12.3 the ML–estimates become

$$\hat{\epsilon}_A = 1.136$$

$$\hat{\epsilon}_B = 1.426$$

$$\hat{\epsilon}_C = -0.505$$

$$\hat{\epsilon}_D = -0.866$$

$$\hat{\epsilon}_E = -1.191$$

$$\hat{\mu} = -2.983$$

$$\hat{\sigma}^2 = 2.544$$

In accordance with fig. 12.3 the estimates $\hat{\epsilon}_A$ to $\hat{\epsilon}_E$ suggest that there is a relatively higher probability of an exposure to exhausting physical work and noise than to smoke and in particular to chemicals and dissolutors whatever the level of the latent variable.

This is in accordance with the interpretation of the estimates in table 12.4 for the latent class model. The very low estimate −2.983 of μ indicate that most of the population is concentrated in the end of the θ–scale, where there is a low probability of an exposure to work hazards. Also this is in agreement with what was found in the analysis by a latent class model. Fig. 12.4 show the item characteristic curves for all five items A to E. △.

The fact that the score t for a given individual is a sufficient statistic for θ can be used to derive an alternative way of estimating the item parameters ϵ_{Ai} to ϵ_{Dl}. Consider the conditional probability $p_{ijkl}(t)$ of response (i,j,k,l) given that the score is $w_i + \ldots + w_l = t$. The probability of observing t is

$$p(t) = \sum_{w_i + \ldots + w_l = t} \sum p_{ijkl}.$$

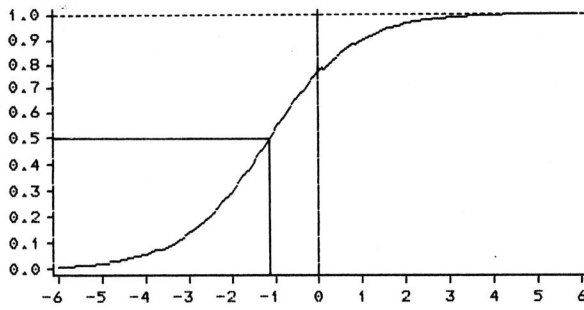

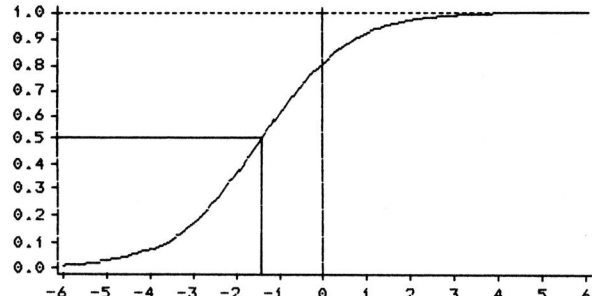

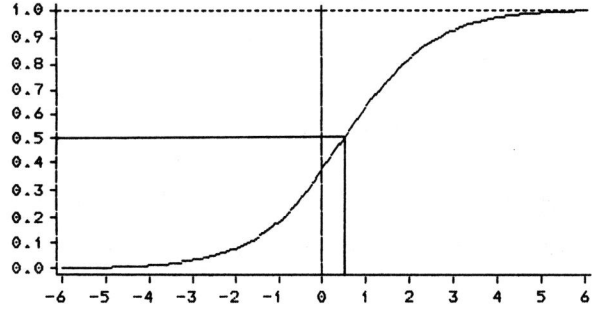

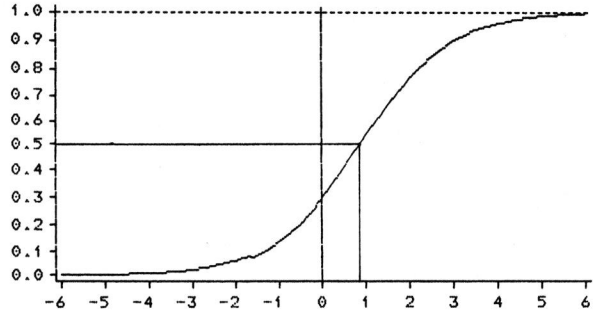

Fig.12.4. Item characteristic curves for four items of the work hazard data.

444

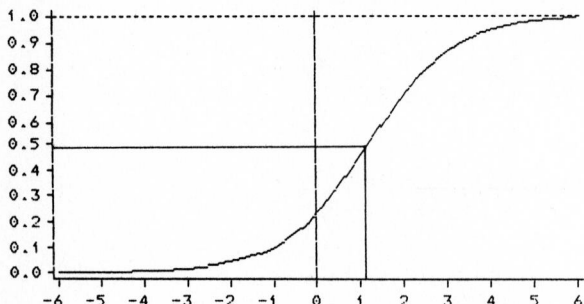

Fig.12.4. Item characteristic curves for item five of the work hazard data.

Hence it follows from (12.15) that

(12.52)
$$p_{ijkl}(t) = \exp(\epsilon_{Ai}+...+\epsilon_{Dl})/g(t;\epsilon)$$

where

(12.53)
$$g(t;\epsilon) = \sum_{w_i+...w_l=t} \sum \exp(\epsilon_{Ai}+...+\epsilon_{Dl}).$$

If the conditional likelihood is defined as the probability distribution of the responses given the observed values of the scores, the conditional log–likelihood function then become

(12.54)
$$\ln L_c(\epsilon) = \sum_i \epsilon_{Ai} x_{i...} +...+ \sum_l \epsilon_{Dl} x_{...l} - \sum_t n_t \ln g(t;\epsilon).$$

Here n_t is as before the number of individuals with score t.

The log–likelihood function (12.54) does not depend on the density $\varphi(\theta)$. Consequently the ϵ's can be estimated independently of the form of $\varphi(\theta)$. Since (12.54) is log–

linear with the ϵ's as canonical parameters and $x_{i...},...,x_{...1}$ as sufficient statistics, the likelihood equations are

$$(12.55) \quad \begin{cases} x_{i..} = E[X_{i...}|n_t \text{ for all } t] \\ \cdots \\ x_{...1} = E[X_{...1}|n_t \text{ for all } t] \end{cases}$$

The conditional expected values in (12.55) are derived from the g–functions as

$$(12.56) \qquad E[X_{i...}|n_t \text{ for all } t] = \Sigma n_t \frac{\partial g(t;\epsilon)}{\partial \epsilon_{Ai}} / g(t;\epsilon)$$

and similarly for the other three equations.

In the binary case, i.e. when $I=J=K=L=2$, the weights can without loss of generality be chosen as $w_1=1$ and $w_2=0$. The possible values of t are then $0,...,4$ and the g–functions become

$$(12.57) \qquad g(t;\epsilon_A,...,\epsilon_D) = \underset{z_1+...+z_4=t}{\Sigma} \exp(\epsilon_A z_1+...\epsilon_D z_4), \ t=0,...,4.$$

For $t=2$, we get for example

$$g(2;\epsilon_A,...,\epsilon_D) = \delta_A\delta_B+\delta_A\delta_C+\delta_A\delta_D+\delta_B\delta_C+\delta_B\delta_D+\delta_C\delta_D,$$

where $\delta_A=e^{\epsilon_A},...,\delta_D=e^{\epsilon_D}$. The derivative of $\varphi(2;\epsilon_A,...,\epsilon_D)$ with respect to ϵ_A is

$$\frac{\partial g(2;\epsilon_A,...,\epsilon_D)}{\partial \epsilon_A} = \delta_A(\delta_B+\delta_C+\delta_D) = \delta_A g(1;\epsilon_B,\epsilon_C,\epsilon_D).$$

In general (12.57) yields

$$\frac{\partial g(t; \epsilon_A, \ldots, \epsilon_D)}{\partial \epsilon_A} = e^{\epsilon_A} g(t-1; \epsilon_B, \ldots, \epsilon_D)$$

This means that the likelihood equations in the binary case have the simple forms

(12.58)
$$\begin{cases} x_{1\ldots} = e^{\epsilon_A} \sum_{t=1}^{4} n_t \dfrac{g(t-1; \epsilon_B, \ldots, \epsilon_D)}{g(t; \epsilon_A, \ldots, \epsilon_D)} \\ \quad \ldots \\ x_{\ldots 1} = e^{\epsilon_D} \sum_{t=1}^{4} n_t \dfrac{g(t-1; \epsilon_A, \ldots, \epsilon_C)}{g(t; \epsilon_A, \ldots, \epsilon_D)} \end{cases}$$

The g–functions necessary for the numerical solution of (12.58) are obtained by the iterative procedure

$$g(t; \epsilon_D) = \begin{cases} e^{\epsilon_D} & \text{for } t=1 \\ 1 & \text{for } t=0 \end{cases}$$

$$g(t; \epsilon_C, \epsilon_D) = g(t; \epsilon_D) + e^{\epsilon_C} g(t-1; \epsilon_D)$$

$$g(t; \epsilon_B, \epsilon_C, \epsilon_D) = g(t; \epsilon_C, \epsilon_D) + e^{\epsilon_B} g(t-1; \epsilon_C, \epsilon_D)$$

$$g(t; \epsilon_A, \epsilon_B, \epsilon_C, \epsilon_D) = g(t; \epsilon_B, \epsilon_C, \epsilon_D) + e^{\epsilon_A} g(t-1; \epsilon_B, \epsilon_C, \epsilon_D),$$

which produces $g(t-1; \epsilon_B, \epsilon_C, \epsilon_D)$ and $g(t; \epsilon_A, \epsilon_B, \epsilon_C, \epsilon_D)$ of the first equation in (12.58). The remaining g–functions are obtained by exchanging the sequence of variables in the procedure.

The estimates to be shown below were obtained by a Newton Raphson procedure, which also provided standard errors of the estimators. Such a procedure usually requires only 3 to 4 iterations and is relatively fast due to the simple iterative procedure for computing the g–functions.

The properties of conditional ML–estimators in this context was studied by Andersen (1970). The asymptotic properties are not covered by the theorems of chapter 3 since the reference probability space changes with increasing sample size. Theorem 3.6 is, however, still valid. Because the observed distribution of t–values $n_0,...,n_4$ are assumed to be known constants, the interior of the convex extension of the support is, however, a bit complicated. The following result is due to Fischer (1981).

Theorem 12.2:

For a four–way table with binary variables the likelihood equations have a unique set of solutions if (assuming that the marginals are ordered $x_{1...} \geq x_{.1..} \geq x_{..1.} \geq x_{...1}$) **the marginals satisfy**

$$0 < x_{1...} < n_1 + n_2 + n_3 + n_4$$

$$x_{1...} + x_{.1..} < n_1 + 2n_2 + 2n_3 + 2n_4$$

$$...$$

$$x_{1...} + x_{.1..} + x_{..1.} < n_1 + 2n_2 + 3n_3 + 3n_4$$

Proof:

The support is from below

$$x_{1...} \geq 0$$

$$...$$

$$x_{...1} \geq 0$$

In addition the item totals can not exceed the marginals obtained in the extreme situation where the only observed responses are (1,1,1,1), (1,1,1,2), (1,1,2,2), (1,2,2,2,) and (2,2,2,2). For this case we obviously have

$$x_{1...} = n_1 + n_2 + n_3 + n_4$$

$$x_{.1..} = n_2 + n_3 + n_4$$

$$x_{..1.} = n_3 + n_4$$

$$x_{...1} = n_4.$$

As is easily seen any deviation from this extreme situation will diminish $x_{1...}, x_{.1..}$ and

$x_{..1.}$, provided $x_{1...} \geq x_{.1..} \geq x_{..1.} \geq x_{...1}$. Hence

(12.59) $$x_{1...} \leq n_1 + n_2 + n_3 + n_4,$$

(12.60) $$x_{1...} + x_{.1..} \leq n_1 + 2n_2 + 2n_3 + 2n_4,$$

and

(12.61) $$x_{1...} + x_{.1..} + x_{..1.} \leq n_1 + 2n_2 + 3n_3 + 3n_4$$

while the sum of all four marginals satisfy

$$x_{1...} + x_{.1..} + x_{..1.} + x_{...1} = n_1 + 2n_2 + 3n_3 + 4n_4.$$

The inequalities (12.59), (12.60) and (12.61) are the constraints on the support due to the given values of $n_0, ..., n_4$. They on the other hand form the boundary of a convex set in the three–dimensional space spanning $x_{1...}$, $x_{.1..}$ and $x_{..1.}$. Hence the likelihood equations have a unique set of solutions, if none of the inequalities (12.59), (12.60) and (12.61) are sharp. ▢.

If the scoring of the response categories is equidistant, i.e. $w_1 = I-1$, $w_2 = I-2, ..., w_I = 0$, the conditional likelihood equations are almost as simple as in the binary case. The derivatives in (12.56) can then be expressed by the g–functions due to the formula

$$\frac{\partial g(t;\epsilon)}{\partial \epsilon_{Ai}} = \sum \cdots \sum_{w_j + ... + w_l = t - w_i} \exp(\epsilon_{Bj} + ... + \epsilon_{Dl}) = e^{\epsilon_{Ai}} g(t - w_i; \epsilon_B, ..., \epsilon_D)$$

which follows directly from (12.53).

Hence the likelihood equations (12.55) become

$$(12.62) \qquad \begin{cases} x_{i...} = \sum_t n_t e^{\epsilon_{Ai}} g(t-w_i; \epsilon_B, \dots, \epsilon_D)/g(t;\epsilon) \\ \dots \\ x_{...1} = \sum_t n_t e^{\epsilon_{Dl}} g(t-w_i; \epsilon_A, \dots, \epsilon_C)/g(t;\epsilon) \end{cases}$$

If the item parameters are estimated by the conditional ML–method, it is normal practice to obtain estimates for μ and σ^2 from equations (12.49) and (12.50) with the CML–estimates inserted for the ϵ's.

Example 12.1. (Continued):

For the complain data the CML–estimates of the ϵ's and the solutions to (12.49) and (12.50) are compared to the direct ML–estimates in table 12.8.

Table 12.8. Comparison of CML–estimates and ML–estimates for the data in table 12.1.

	CML–estimates	ML–estimates
$\hat{\epsilon}_A$	2.106	2.012
$\hat{\epsilon}_B$	0.227	0.281
$\hat{\epsilon}_C$	0.055	0.107
$\hat{\epsilon}_D$	−2.388	−2.400
$\hat{\mu}$	1.854	1.843
$\hat{\sigma}^2$	1.536	1.567

The differences between the two sets of estimates is obviously minimal. △.

Example 12.2. (Continued)

For the work hazard data in table 12.3 the CML–estimates and the direct ML–estimates are compared in table 12.9.

Table 12.9. CML–estimates and direct ML–estimates compared for the work hazard data in table 12.3.

	CML–estimates	ML–estimates
$\hat{\epsilon}_A$	1.132	1.136
$\hat{\epsilon}_B$	1.422	1.426
$\hat{\epsilon}_C$	−0.494	−0.505
$\hat{\epsilon}_D$	−0.861	−0.866
$\hat{\epsilon}_F$	−1.199	−1.191
$\hat{\mu}$	−2.979	−2.983
$\hat{\sigma}^2$	2.536	2.544

For this data set the differences are even smaller than for the complain data. In general the differences tend to diminish when the number of items increase. △.

The numerical problems connected with the joint ML–estimation of the item parameters and the parameters of the latent density was discussed by Bock and Aitkin (1981). The Bock–Aitkin paper deals with the two–parameter logistic model (12.17). Thissen (1982) shows how the results apply to the one–parameter logistic model, i.e. to the solution of the likelihood equations (12.49) to (12.51). Bock and Aitkin suggested to reparametrize the model by the transformation $\theta^* = (\theta - \mu)/\sigma$ of the latent variable, such that (12.25) becomes

$$(12.63) \qquad p_{ijkl} = \exp\{\epsilon_A^* z(i) + ... + \epsilon_D^* z(l)\} \int e^{\theta^* \sigma t} H^{-1}(\theta^*) \varphi(\theta^*) d\theta^*,$$

where $\varphi(\theta)$ is the standard normal density, $\epsilon_A^* = \epsilon_A + \mu, ..., \epsilon_D^* = \epsilon_D + \mu$ and

$$H(\theta^*) = (1 + e^{\epsilon_A^* + \sigma\theta^*})...(1 + e^{\epsilon_D^* + \sigma\theta^*}).$$

The likelihood derived from (12.63) has of course the same solutions as those derived from (12.25), but now

$$\tfrac{1}{4}(\epsilon_A^* + ... + \epsilon_D^*) = \mu.$$

Bock and Aitkin computed the integrals appearing in the likelihood equations (12.49) to (12.51) by **Gauss–Hermite quadrature** using only a few nodes. Thissen (1982) reports that eleven nodes seem to perform quite well. In addition Bock and Aitkin recommended an EM–type algorithm for obtaining the solution of the likelihood equations. They finally suggested to characterize the density distribution of the latent variabel by computing the posterior density of θ given the data, derived above as (12.47) at the nodes of the quadrature formula. Thus if $\theta_1, ..., \theta_Q$ are the nodes, the density $\varphi(\theta)$ is estimated as

$$(12.64) \qquad \hat{\varphi}(\theta_q) = e^{\theta_q H^{-1}(\theta_q)} \overline{\varphi}(\theta_q) / \sum_{q=1}^{Q} e^{\theta_q H^{-1}(\theta_q)} \overline{\varphi}(\theta_q), \quad q=1,...,Q,$$

where $\overline{\varphi}(\theta_q)$ is the weight used at θ_q in the quadrature formula. The procedure can be repeated by estimating the parameters $\epsilon_A^*, ..., \epsilon_D^*$ and σ using a quadrature formula based on the empirical density (12.64) and use these estimates to obtain a new empirical density (12.64), until it stabilizes. The obtained values (12.64) then give an empirical characterization of the latent density $\varphi(\theta)$. Similar ideas are contained in Mislevy (1984), who also gives a useful survey of various methods for estimating the ϵ's and φ jointly. In Mislevy (1987) it is shown how to incorporate auxiliary information into the latent density $\varphi(\theta)$.

In some situations one would like to estimate the item parameters and investigate the fit of the model without specifying the latent density. For the score model (12.16) this can be done by observing that the score

$$t = w_i + ... + w_1$$

is a sufficient statistic for the latent parameter θ. This means that the observed distribu–

tion of individuals over score–values contains the necessary information about the density $\varphi(\theta)$.

Without specifying the density $\varphi(\theta)$, the logarithm of the likelihood (12.27) is

$$(12.65) \qquad \ln L(\epsilon_A,...,\epsilon_D,\varphi_0,...,\varphi_T) = \sum_i \epsilon_{Ai} x_{i...} +...+ \sum_l \epsilon_{Dl} x_{...l} + \sum_t n_t \ln \varphi_t.$$

where

$$\varphi_t = \int e^{\theta t} H^{-1}(\theta) \varphi(\theta) d\theta$$

and T is the maximal score value.

Hence the canonical parameters in a log–linear representation of (12.65) are the ϵ's and the logarithms of the φ_t's.

The ML–estimates for the φ_t's regarded as parameters are

$$(12.66) \qquad\qquad\qquad \hat{\varphi}_t = n_t/n,$$

such that the $\hat{\epsilon}$'s that maximize (12.65) together with the estimates (12.66) can be regarded as ML–estimates for a model, where the form of $\varphi(\theta)$ is not specified.

In this formulation it is assumed that the variation of the ϵ's in $H(\theta)$ allow for all possible vectors φ_t for which $\sum \varphi_t = 1$. We return to this question in the next section.

It should be noted that (12.65) is not a latent class model since the assignment of an individual to a latent class is determined by the response.

If the contingency table formed by variables A,B,C,D is extended by a categorical variable with observed values t, the log–likelihood function (12.65) obviously forms a log–linear model. This five–dimensional table is incomplete, but applying the theory in chapter 7, the ϵ's and $w_t = \ln \varphi_t$ can be estimated. This approach was discussed by Kelderman (1984), cf. also Fienberg (1980), chapter 8.

12.7. Testing the goodness of fit

Since all models pertain to an observed contingency table, the fit of a given data set is in general obtained through a goodness of fit test based on the test statistic

$$(12.67) \qquad Z = 2 \sum_i \sum_j \sum_k \sum_l X_{ijkl} \ln(X_{ijkl}/\hat{\mu}_{ijkl})$$

where

$$(12.68) \qquad \hat{\mu}_{ijkl} = n\hat{p}_{ijkl}$$

and $\hat{p}_{ijkl}$ is the cell probability estimated under the given model. Thus for the latent class model (12.2), the goodness of fit test statistic is (12.67) with

$$(12.69) \qquad \hat{p}_{ijkl} = \sum_{m=1}^{M} \hat{\pi}^A_{im} \ldots \hat{\pi}^D_{lm} \hat{\varphi}_m.$$

Since there are $(I+J+K+L-4)M+M-1$ estimated parameters, the number of degrees of freedom for the approximating χ^2–distribution is

$$(12.70) \qquad df = IJKL - (I+J+K+L-3)M.$$

Example 12.3. (Continued)

For the Coleman data in table 12.5 consider the following three latent class models.

Model I: Two latent classes, no constraints.

Model II: Four latent classes and constraints correspondding to two binary latent variable, θ_1 affecting the membership items and θ_2 affecting the attitude items, i.e. constraints (12.40) and (12.41).

Model III: Same as model II, but with same response probabilities on the membership items at both occassions and with the same response probabilities on the attitude items at both occasions, i.e. constraints (12.40), (12.41) and (12.42).

The observed values of the test statistic (12.67) are shown in table 12.10 with the appropriate degrees of freedom and levels of significance.

Table 12.10. Observed value of the goodness of fit test statistic for three latent class models fitted to the Coleman data.

Model	Test statistic	Degrees of freedom	Level of significance
I	249.50	6	<0.0005
II	1.27	4	0.866
III	40.74	8	<0.0005

In table 12.11 the expected values are shown for each of the three models together with the observed values.

Table 12.11. Observed numbers for each response pattern and expected numbers for each of the three investigated models.

Response pattern	Observed numbers	Expected numbers		
		Model I	Model II	Model III
1 1 1 1	458	408.30	454.87	455.72
1 1 1 2	140	199.73	144.20	157.78
1 1 2 1	110	94.34	109.14	148.33
1 1 2 2	49	68.49	48.85	64.50
1 2 1 2	171	227.71	172.27	157.78
1 2 1 2	182	112.71	179.56	179.88
1 2 2 1	56	77.92	58.25	64.50
1 2 2 2	87	63.80	85.72	94.14
2 1 1 1	184	159.64	188.64	148.33
2 1 1 2	75	97.12	68.80	64.50
2 1 2 1	531	403.04	530.68	530.59
2 1 2 2	281	397.34	283.11	310.40
2 2 1 1	85	110.58	82.12	64.50
2 2 1 2	97	76.22	101.38	94.14
2 2 2 1	338	451.47	337.32	310.40
2 2 2 2	554	449.60	553.06	552.51

In accordance with the results in Goodman (1974) a four–class latent class model thus fits the data, where the latent classes are formed by two binary latent variables θ_1 affecting the membership items and θ_2 affecting the attitude items. △.

For the continuous latent structure model, (12.16), the expected values are $n\hat{p}_{ijkl}$, with

(12.71)
$$\hat{p}_{ijkl} = \exp(\hat{\epsilon}_{Ai} + \ldots + \hat{\epsilon}_{Dl}) \cdot \int e^{\theta t} \hat{H}^{-1}(\theta) \varphi(\theta | \hat{\alpha}, \hat{\beta}) d\theta,$$

and

$$\hat{H}(\theta) = \sum_i \exp(\hat{\epsilon}_{Ai} + w_i \theta) \ldots \sum_l \exp(\hat{\epsilon}_{Dl} + w_l \theta).$$

Due to the constraints imposed, there are $I+J+K+L-5$ free ϵ parameters. Hence the approximating χ^2-distribution to (10.67) has

$$df = IJKL - 1 - (I+J+K+L-5)-2 = IJKL - (I+J+K+L)+2$$

degrees of freedom.

For four binary items the number of degrees of freedom is thus df=16−8+2=10.

As an alternative to the test statistic (12.67), Andersen (1973a) and Andersen and Madsen (1977) suggested **a two stage procedure** for testing the goodness of fit.

Since (12.54) is valid for any set of individuals, it is in particular valid for a subset formed by all individuals in score group t, i.e. all individuals for with score $w_i + \ldots + w_l = t$. Let

$$x_{it}^A, \ldots, x_{lt}^D$$

be the marginals $x_{i\ldots}, \ldots, x_{\ldots l}$ if summation is only over individuals in score group t. $L_c(\epsilon)$ then factorizes as

(12.72)
$$L_c(\epsilon) = \prod_t L_c^{(t)}(\epsilon),$$

where

(12.73)
$$L_c^{(t)}(\epsilon) = \exp(\sum_i \epsilon_{iA} x_{it}^A + \ldots + \sum_l \epsilon_{Dl} x_{lt}^D)/g(t;\epsilon)^{n_t}.$$

From this likelihood function the ϵ's can estimated except for extreme score groups e.g. $t=0$ and $t=4w_1$, where $L_c^{(t)}(\epsilon)=1$ for all ϵ's because only one response (i,j,k,l) can be observed. If $\hat{\epsilon}$ is the set of over–all CML–estimates obtained from (12.55) and $\hat{\epsilon}^{(t)}$ is the set of score group estimates obtained from maximizing (12.73), it was proved in Andersen (1973a), for the binary case, and in Andersen (1973b), for the general case, that the test statistic

$$(12.74) \qquad\qquad Z_c = 2 \sum_t \ln L_c^{(t)}(\hat{\epsilon}^{(t)}) - 2\ln L_c(\hat{\epsilon})$$

is approximately χ^2–distributed with

$$df = (T{-}1)(I{+}J{+}K{+}L{-}5),$$

degrees of freedom, when T is the number of score groups for which estimates $\hat{\epsilon}^{(t)}$ are obtained.

If the model fits the data the score group estimates $(\hat{\epsilon}_{Ai}^{(t)},...,\hat{\epsilon}_{Dl}^{(t)})$ approximates the true values of the ϵ's as wells as the over–all estimates $(\hat{\epsilon}_{Ai},...,\hat{\epsilon}_{Dl})$. If the model does not fit the data, however, the likelihood (12.73) will depend on the latent parameters of the individuals in score group t. Since the observed value z_c of (12.74) is 0 if $\hat{\epsilon}_{Ai}^{(t)}=\hat{\epsilon}_{Ai},...,$ $\hat{\epsilon}_{Dl} = \epsilon_{Dl}^{(t)}$ for all i,...,l and t, the observed value of Z_c is an indicator of discrepancies between model and data. The larger the observed value z_c of Z_c, the less is the fit of the model. Hence the fit can be judge from the significance level

$$P(Q \geq z_c),$$

where $Q \sim \chi^2((T{-}1)(I{+}J{+}K{+}L{-}5))$.

The goodness of fit test statistic Z_c can be supplemented with a graphical check where for each score t, $\hat{\epsilon}_{Ai}^{(t)}$ is plotted against $\hat{\epsilon}_{Ai}$, $\hat{\epsilon}_{Bj}^{(t)}$ against $\hat{\epsilon}_{Bj}$ etc.

Gustavsson (1980), van den Wollenberg (1982) and Molenaar (1983) has criticized the test statistic (12.74) for being insensitive to certain departures from the model assumptions. Thus van den Wollenberg suggests to use tests statistics that are more sensitive to violations of (a) the one–dimensionality of the latent variable θ and (b) the conditional independence of the responses given the latent variable.

If a goodness of fit test of the model is only based on the conditional likelihood (12.72), the remaining factor of the compete likelihood, is

$$(12.75) \qquad L_T = \underset{t}{\Pi} p(t)^{n_t},$$

where

$$(12.76) \qquad p(t) = g(t;\epsilon) \int e^{\theta t} H^{-1}(\theta) \varphi(\theta | \alpha, \beta) d\theta.$$

If $\hat{\alpha}$ and $\hat{\beta}$ are the ML–estimates of α and β the test statistic (12.74) can then be supplemented by the statistic

$$(12.77) \qquad Z_T = 2 \underset{t}{\Sigma} n_t (\ln(n_t) - \ln(n\hat{p}(t)))$$

where

$$(12.78) \qquad \hat{p}(t) = g(t;\epsilon) \int e^{\theta t} H^{-1}(\theta) \varphi(\theta | \hat{\alpha}, \hat{\beta}) d\theta.$$

For known values of the ϵ's, Z_T is approximately χ^2–distributed with $T{-}1{-}2$ degrees of freedom. The expression $\hat{p}(t)$ is, however, relatively insensitive to changes in the ϵ's. Hence if the ϵ's in (12.76) are replaced by the CML–estimates one can still with caution apply the χ^2–approximation.

The significance level of Z_c primarily measure if the data supports that the response probability $p_{ijkl}(\theta)$ given the latent variable has the structural form (12.14) required in the score model. The significance level of Z_T on the other hand measure the extent to which the latent density $\varphi(\theta | \alpha, \beta)$ describes the variation of θ in the population.

A two–stage procedure based on applying first Z_c and then Z_T thus makes it possible the check two in nature very different aspects of the model structure.

One may decide to disregard any prior assumptions on the form of $\varphi(\theta)$. In this case the probabilities $p(t)$ are estimated by the frequencies n_t/n. The estimated cell frequencies (12.71) are then computed as

$$\hat{p}_{ijkl} = \hat{p}_{ijkl}(t)n_t/n.$$

where $\hat{p}_{ijkl}(t)$ is the estimated conditional probability (12.52) of being in cell (i,j,k,l) given score t, and the estimated expected numbers $n\hat{p}_{ijkl}$ become

$$(12.79) \qquad n\hat{p}_{ijkl} = n\exp(\hat{\epsilon}_{Ai}+...+\hat{\epsilon}_{Dl})/g(t;\hat{\epsilon})\left[\frac{n_t}{n}\right]$$

If (12.79) are the expected numbers in (12.67), Z is approximately χ^2-distributed with

$$df = IJKL-1 - (I+J+K+L-5)-(T-1) = IJKL-(I+J+K+L)-T+5,$$

degrees of freedom, where T is the number of score groups.

The question arises, of course, if there exists a density $\varphi(\theta)$ such that the probabilities $\hat{p}(t)$ can be expressed as

$$\hat{p}(t) = g(t;\hat{\epsilon}) \int e^{\theta t}H^{-1}(\theta)\varphi(\theta)d\theta.$$

In Tjur (1982) it was shown that if such a density exists, then $\varphi(\theta)$ and the CML–estimates for the ϵ's jointly maximizes the log–likelihood function

$$(12.80) \qquad \ln L = \Sigma_i \epsilon_{Ai}x_{i...} +...+\Sigma_l \epsilon_{Dl}x_{...l}+ \Sigma_t n_t\ln\left[\int e^{\theta t}H^{-1}(\theta)\varphi(\theta)d\theta\right]$$

corresponding to (12.27) for an arbitrary density φ.

The problem of a non–parametric maximalization of lnL has been further discussed by de Leeuw and Verhelst (1986), by Lindsay (1987), by Lindsay, Clogg and Gregg (1989) and by Engelen (1987).

Cressie and Holland (1983) discuss the fact that (12.80) is a log–linear model, if the expressions

$$\lambda_t = \ln \left[\int e^{\theta t}H^{-1}(\theta)\varphi(\theta)d\theta\right]$$

are regarded as parameters to be estimated.

Example 12.1. (Continued)

We shall now compare the application of three latent structure models to the complain data in table 12.1. The models are:

Model I: A two class unrestricted latent class model.

Model II: A latent structure model with an unrestricted continuous latent density.

Model III: A latent structure model with a normal latent density.

The estimated parameters of model I are shown in table 12.12, while the estimated parameters of model III and the $\hat{\epsilon}$'s of model II are shown in table 12.9. The estimates $\hat{p}(t)$ of the score group probabilities are shown in table 12.14. Before proceeding to compare all three models, we shall discuss the two–stage procedure for testing the goodness of fit of model III.

With four binary items and the score weights chosen as $w_1=1$ and $w_2=0$, the possible values of the score t are 0,1,2,3 and 4. The scores t=0 and t=4 are extreme scores, since t=0 can only be obtained for i=j=k=l=2 and t=4 can only be obtained for i=j=k=l=1. The $\hat{\epsilon}^{(t)}$'s that maximize (12.73) are shown in table 12.13 together with the over–all CML–estimates of the ϵ's. Score groups 1 and 2 are merged because one of the item totals is zero in score group 1.

Table 12.12. The estimated parameters of an unrestricted two–class latent class model applied to the data in table 12.1.

	Latent class	
	m=1	2
$\hat{\pi}_{1m}^{A}$	0.983	0.864
$\hat{\pi}_{1m}^{B}$	0.894	0.619
$\hat{\pi}_{1m}^{C}$	0.940	0.316
$\hat{\pi}_{1m}^{D}$	0.472	0.066
$\hat{\varphi}_{m}$	0.810	0.190

Table 12.13. Score group CML–estimates and over–all CML–estimates (with standard errors in parantheses) of the item parameters for the data in table 12.1.

	Score group t=1–2	3	Over–all
$\hat{\epsilon}_{1A}$	2.273	1.552	2.106(0.200)
$\hat{\epsilon}_{1B}$	0.273	0.116	0.227(0.120)
$\hat{\epsilon}_{1C}$	−0.119	0.596	0.055(0.117)
$\hat{\epsilon}_{1D}$	−2.427	−2.264	−2.388(0.132)

Apart from $\hat{\epsilon}_{1C}^{(3)}$, the score group estimates match the over–all estimates well. The observed value of the test statistic Z_c is

$$z_c = 7.31.$$

With df=(2–1)(8–5)=3 degrees of freedom, the significance level is

$$p = P(Q{\geq}z_c) = 0.063.$$

The Rasch–model thus seems to fit the data well.

Table 12.14 shows the observed distributed of persons over the five score groups together with the frequencies n_t/n and the estimated expected numbers $\hat{np}(t)$ under the assumption of normal latent density, i.e.

$$\hat{p}(t) = g(t;\hat{\epsilon})\int e^{\theta t}\hat{H}^{-1}(\theta)\varphi(\theta|\hat{\mu},\hat{\sigma}^2)d\theta,$$

where $\varphi(\theta|\mu,\sigma^2)$ is the density in the normal distribution with mean μ and variance σ^2. The estimates $\hat{\mu}$ and σ^2 are shown in table 12.8.

Table 12.14. Observed numbers and frequencies for all score groups and expected numbers given a normal latent density with mean $\hat{\mu}=1.854$ and variance $\hat{\sigma}^2=1.536$.

Score	Observed numbers n_t	Frequencies n_t/n	Expected numbers $\hat{np}(t)$
t=0	8	0.013	3.88
1	25	0.042	32.00
2	108	0.180	109.73
3	266	0.443	258.18
4	193	0.322	196.22
Total	600	1.000	600.01

The observed value of the test statistic Z_T is

$$Z_T = 5.32.$$

With df=5-1-2=2 degrees of freedom the significance level is

$$p = 0.070,$$

which indicate that a normal latent density describe the data well. Together the two tests thus show that a Rasch model fits the data.

The test statistic (12.67) based on the observed distribution over response pattern allow us, however, to compare all three model. Under model I, the expected numbers are computed as (12.68) with estimated response probabilities (12.69). Under model II, the expected numbers are computed as (12.79) and under model III the expected numbers are computed as (12.68) with the estimated response probabilities given by (12.71), where $\varphi(\theta|\hat{\alpha},\hat{\beta})=\varphi(\theta|\hat{\mu},\hat{\sigma}^2)$ is a normal density function. Table 12.15 exhibit the observed numbers for each response pattern and the expected numbers under each of the three models. In all three cases the fit seems reasonable good.

Table 12.15. Observed numbers for each response pattern and expected numbers under three latent structure models.

Response pattern	Observed numbers	Expected numbers		
		Model I	Model II	Model III
1 1 1 1	193	190.63	193.00	195.60
1 1 1 2	227	229.96	227.12	223.61
1 1 2 1	13	14.86	19.74	18.23
1 2 1 1	21	23.36	16.61	15.32
2 1 1 1	5	3.57	2.54	2.71
1 1 2 2	58	52.36	52.34	52.21
1 2 1 2	40	36.20	44.03	43.86
1 2 2 1	4	3.12	3.83	3.58
2 1 1 2	5	6.58	6.73	7.76
2 1 2 1	1	0.65	0.58	0.63
2 2 1 1	0	0.52	0.49	0.53
1 2 2 2	20	25.51	19.34	23.72
2 1 2 2	3	6.39	2.96	4.20
2 2 1 2	2	2.18	2.49	3.53
2 2 2 1	0	0.29	0.22	0.29
2 2 2 2	8	3.82	8.00	4.22

The observed values of the test statistic (12.67), the degrees of freedom and the significance levels are shown in table 12.16.

Table 12.16. Observed values z of (12.67) and corresponding significance levels for models I, II and III.

Model	z	df[1]	p
I	9.79	4	0.044
II	7.16	6	0.306
III	12.01	8	0.151

1) Because of the small expected numbers in table 12.15, response patterns 2112, 2121 and 2211 are grouped as are response patterns 2212 and 2221.

The levels of significance in table 12.16 show that both the models based on a continuous latent density fits the data well. The latent class model does not fit the data quite well in spite of the fact that as many as 9 parameters are allowed to vary.

12.8. Diagnostics

If a given model fails to fit the data, it is necessary to have tools to detect the directions of the model departures and if possible the courses of the lack of fit. Such tools are called

diagnostics. The literature contain a number of suggestions for such diagnostics applied to latent structure models. For the binary score model, or the Rasch model, of section 12.5, van den Wollenberg (1982), Molenaar (1983) and Glas (1988) contain a number of diagnostics. The most direct form of diagnostics are residuals based on the observed and expected numbers for each response pattern. Let, as in section 12.7, x_{ijkl} be the observed number of persons with response pattern (ijkl) and X_{ijkl} the corresponding random variable. The expected numbers are then given by

$$\mu_{ijkl} = E[X_{ijkl}] = np_{ijkl},$$

where the response probabilties can be determined by a latent class model (12.69) or by a continuous latent structure model (12.71). Let further $\delta_1,...,\delta_R$ denote all the parameters of the model, i.e. $\pi^A_{im},...,\pi^D_{lm}$, φ_m for all i,j,k,l and m in case of a latent class model and $\epsilon_{Ai},...,\epsilon_{Dl}$ for all i,j,k,l and α,β in case of a continuous latent structure model. The expected numbers can then be written on the common form

$$\mu_{ijkl} = np_{ijkl}(\delta_1,...,\delta_R).$$

According to the general formula (3.40) the **standardized residuals** are then given by

(12.81)
$$r_{ijkl} = (X_{ijkl} - \hat{\mu}_{ijkl})/\hat{\sigma}_{ijkl},$$

where

$$\hat{\mu}_{ijkl} = np_{ijkl}(\hat{\delta}_1,...,\hat{\delta}_R) = n\hat{p}_{ijkl}$$

and

$$\hat{\sigma}^2_{ijkl} = \hat{\mu}_{ijkl}(1 - \hat{p}_{ijkl} - \sum_r \sum_s \hat{p}^{-1}_{ijkl} \frac{\partial \hat{p}_{ijkl}}{\partial \delta_r} \frac{\partial \hat{p}_{ijkl}}{\partial \delta_s} i^{rs}),$$

with i^{rs} being the element in cell (r,s) of the information matrix. If the ML–estimates $\hat{\delta}_1,...,\hat{\delta}_R$ are computed using a Newton–Raphson procedure, the residuals (12.81) are obtained as by–products of the procedure. Glas (1988) also derived standardized residuals.

Example 12.4.

Clogg (1979) applied a latent class model to the 3–dimension contingency table shown in table 12.17. The three variables represent aspects of life satisfaction. Variable A is "satisfaction with hobbies", variable B "satisfaction with residence" and variable C "satisfaction with the family". The original 7 response categories were recoded, such that the responses 1,2 and 3 in table 10.17, below are

1. "a very great deal" or "a great deal"
2. "quite a bit" or "a fair amount"
3. "some", "a little" or "none".

The data shown in table 12.17 are the observed responses from a sample of 1472 individuals obtained as part of the 1975 US General Social Survey.

Table 12.17. The observed responses for 1472 individuals in the 1975 US General Social Survey on three trichotomous variables concerning life satisfaction.

A	B	C 1	C 2	C 3
1	1	466	27	16
1	2	191	38	14
1	3	64	18	5
2	1	126	31	5
2	2	117	58	12
2	3	45	23	3
3	1	54	12	7
3	2	49	26	11
3	3	23	16	15

Source: Clogg (1979), table 1.

Clogg applied various latent class models to the data in table 12.17. We shall now try to fit a score model (12.16) to the data, assuming that the scoring of the three response categories is equidistant, i.e. $w_1=2$, $w_2=1$ and $w_3=0$. With this scoring the possible values of the score t are 0,1,2,3,4,5 and 6. The ML–estimates of the parameters are shown in table 12.18 together with the item totals.

Table 12.18. Item totals and ML–estimates of the parameters for a score model applied to the data in table 12.17.

Item	Response category	Item total	Item parameter
A	1	839	–0.666
	2	420	–0.013
	3	213	0.000
B	1	744	–0.877
	2	516	+0.181
	3	212	0.000
C	1	1135	+0.873
	2	249	+0.504
	3	88	0.000
$\hat{\mu}$			1.308
$\hat{\sigma}^2$			0.859

The estimated numbers are shown in table 12.19 together with the observed numbers and the standardized residuals.

The goodnes of fit test statistic (12.67) based on the observed and expected numbers in table 12.19 has observed value

$$z = 60.46,$$

which with 19 degrees of freedom has a significance level of

$$p < 0.0005.$$

The score model does not accordingly fit the data.

The standardized residuals in table 12.19 show that the fit is relatively good as a whole, since most residuals are small. In fact only 7 out of 27 residuals are larger than 2. These residuals correspond to cells 111,112,113,121,211,222, and 233. The lack of fit in cell 222 is most likely due to the well known fact that many individuals choose to use a middle category whatever the issue. The significant residuals in cells 111,112,121 and 211 with an overrepresentation in cell 111 and underrepresentations in cells 112,121 and 211 are due to a tendency to go all the way to score t=6 if the score is already t=5. This shows that individuals, who are satisfied except with one of the items, will tend to claim that they are satisfied also with the last item. The last two cells with significant residuals are not so easy to explain, but with 27 cells in the table a few residuals are allowed to be significant and the values of the residuals for cells 113 and 233 are at any rate not among the largest residuals.

Table 12.19. Observed numbers, expected numbers and standardized residuals under a score model with equidistant weights

Response pattern Variabel A B C	Observed numbers	Expected numbers	Standardized residuals
1 1 1	466	436.47	4.502
1 1 2	27	53.53	−4.490
1 1 3	16	8.94	2.658
1 2 1	191	213.14	−2.408
1 2 2	38	41.74	−0.676
1 2 3	14	10.61	1.151
1 3 1	64	53.44	1.976
1 3 2	18	15.93	0.566
1 3 3	5	5.97	−0.420
2 1 1	126	147.43	−2.562
2 1 2	31	28.87	0.444
2 1 3	5	7.34	−0.929
2 2 1	117	114.96	0.225
2 2 2	58	34.26	4.669
2 2 3	12	12.84	−0.260
2 3 1	45	43.86	0.206
2 3 2	23	19.27	0.948
2 3 3	3	10.48	−2.554
3 1 1	54	44.08	1.973
3 1 2	12	13.14	−0.338
3 1 3	7	4.92	0.982
3 2 1	49	52.31	−0.568
3 2 2	26	22.99	0.713
3 2 3	11	12.50	−0.476
3 3 1	23	29.43	−1.374
3 3 2	16	18.76	−0.746
3 3 3	15	14.76	0.077

The conclusion of the analysis seems to be that a score model with equidistant scores and a normal latent density fits the data relatively well. The main model departures seem to be associated with a well known effect for questionnaire data, where the "diagonal" responses 111,222, and 333 tend to be overrepresented. △.

Standardized residuals can also be used in connection with the two–stage procedure suggested in section 12.7. The conditional likelihood is composed of a product of independent likelihood functions (12.73), where each factor covers a section of the list of possible response patterns, namely those patterns (ijkl) for which

$$w_i + w_j + w_k + w_l = t.$$

Within this section the expected numbers are computed as

$$\mu_{ijkl}(t) = E[X_{ijkl}|t] = n_t p_{ijkl}(t),$$

where $p_{ijkl}(t)$ is the conditional probability of the response ijkl given $w_i+w_j+w_k+w_l=t$. For the score model, the conditional probability $p_{ijkl}(t)$ is given by (12.52). The estimated expected numbers are obtained by inserting the CML–estimates of the ϵ's in (12.52) yielding

$$\hat{\mu}_{ijkl}(t) = n_t \hat{p}_{ijkl}(t).$$

Standardized residuals corresponding to the test statistic Z_c are then derived as

(12.82) $$r_{ijkl}(t) = (x_{ijkl} - \hat{\mu}_{ijkl}(t))/\hat{\sigma}_{ijkl}(t),$$

where $\hat{\sigma}^2_{ijkl}(t)$ is an estimate of the variance of $X_{ijkl} - \hat{\mu}_{ijkl}(t)$. The exact form of $\hat{\sigma}^2_{ijkl}(t)$ can be obtained from (3.40), section 3.5. Note that the residuals (12.82) are independent of the form of the latent density. An inspection of these residuals can, therefore, be used to evaluate if a lack of fit of the model is in the latent structure of the model, rather than in the form of the latent density.

Residuals to evaluate if the lack of fit is due to the form of the latent density can be derived from the distribution over scores t=0,...,T. The observed number of individuals with score t is n_t, while the expected number is equal to

$$\mu(t) = np(t),$$

where p(t) is the probability (12.76). Hence standardized residuals are obtained as

(12.83)
$$r_t = (n_t - \hat{\mu}(t))/\hat{\sigma}(t),$$

where $\hat{\sigma}^2(t)$ is an estimate of $\text{var}[n_t - \hat{\mu}(t)]$ and $\hat{\mu}(t) = n\hat{p}(t)$ with $\hat{p}(t)$ given by (12.78).

As the score is an estimate of the latent variable θ, an inspection of the residuals r_t, t=0,...,T, can provide information about the range of θ's values where the lack of fit occurs.

Example 12.4 (Continued)

The joint use of the residuals (12.82) and (12.83) can be illustrated by the data in table 12.17. Table 12.20 show the observed numbers for each response pattern together with the estimated expected number $\hat{\mu}_{ijkl}(t)$ for response pattern (ijkl) given score t and the standardized residual $r_{ijkl}(t)$. In the table the response pattern are ordered according to score group. Note that the $\hat{\mu}_{ijkl}(t)$ add up to n_t within each score group. Below the section of response patterns with score t in table 12.20 is shown the observed number n_t, the expected number $\hat{\mu}(t)$ and the standardized residual r_t.

Table 12.19 reveals that the lack of fit is mainly in the form of the latent density apart from the overrepresentation of pattern 222. The significant residuals are concentrated at the upper end of the score scale. △ .

Table 12.20. Observed numbers, expected numbers and standardized residuals for the two-stage procedure applied to the data in table 12.17.

Response pattern	Score	Observed number	Expected number	Standardized residual
1 1 1		466	466.0	–
	6	466	435.33	4.581
1 1 2		27	43.31	–3.178
1 2 1		191	180.35	1.833
2 1 1		126	120.34	0.976
	5	344	416.76	–4.670
1 1 3		16	9.82	2.258
1 2 2		38	46.77	–1.649
1 3 1		64	56.47	1.503
2 1 2		31	31.21	–0.044
2 2 1		117	129.98	–2.182
3 1 1		54	45.75	1.740
	4	320	291.33	2.204
1 2 3		14	12.68	0.433
1 3 2		18	17.52	0.130
2 1 3		5	8.46	–1.323
2 2 2		58	40.33	3.637
2 3 1		45	48.69	–0.750
3 1 2		12	14.19	–0.648
3 2 1		49	59.12	–2.005
	3	201	176.64	2.173
1 3 3		5	6.07	–0.474
2 2 3		12	13.97	–0.648
2 3 2		23	19.30	1.029
3 1 3		7	4.92	1.006
3 2 2		26	23.44	0.676
3 3 1		23	28.30	–1.328
	2	96	94.73	0.154
2 3 3		3	7.71	–2.057
3 2 3		11	9.36	0.680
3 3 2		16	12.93	1.234
	1	30	42.15	–2.328
3 3 3		15	15.00	–
	0	15	15.07	–0.021

The test statistic (12.74) was based on a comparison of score group estimates $\hat{\epsilon}_{iA}^{(t)},...,\hat{\epsilon}_{lD}^{(t)}$ for each score group t and the over–all CML–estimates $\hat{\epsilon}_{iA},...,\hat{\epsilon}_{lD}$. As mentioned one can supplement the test statistic with a graphical check if the score group estimates are plotted against the over–all estimates with one plot for each score group. On such plots the points should cluster randomly around the identity line. These plots are helpful in determining if the lack of fit is due to certain score groups or to certain of the variables.

Example 12.5.

In psychiatrics so called rating scales are often used to measure the degree of psychiatric disturbance. Such ratings are based on batteries of items, each reflecting an aspect of a persons vulnerability. On each item a person can score positively or negatively. The rating is the number of negative scores. We consider data from a battery of six such items. If the items are labeled A,B,C,D,E and F, the data consists of six binary measurements for each person thus forming a 6–dimensional contingency table. With 6 binary variables the number of response patterns is 64 and we shall not show the complete data set. In table 12.21 the observed numbers are shown for a selection of response patterns. Table 12.22 summarize the within score groups and over–all estimates of the item parameters.

In fig.12.5 the score group estimates are plotted against the over–all estimates. On all five plots the points corresponding to the variables cluster to some extend around identity lines. The standard error is about 0.1 for all six over–all estimates. Hence we should allow deviations of the magnitude $\pm$ 0.2 around the identity lines.

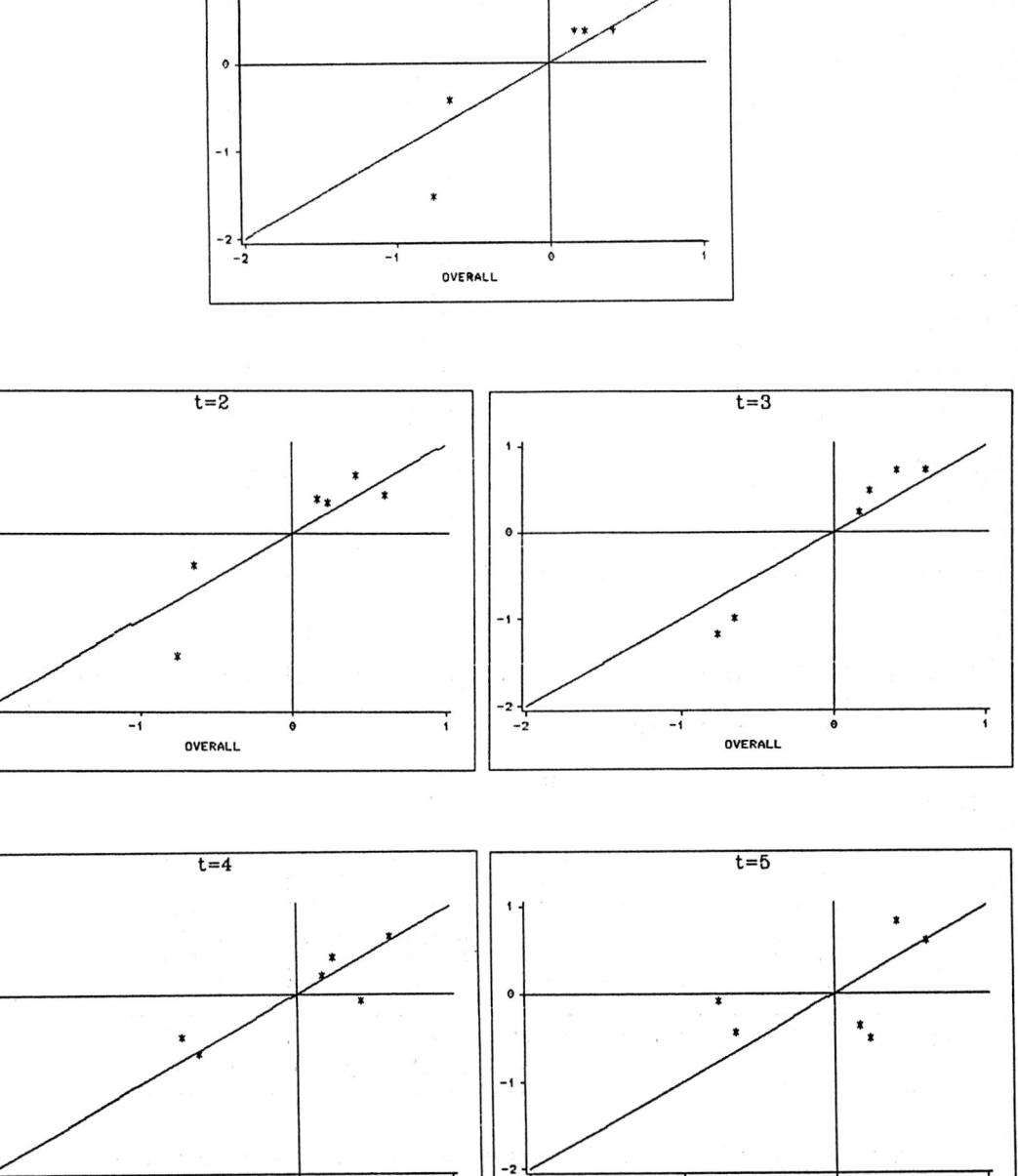

Fig.12.5. Score group estimates of the item parameter plotted against over–all CML–estimates for all five non–trivial score groups.

Table 12.21. Observed and expected numbers for a Rasch and a Birnbaum model for selected response patterns.

Response patterns	Observed numbers	Expected numbers Rasch model	Birnbaum model
111111	50	50.64	53.78
111211	15	6.99	10.31
112111	13	7.47	11.14
121111	14	16.81	14.72
211111	10	18.84	9.63
221111	22	14.09	10.87
221112	9	5.91	6.44
221121	9	7.13	7.74
221122	11	5.64	7.70
221211	9	8.58	9.17
221212	9	6.79	8.47
221221	9	8.18	9.86
221222	13	12.81	16.86
222111	11	9.16	9.72
222112	9	7.25	8.92
222122	13	13.69	17.63
222211	10	10.52	11.63
222212	13	16.47	17.98
222221	22	19.85	20.18
222222	74	72.18	68.07

Source: Data obtained from the Institute for Psychiatry Demography. University of Aarhus.

Table 12.22. CML–estimates for the item parameters both over–all and within scoregroups. For the overall estimates, standard errors are shown in parantheses.

Score group	A	B	Variable C	D	E	F
t=1	−1.519	−0.420	0.353	0.353	0.353	0.879
2	−1.419	−0.372	0.383	0.338	0.644	0.427
3	−1.169	−0.995	0.227	0.488	0.724	0.724
4	−0.491	−0.683	0.201	0.409	−0.092	0.656
5	−0.101	−0.437	−0.363	−0.506	0.815	0.592
Over–all	−0.761 (0.111)	−0.648 (0.109)	0.163 (0.105)	0.229 (0.106)	0.415 (0.106)	0.602 (0.107)

There are obviously some marked discrepancies on fig. 12.5. A closer inspection reveals that the most obvious ones are connected with variable A and with score group 5. It is thus worth checking the nature of variable A, and what is characteristic otherwise of persons with a high rating. In the next section we shall return to the role of variable A. △.

12.9. Score models with varying discriminating powers.

We now return to the **two–parameter logistic model** (12.17), where the binary probabilities have the forms

$$\pi_1^A(\theta) = \exp\{(\theta+\epsilon_A)\delta_A\}/[1 + \exp\{(\theta+\epsilon_A)\delta_A\}]$$

$$\ldots$$

$$\pi_1^D(\theta) = \exp\{(\theta+\epsilon_D)\delta_D\}/[1 + \exp\{(\theta+\epsilon_D)\delta_D\}]$$

The probability of response (ijkl) then becomes

$$p_{ijkl} = \int p_{ijkl}(\theta)\varphi(\theta|\mu,\sigma^2)d\theta$$

with

$$p_{ijkl}(\theta) = [\pi_1^A(\theta)]^{z_i}[1-\pi_1^A(\theta)]^{1-z_i}\ldots[\pi_1^D(\theta)]^{z_1}[1-\pi_1^D(\theta)]^{1-z_1},$$

where $z_i=1$ for response $i=1$ and $z_i=0$ for response $i=2$. After some algebra it turns out that the likelihood function has the form

$$(12.84) \qquad \ln L = \epsilon_A\delta_A x_{1\ldots} +\ldots+\epsilon_D\delta_D x_{\ldots 1}+ \sum_t n_t \int e^{t\theta}H^{-1}(\theta)\varphi(\theta|\mu,\sigma^2)d\theta,$$

where

$$(12.85) \qquad t = \delta_A z_i+\ldots+\delta_D z_1$$

and

$$(12.86) \qquad H(\theta) = (1+e^{(\theta+\epsilon_A)\delta_A})\ldots(1+e^{(\theta+\epsilon_D)\delta_D})$$

The score is thus the weighted average of the 1–answers with the item discriminating powers as weights. If the δ's are not multipla of a few integers, there will almost certainly be only one response pattern for each score value. In this case the sum over t in (12.84) has as many terms as there are response patterns.

It is straight forward to maximize (12.84) with respect to all the parameters al–

though the integral

$$\int e^{t\theta} H^{-1}(\theta) \varphi(\theta | \mu, \sigma^2) d\theta$$

must be evaluated numerically for each score value.

In order to remove the obvious arbitrarinesses in the parametrization, we introduce the constraints

$$\delta_A \delta_B \delta_C \delta_D = 1$$

and

$$\epsilon_A \delta_A + ... + \epsilon_D \delta_D = 0.$$

Including μ and σ^2, there are accordingly 2+6=8 parameters in the model.

Given ML–estimates $\hat{\epsilon}_A, ..., \hat{\epsilon}_D$, $\hat{\delta}_A, ..., \hat{\delta}_D$, $\hat{\mu}$ and $\hat{\sigma}^2$ for the parameters, the fit of the model can be checked through the test statistic (12.67), which in the binary case I=J=K=L=2, considered here, has

$$df = 2^4 - 1 - 8 = 7$$

degrees of freedom.

If the variables are not binary there are I+J+K+L–5 unconstrained ϵ's and 3 unconstrained δ's, such that (12.67) has

$$df = IJKL - 1 - (I+J+K+L-5) - 3 - 2 = IJKL - (I+J+K+L) - 1$$

degrees of freedom.

If there are more than four variables, say M, and the M variables are binary, there are M–1 unconstrained ϵ's and M–1 unconstrained δ's. Hence the degrees of freedom for (12.67) is

$$df = 2^M - 1 - 2(M-1) - 2 = 2^M - 2M - 1.$$

Example 12.5. (Continued)

Table 12.23 show the estimates of the parameters of a one–parameter logistic or Rasch model applied to the data in example 12.5. Note that there are six variables denoted A, B,C,D,E and F. As suggested by fig.12.5 the Rasch model does not fit the data. As an alternative consider the two–parameter logistic or Birnbaum model (12.17). The last two columns of table 12.23 show the estimated parameters of this model. The estimates of the item discriminating powers suggest that an assumption of equal item discriminating powers is not likely to hold.

Table 12.23. Parameter estimates for the Rasch model and the Birnbaum model applied to six variables from a psychiatric rating battery.

Variable	Rasch model $\hat{\epsilon}$	Birnbaum model $\hat{\epsilon}$	$\hat{\delta}$
A	−0.762	−0.504	1.919
B	−0.648	−0.535	1.111
C	0.164	0.247	0.750
D	0.230	0.325	0.760
E	0.415	0.524	0.878
F	0.602	0.712	0.938
$\hat{\mu}$	−0.219	−0.316	
σ^2	2.943	3.159	

The fit of Birnbaum model can be tested by the test quantity (12.67), but due to the large number possible response patterns many expected numbers are small. In order to make the χ^2–approximation valid all together 27 response patterns were merged into 4 groups of response patterns. The test statistic (12.67) then has observed value

$$z = 50.58$$

with df=64–23–1–5–5–2=28 degrees of freedom. The significance level is

$$p = P(Q \geq 50.58) = 0.006.$$

The fit is clearly better than for the Rasch model where, after 25 response patterns have been merged into 3 groups, (12.67) has observed value

$$z = 72.10$$

with 64–22–1–5–2=34 degrees of freedom. The significance level for the for of the Rasch model is thus

$$P(Q \geq 72.10) = 0.0001.$$

Table 12.21 shows the expected numbers under the Birnbaum model.

In cases like this it is tempting to exclude variable A as an item for the rating of psychiatric disturbance. Practitioners claim, however, that variable A in this particular case covers a very important psychiatric aspect and hence must be included. The estimates in table 12.23 then strongly suggest to use a rating, where variable A is assigned weight 2, while variables B,C,D,E and F are assigned weights 1. $\triangle$.

12.10. Comparison of latent structure models

In several of the examples we have applied a latent class model as well as a latent structure model with a continuous latent density to the data. Often both types of models fit the data well and a comparison is called for. There are two main differences between the models

(i)   In the continuous latent structure model the latent variable can vary freely over the real axes, while for a latent class model it is assumed that all persons in the sample can be classified into a few groups with the same value of the latent variable within the group.

(ii) In a latent class model there are no constraints on the conditional distribution of the response given the value of the latent variable, while in a continuous latent structure model it is necessary to assume a certain structure in the conditional probability of the response given the latent variable.

These differences mean that it is quite possible that for some data sets one model fits best, while for other data sets the other model fits best. It turns out on the other hand that in case the estimated conditional probabilities in a latent class model exhibit the same structure as the estimated conditional probabilities for a continuous latent structure model, then the two sets of parameters are directly comparable and can be given the same type of interpretation. It is even possible to formulate a latent class model, where the conditional probabilities have the same structure as in the Rasch model, i.e.

$$(12.87) \qquad \pi^A_{im} = \exp(\theta_m + \epsilon^A_i)/\sum_i \exp(\theta_m + \epsilon^A_i)$$

and similar expressions for the other items. In formula (12.87) $\theta_1,...,\theta_M$ are parameters connected with the M latent classes, which represent average values of the latent variable within the latent classes. Such models were considered by Formann (1985) and by Rost (1988).

Example 12.2. (Continued)

Consider again the data in table 12.3. The test statistic (12.67) has observed value

$$z = 8.91$$

for a three–class latent class model. Since several expected numbers are very small 13 response patterns has been merged into 5 groups, leaving 24 observed and expected numbers in the test statistic. The degrees of freedom are accordingly

$$df = 24-1-3\cdot 5-2 = 6.$$

The significance level is thus

$$p = 0.179.$$

For a Rasch model the test statistic (12.67) with the same grouping as for the latent class model has observed value

$$z = 43.41.$$

The degress of freedom are here

$$df = 24-1-4-2 = 17,$$

and the significance level

$$p = 0.0004.$$

A latent class model thus fits the data better than a Rasch model.

In order to compare the parameters of the latent class model with the parameters of the Rasch model, one can obtain average values θ_1, θ_2 and θ_3 for the latent classes by minimizing the sum

$$(\hat{\pi}^A_{1m} - \pi^A_{1m}(\theta_m))^2 + ... + (\hat{\pi}^E_{1m} - \pi^E_{1m}(\theta_m))^2,$$

where $\pi^A_{1m}(\theta)$ is the right hand side of (12.87) and $\hat{\pi}^A_{1m}$ the estimate from the latent class analysis. This yields the values $\theta_1=-4.170$, $\theta_2=-1.264$ and $\theta_3=0.260$.

For these values of θ_m it is possible to evaluate the extent to which the estimates based on a Rasch model and a latent class model are in agreement. In table 12.24, $\hat{\pi}^A_{1m},\ldots,\pi^E_{1m}$ and $\pi^A_1(\theta_m),\ldots,\pi^E_1(\theta_m)$ are compared for all three latent classes. The amount of agreement is substantial but there are also clear disagreements, especially in the smallest latent class, where the conditional probabilities of the latent class model does not show the variation to be expected under a Rasch model structure. $\triangle$.

Table 12.24. The item response probabilities $\pi^A_1(\theta),\ldots,\pi^E_1(\theta)$ evaluated at $\theta_1=-4.170$, $\theta_2=-1.264$ and $\theta_3=0.260$ compared with the latent class response probabilities.

Item S:	Latent class m	θ_m	Response probability Latent class $\hat{\pi}^S_{1m}$	Latent structure $\pi^S_1(\theta_m)$
A	1	−4.170	0.122	0.046
	2	−1.264	0.403	0.468
	3	0.260	0.619	0.802
B	1	−4.170	0.052	0.060
	2	−1.264	0.741	0.541
	3	0.260	0.727	0.844
C	1	−4.170	0.013	0.025
	2	−1.264	0.197	0.319
	3	0.260	0.290	0.682
D	1	−4.170	0.018	0.006
	2	−1.264	0.033	0.106
	3	0.260	0.652	0.353
E	1	−4.170	0.009	0.005
	2	−1.264	0.021	0.079
	3	0.260	0.583	0.282

Another example of an empirical comparison of a Rasch model and a latent class model is due to Masters (1985).

12.11. Estimation of the latent variable

If a latent class model or a latent structure model with a continuous latent variable is accepted, the value of the latent variable can be estimated for each individual in the sample.

For a latent class model the probability of belonging to latent class m given the response is according to (12.1)

$$(12.88) \qquad \pi_{m|ijkl} = \pi_{im}^A \pi_{jm}^B \pi_{km}^C \pi_{lm}^D \varphi_m / \sum_{m=1}^{M} \pi_{im}^A \pi_{jm}^B \pi_{km}^C \pi_{lm}^D \varphi_m.$$

For an individual with response (ijkl) the class m, which maximizes (12.88), is the most likely class for the individual to belong to. Thus each individual can based on the response be assigned to a latent class, he or she most likely belongs to. This assignment to latent classes represents an estimation of the value of the latent variable for each individual in the sample.

For a latent structure model, the value $\hat{\theta}$ of the latent variable for a given individual can be estimated with or without knowledge of the latent density φ. A priori without knowledge of the functional form of $\varphi(\theta)$, θ can be estimated as the value, which maximizes the conditional probability of the response (ijkl) given θ, i.e. according to (12.7) the probability

$$P_{ijkl}(\theta) = \pi_i^A(\theta) \pi_j^B(\theta) \pi_k^C(\theta) \pi_l^D(\theta).$$

For the Rasch model (12.11) this probability is equal to

$$(12.89) \qquad P_{ijkl}(\theta) = e^{t\theta} / H(\theta),$$

such that there is one estimate $\hat{\theta}$ for each value of the score t. The expression (12.89) is easily maximized by numerical methods.

If the form of φ is known and the parameters of $\varphi(\theta)$ estimated, θ can be estimated **a posteriori**. The a posteriori distribution of θ, given the response (ijkl), is

$$\varphi(\theta|ijkl) = \pi_i^A(\theta)\pi_j^B(\theta)\pi_k^C(\theta)\pi_l^D(\theta)\varphi(\theta)/\int \pi_i^A(\theta)\pi_j^B(\theta)\pi_k^C(\theta)\pi_l^D(\theta)\varphi(\theta)d\theta.$$

Hence an estimate of θ can be obtained as the expected value of θ given the response, or

$$(12.90) \qquad\qquad \hat{\theta} = E[\theta|ijkl] = \int \theta\varphi(\theta|ijkl)d\theta.$$

For the Rasch model, (12.90) is equal to

$$\hat{\theta} = E[\theta|ijkl] = E[\theta|t] = \int \theta e^{t\theta}H^{-1}(\theta)\varphi(\theta)d\theta / \int e^{t\theta}H^{-1}(\theta)\varphi(\theta)d\theta.$$

Example 12.2. (Continued)

For the data on work hazards in table 12.3, the assigned latent class based on maximizing (12.88) is for each response pattern shown in table 12.25. The table also shows for each score t the a priori estimate of θ based on maximizing (12.89) and the a posteriori estimate of θ based on (12.90). Note that the extreme scores t=0 and t=5 does lead to a priori estimates of θ, since (12.89) in these cases is independent of θ.

Table 12.25. Assigned latent class for each response pattern and a priori and a posteriori estimates of θ for each score t.

Response pattern	Assigned latent class	Score	Estimate of θ A priori	A posteriori
11111	3	0	–	–3.763
11112	3	1	–1.725	–2.381
11121	3	2	–0.497	–1.362
11122	2	3	0.549	–0.515
11211	3	4	1.710	0.269
11212	3	5	–	1.049
11221	3			
11222	2			
12111	3			
12112	3			
12121	3			
12122	2			
12211	3			
12212	3			
12221	3			
12222	1			
21111	3			
21112	3			
21121	3			
21122	2			
21211	3			
21212	3			
21221	3			
21222	2			
22111	3			
22112	3			
22121	3			
22122	1			
22211	3			
22212	1			
22221	1			
22222	1			

The fact that the values of θ estimated a posteriori are considerable lower than the a priori estimated values is a reflection of the negative value of the latent mean value. $\triangle$.

12.12. Exercises

12.1. The questionnaire of the Danish Welfare Study contain a number of items related to the same background variable. Three examples are work hazards, psychic inconveniences and physical inconveniences. Among the items related to work hazards the following five are selected here: "At your work are you often exposed to (1) noise, (2) bad light,

(3) toxic substances, (4) heat, (5) dust. The five selected items for psychic inconveniences are: "Do you suffer from (1) neuroses,, (2) sensitivity to noise, (3) sensitivity towards petitessen, (4) troubling thoughts, (5) shyness. The five selecteditems for physical inconveniences are: "Are you often troubled by (1) diarrhoeas, (2) pains in your back, (3) colds, (4) coughing, (5) headaches.

For each variable the item is scored in a dichotomous way. For work hazards as

Yes, always =1,

Yes, sometimes = 1,

No = 2,

for psychic inconveniences as

Yes = 1,   No = 2,

and for physical inconveniences as

Yes =1,   No = 2.

The table below show for each of the three variables the observed number of respondents in the welfare study for each response pattern.

(a)   Try to fit a latent class model with three latent classes to each of the three data sets.

(b)   If a latent class model fits, interpret the parameters.

(c)   Try to fit a Rasch model to the three data sets both with and without assuming a normal latent density.

(d)   If the models in (c) do not fit the data, use standardized residuals to describe which response patterns contribute most to the model departures.

Response pattern	Observed numbers Work hazards	Psychic in- conveniences	Physical in- conveniences
11111	70	34	16
11112	15	10	5
11121	34	21	24
11122	6	17	16
11211	39	17	11
11212	21	4	12
11221	38	15	79
11222	49	20	98
12111	103	63	18
12112	39	21	11
12121	129	45	19
12122	66	38	15
12211	115	42	9
12212	116	29	9
12221	217	65	54
12222	409	92	97
21111	4	14	37
21112	8	4	38
21121	7	35	73
21122	3	32	82
21211	12	21	55
21212	22	12	70
21221	16	99	365
21222	60	172	689
22111	24	46	30
22112	27	27	61
22121	54	115	84
22122	99	176	178
22211	63	98	44
22212	193	107	131
22221	168	776	454
22222	1499	2746	2267

12.2. Goodman (1975) used three famous data sets to illustrate latent class analysis. The table below is the data as presented by Goodman. For further information on the background of the data sets, the reader should consult Goodman's article.

(a) If there are no constraints on the parameters, how many latent classes can be estimated from these data sets.

(b) Estimate the parameters of a latent class model with the highest number of classes from (a) and interpret the parameters.

(c) Evaluate the fit of a latent class model for each of the data sets.

Response pattern	Data set		
ABCD	Stouffer–Toby questionnaire	McHugh test	Lazarsfeld–Stouffer questionnaire
1111	42	23	75
1112	23	5	69
1121	6	5	55
1122	25	14	96
1211	6	8	42
1212	24	2	60
1221	7	3	45
1222	38	8	199
2111	1	6	3
2112	4	3	16
2121	1	2	8
2122	6	4	52
2211	2	9	10
2212	9	3	25
2221	2	8	16
2222	20	34	229

12.3. Consider again the three data sets in exercise 12.2.

(a) Fit a Rasch model to each of the data sets and estimate the parameters of the models.

(b) If the model fit the data interprete the parameters.

(c) If the model does not fit the data, use standardized residuals to identify reasons for the lack of fit.

12.4. The data below is from the Danish Welfare Study. The respondents were asked if they possessed (1) a freezer, (2) a dish washer, (3) a black and white TV–set, (4) a colour TV–set and (5) a swimming pool. The table below show with Yes=1 and No=2 the number of respondents for each response pattern. The purpose of analysing this data set was to study if an index for wealth could be derived from the responses.

An analysis by a latent structure model with a normal latent density reveals that the best fit is by a normal distribution with variance equal to zero.

The item parameters of a Rasch model are estimated to $(\hat{\epsilon}_1,....\hat{\epsilon}_5)=(1.997, -0.508,$ 1.166, 0.566, −3.220). The mean value of the best fitting normal density (with $\hat{\sigma}^2=0$) was estimated to $\hat{\mu}=-0.915$.

Response pattern	Number of respondent	Score	Number of respondent
11111	5	t=0	182
11112	63	1	1048
11121	12	2	2980
11122	328	3	849
11211	22	4	100
11212	398	5	5
11221	0		
11222	23		
12111	2		
12112	97		
12121	10		
12122	1739		
12211	12		
12212	1149		
12221	2		
12222	94		
21111	1		
21112	1		
21121	1		
21122	19		
21211	2		
21212	26		
21221	0		
21222	5		
22111	0		
22112	18		
22121	2		
22122	670		
22211	2		
22212	279		
22221	0		
22222	182		

(a) Show that the probability of obtaining score t in this case become

$$(t) = g(t;\epsilon_1,...,\epsilon_5)e^{t\mu} / \prod_{j=1}^{5} (1+e^{\epsilon_j+\mu})$$

(b) Using (a) find the expected number of respondent for each score value.

(c) Evaluate the fit of the distribution derived in (a).

(d) Describe the way the observed score distribution differ from the expected distribution.

12.5. Formann (1985) presented the second part of the Coleman–data, which was analyzed in example 12.3, and which only include the boys in the school classes, The table below, taken from Formann's paper, show the corresponding data for girls.

Response pattern	Number of girls
1111	484
1112	93
1121	107
1122	32
1211	112
1212	110
1221	30
1222	46
2111	126
2112	40
2121	768
2122	321
2211	74
2212	75
2221	303
2222	536

(a)   Analyze these data by a suitable latent class model.

(b)   Compare with the results obtained in sections 12.5 and 12.7.

12.6 Formann (1988) presented the responses to five items concerning the attitude toward nuclear energy. The sample consisted of 600 germans, who could answer each item with "agree"=1 and " do not agree"=2. The formulations of the five items were: (1) "In the near future, alternate sources of energy will not be able to substitute nuclear energy." (2) "It is difficult to decide between the different types of power stations if one carefully considers all their pros and cons." (3) "Nuclear power stations should not be put into operation before the problems of radioactive waste have been solved." (4) "Nuclear power stations should not be put in operation before it is proven that the radiation caused by them is harmless." (5) "The foreign power stations now in operation should be closed down."

(a)   Fit a latent class model to the data.

(b)   Interpret the parameters of the model.

Response pattern	Number of respondents
11111	24
11112	52
11121	2
11122	18
11211	2
11212	2
11221	0
11222	3
12111	15
12112	37
12121	2
12122	22
12211	5
12212	3
12221	0
12222	6
21111	61
21112	65
21121	1
21122	22
21211	16
21212	2
21221	2
21222	6
22111	118
22112	41
22121	1
22122	14
22211	39
22212	3
22221	11
22222	3

12.7. Clogg and Goodman (1984) presented a data set originally due to Solomon, based on the responses to four items concerning the attitude towards science and a scientific career. Responses were obtained from a group of students with high IQ and from a group with low IQ. The data set is shown below.

(a)   Analyse the responses from each of the two groups of students by a latent class model.

(b)   Compare the parameters estimated from the two data sets.

(c)   Analyse the data set obtained by combining the two groups of students to one group.

(d)   Compare the results of (a), (b) and (c).

Number of students with

Response pattern	High IQ	Low IQ
1111	122	62
1112	68	70
1121	33	31
1122	25	41
1211	329	283
1212	247	253
1221	172	200
1222	217	305
2111	20	14
2112	10	11
2121	11	11
2122	9	14
2211	56	31
2212	55	46
2221	64	37
2222	53	82

12.8. In an American study from 1947, a sample of 1729 individuals were cross–classified according to four items regarding general knowledge. The four items were: (1) read newspapers, (2) listen to radio, (3) read books and magazines, (4) attend lectures. The responses were Yes=1 and No=2. The number of respondents for each response pattern is shown below.

Response pattern	Number of respondents
1111	31
1112	169
1121	12
1122	94
1211	4
1212	32
1221	7
1222	63
2111	45
2112	378
2121	13
2122	231
2211	11
2212	150
2221	12
2222	477

(a)   Try to describe the data by a log–linear model for a four way contingency table with as few as possible interactions.

(b)   Analyse then the data by a latent class model or a latent structure model with a continuous latent variable.

(c)   Compare the results of (a) and (b).

12.9. The Law School Admission Test (LSAT) used extensively in the United States consists of a number of dichotomous items, which can be answered correctly or incorrectly. The table below show the observed distribution over response patterns for two sections of the LSAT, each compromising five items. The purpose of the LSAT is to produce a score for each respondent, which estimate his or her general ability to solve the problems presented as items in the test. These problems are constructed such that the higher the likelihood of solving them, the more probable it should be that the student taken the test will complete a law school with a good result.

(a)   For each of the sections formulate a latent structure model. Explain the parameters of the chosen model and relate the parameters and their estimates to the purpose of the LSAT.

(b)   Check the fit of the chosen model and estimate the parameters for each section.

(c)   Make a recommendation as to which of the 10 items to include in the test.

(d)   For the five items in section 6 estimate for each of the possible score values the value of the latent parameter.

| | Number of students | |
Response pattern	Section 6	Section 7
11111	298	308
11112	28	32
11121	61	136
11122	11	18
11211	173	35
11212	21	7
11221	56	25
11222	16	6
12111	80	90
12112	15	15
12121	28	51
12122	3	14
12211	81	34
12212	14	11
12221	29	39
12222	10	7
21111	15	28
21112	2	8
21121	3	23
21122	0	7
21211	16	7
21212	0	3
21221	8	5
21222	1	10
22111	4	17
22112	3	3
22121	1	19
22122	1	3
22211	11	7
22212	2	1
22221	6	19
22222	3	12

12.10. Engelen (1987) used the responses from 395 Dutch students on five items from a test in mathematics, called the IEA, to illustrate the use of latent structure models. He claimed that a Rasch model fitted the data well.

(a) Formulate a Rasch model for the data and check if Engelens claim is correct.

(b) Assume a normal latent density, estimate its parameters and interpret their meaning in the given context.

(c) Make a comparison applying a Rasch model and applying a latent class model to the data set.

Response pattern	Number of students
11111	74
11112	46
11121	1
11122	1
11211	20
11212	21
11221	0
11222	4
12111	2
12112	7
12121	0
12122	0
12211	2
12212	9
12221	3
12222	0
21111	18
21112	28
21121	2
21122	3
21211	22
21212	62
21221	2
21222	10
22111	0
22112	11
22121	0
22122	4
22211	3
22212	26
22221	0
22222	14

12.11. Cox, Przepiora and Plackett (1982) analysed the data shown below by various models. The data is described in the paper by Cox et al. as follows.

"Seven pathologists independently classified histological slides made from cold knife cone biopsies of the uterine cervix. Lesions on each of 118 slides were classified into five broad histological categories: 1, negative; 2, atypical squamous hyperplasia; 3, carcinoma in situ; 4, squamous carcinoma with early stromal invasions; 5, invasive carcinoma. These categories represent an ordering of the "involvement" of the lesion, 1 being the least involved and 5 the most involved. A complete slide is classified into one of the five categories on the basis of the most involved lesion identified. The study was designed to obtain information about variability in classification."

Slide No.	A	B	Pathologist C	D	E	F	G	Slide No.	A	B	Pathologist C	D	E	F	G
1	4	3	4	2	3	3	3	60	1	1	2	1	1	1	1
2	1	1	1	1	1	1	1	61	1	3	2	1	2	1	1
3	3	3	3	3	3	3	3	62	4	3	3	3	3	2	3
4	4	3	3	4	3	3	3	63	1	3	2	2	2	1	2
5	3	3	3	3	3	3	3	64	2	3	2	2	3	2	3
6	2	1	2	1	1	1	1	65	4	3	3	3	3	3	3
7	1	1	1	1	2	1	1	66	3	3	3	4	3	2	4
8	3	3	2	3	2	2	3	67	1	1	1	1	1	1	1
9	2	2	2	2	3	1	2	68	2	3	2	2	3	2	2
10	1	1	1	1	2	1	1	69	3	3	2	3	3	1	3
11	5	5	5	4	5	5	5	70	1	1	1	1	1	1	1
12	1	1	1	1	2	1	1	71	4	3	3	3	3	3	3
13	3	3	3	2	3	3	3	72	3	3	3	2	3	1	3
15	2	2	2	1	1	1	2	73	3	3	3	3	3	2	3
16	4	3	3	2	3	2	3	74	4	3	1	3	3	2	3
17	3	3	2	3	3	3	3	76	1	2	1	1	1	1	1
18	2	3	2	2	3	2	3	77	2	2	1	2	2	1	2
19	2	1	2	1	2	1	1	78	2	3	2	1	3	2	2
22	2	3	2	2	2	1	3	79	2	1	1	2	1	1	1
23	1	1	2	1	1	1	1	80	4	4	3	2	4	1	3
24	4	3	3	4	3	3	3	81	1	1	1	1	1	1	1
25	1	1	2	1	2	1	1	82	4	4	3	3	4	3	3
26	1	1	1	1	1	1	1	83	5	5	1	4	5	5	4
27	2	1	2	2	2	1	2	84	2	3	2	2	2	1	2
28	4	4	4	2	4	3	3	85	4	4	4	2	5	1	3
29	3	3	3	2	3	2	3	86	3	3	2	3	3	3	3
30	3	3	3	3	3	2	3	87	4	3	3	3	3	3	3
31	1	1	1	1	1	1	1	88	4	2	3	2	3	2	3
32	4	3	3	3	3	2	3	89	2	3	2	2	4	1	3
33	3	3	3	3	3	3	3	90	3	3	3	2	4	2	3
34	1	1	1	1	1	1	1	91	3	3	2	1	3	2	2
35	3	3	3	2	3	1	3	92	4	4	3	2	4	1	3
36	2	2	2	2	3	1	2	93	3	3	2	2	3	2	2
37	3	3	2	2	3	1	3	94	1	1	2	1	2	1	1
38	5	3	3	3	4	1	3	95	3	3	3	2	4	3	3
39	2	1	1	1	2	1	1	96	4	3	1	1	2	1	2
40	3	3	2	2	3	1	3	98	4	3	3	4	4	3	3
41	3	3	3	3	3	2	3	99	1	2	2	1	2	1	2
42	5	5	5	5	5	5	5	100	3	3	3	2	4	2	3
43	5	3	3	2	3	2	3	101	4	4	3	4	4	3	4
44	3	2	2	2	2	1	2	102	3	3	2	2	3	3	3
45	1	1	1	1	2	1	1	103	1	1	1	1	1	1	1
46	2	3	1	2	3	1	3	104	2	3	2	2	4	1	2
47	4	4	4	3	3	3	3	105	3	3	3	3	3	2	3
48	3	3	3	2	3	2	3	106	2	3	1	1	3	1	1
49	3	2	2	2	2	1	1	107	3	3	2	2	3	2	3
51	2	3	2	2	2	2	2	108	3	3	2	2	3	1	3
52	3	3	3	4	3	2	3	110	2	2	1	1	2	1	1
53	4	3	3	3	3	5	3	111	1	1	1	1	2	1	1
54	3	3	2	2	4	2	3	112	3	3	2	2	2	2	3
55	3	3	3	3	3	2	3	113	3	3	2	2	2	1	2
56	2	2	2	1	2	2	2	114	2	3	1	1	2	1	1
57	2	3	2	2	3	1	3	115	3	3	2	2	3	2	3
58	1	1	1	1	1	1	1	116	1	1	1	1	2	1	1
59	3	3	3	3	3	3	3	117	3	3	3	2	3	2	3

Cont.	A	B	C	D	E	F	G
118	3	3	2	2	3	1	3
119	1	1	1	1	2	1	1
120	1	1	1	1	1	1	1
121	2	2	1	1	2	1	2
122	5	3	4	2	3	4	3
123	4	3	4	2	4	1	3
124	1	1	1	1	2	1	1
126	2	3	1	1	2	1	2

(a) Your first task is to summarizes the data in two ways.

Table I. Form a two–way contingency table with the seven pathologists as rows and the 5 response categories as columns.

Table II. Merge response categories 1 and 2 into a new response category 1 and response categories 3,4 and 5 into a new category 2. Then form a table of number of observed slides for each response pattern of the seven pathologists.

(b) Analyze table I by a suitable association model from chapter 10 or by corres–pondance analysis (chapter 11).

(c) Although table II is rather sparse, try to analyse it by a latent class model.

(d) Is it possible to compare the results obtained in (b) and (c)?

12.12. In 1982 the Department of Defense in the USA sampled 776 young people in connection with a survey called the Profile of American Youth. The table below show the responses from this sample on four items from the Arithmetic Reasoning Test of the Armed Services Vocational Aptitude Battery, Form 8A. The observed numbers is shown for the complete sample and for the sample broken down into four subsamples based on demographic criteria. In the table a correct answer is indicated by 1 and an incorrect answer by 0.

(a) Try to fit each of the five data sets by a suitable latent structure model.

(b) Compare the results from the four demographic subsamples.

Subsample

Response pattern	1	2	3	4	Total sample
0000	23	20	27	29	99
0001	5	8	5	8	26
0010	12	14	15	7	48
0011	2	2	3	3	10
0100	16	20	16	14	66
0101	3	5	5	5	18
0110	6	11	4	6	27
0111	1	7	3	0	11
1000	22	23	15	14	74
1001	6	8	10	10	34
1010	7	9	8	11	35
1011	19	6	1	2	28
1100	21	18	7	19	65
1101	11	15	9	5	40
1110	23	20	10	8	61
1111	86	42	2	4	134
Total	263	228	140	145	776

REFERENCES

Agresti, A. (1982): Analysis of Ordinal Categorical Data. New York: John Wiley and Sons.

Agresti, A. (1983): A survey of strategies for modeling cross–classifications having ordinal variables. Jour.Amer.Statist. Assoc. 78, 184–198.

Agresti, A., Chiang, C. and Kezouh, A. (1987): Order–restricted score parameters in association models for contingency tables. Jour.Amer.Statist.Assoc., 82, 619–623.

Albert, A. and Andersson, J.A. (1984): On the existence of maximum likelihood estimates in logistic regression models. Biometrika, 71, 1–10.

Amemiya, T. (1981): Qualitative response models. A survey. Jour. Econ.Litterature, 19, 1483–1536.

Andersen, A.H. (1974): Multidimensional contingency tables. Scand.Jour.Stat. 1, 115–127.

Andersen , E.B. (1970): Asymptotic properties of conditional maximum likelihood estimators. Jour.Royal.Statist.Soc. B. 32, 283–301.

Andersen, E.B. (1973a): A goodness of fit test for the Rasch model. Psychometrika. 38, 123–140.

Andersen, E.B. (1973b): Conditional inference and multiple–choice questionnaires. Brit.Jour.Math.Stat.Psych. 26, 31–44.

Andersen, E.B. (1974): A reanalysis of lung cancer cases in Fredericia. (In Danish) Ugeskrift for Læger. 136/48, 2704–2705.

Andersen E.B. and Madsen, M. (1977): Estimating the parameters of the latent population distribution. Psychometrika. 42, 357–374.

Andersen, E.B. (1980a): Discrete Statistical Models with Social Science Applications. Amsterdam: North Holland Publishing Co.

Andersson, T.W. and Goodman, L.A. (1957): Statistical inference about Markov chains. Ann.Math.Stat. 28, 89–109.

Andersson, T.W. (1954): On estimation of parameters in latent structure analysis. Psychometrika, 19, 1–10.

Andersson, J.A. (1984): Regression and ordered categorical variables. Jour.Royal Statist.Soc.B., 46, 1–30.

Andersson, J.A. and Blair, V. (1982): Penalized maximum likelihood estimation in logistic regression and discrimination. Biometrika, 69, 123–136.

Andrich, D. (1978a): A binomial latent trait model for the study of Likert–style attitude questionnaires. Brit.Jour.Math.Statist.Psych. 31, 86–98.

Andrich, D. (1978b): A rating formulation for ordered response categories.Psychometrika. 43, 561–573.

Andrich, D. (1982): An extension of the Rasch model for ratings providing both location and dispersion parameters. Psychometrika. 47, 105–113.

Asmussen, S. and Edwards, D. (1983): Collapsibility and response variables in contingency tables. Biometrika. 70, 567–578.

Baker, R.J. and Nelder (1978): The GLIM System. Oxford: Numerical Algorithm Group.

Barndorff–Nielsen, O. (1978): Information and Exponential Families in Statistical Theory. New York: J. Wiley and Sons.

Bartholomew, D.J. (1980): Factor analysis for categorical data. Jour.Royal Stat.Soc.B. 42, 293–321.

Begg, C.B. and Gray, R. (1984): Calculation of polychotomous logistic regression parameters using individualized regressions, 71, 11–18.

Benzecri, J.–P. (1973): L'Analyse des Données. Tome 2: L'Analyse des Correspondances. Paris: Dunod.

Berkson, J. (1944): Application of logistic functions to bio–assay. Jour.Amer.Statist. Assoc. 39, 357–365.

Berkson, J. (1953): A statistically precise and relatively simple method of estimating the bio–assay with quantal response, based on the logistic function. Jour.Amer. Statist. Assoc. 48, 565–599.

Birch, M.W. (1963): Maximum likelihood in three–way contingency tables. Jour.Royal Stat.Soc. B., 25, 220–233.

Birch, M.W. (1964): A new proof of the Pearson–Fisher theorem. Annals Math. Statist. 35, 817–824.

Birnbaum, A. (1968): Logistic item trait models. In: Lord, F.M. and Novick, M.R.: Statistical Theories of Mental Test Scores. Reading: Addison and Wesley.

Bishop, Y.M.M. (1969): Incomplete thre–dimensional contingency tables. Biometrics. 25, 119–128.

Bishop, Y.M.M. (1970): Effects of collapsing multidimensional contingency tables. Biometrics. 27, 545–562.

Bishop, Y.M.M. and Feinberg, S.E. (1969): Incomplete two–dimensional contingency tables. Biometrics. 25, 119–128.

Bishop, Y.M.M., Feinberg, S.E. and Holland, P.W. (1975): Discrete Multivariate Analysis. Theory and Practice. Cambridge: MIT–press.

Bock, R.D. (1972): Estimating the parameters and latent ability when responses are scored in two or more nominal categories. Psychometrika.37, 29–51.

Bock, R.D. and Aitkin, M. (1981): Marginal maximum likelihood estimation of item parameters: Application of an EM algorithm. Psychometrika. 46, 443–459.

Box, G.E.P. and Cox, D.R. (1964): An analysis of transformations. Jour. Royal Statist. Soc. B, 26, 211–252.

Bowker, A.H. (1948): A test for symmetry in contingency tables.Jour.Amer.Statist.Assoc. 43, 572–574.

Caussinus, H. (1965): Contribution a l'analyse Statistique des tableaux de correlation. Ann.Fac.Sci.Univ. Toulouse. 29, 77–182.

Caussinus, H. (1986): Discussion of L.A. Goodman: Some useful extensions of the usual correspondance analysis approach and the usual log–linear models approach in the analysis of contingency tables. Int.Statist. Review. 54, 243–309.

Chen, T. and Fienberg, S.E. (1986): The analysis of contingency tables with incompletely classified data. Biometrics. 32, 133–144.

Choulakian, V. (1988): Exploratory analysis of contingency tables by log–linear formulation and generalizations of correspondence analysis. Psychometrika. 53, 235–250.

Christofferson, A. (1975): Factor analysis of dichotomyzed variables. Psychometrika. 40, 5–32.

Christofferson A. and Muthen, B. (1981): Simultaneous factor analysis of dichotomous variables in several groups. Psychometrika. 46, 407–419.

Clemmensen, J., Hansen, G., Nielsen, A., Roøje, J., Steensberg, J., Sørensen, S. and Toustrup, J. (1974): Lung cancer and air pollution in Fredericia. (In Danish). Ugeskrift for Læger. 136, 2260–2268.

Clogg, C. (1979): Some latent structure models for the analysis of Likert–type data. Social Science Research, 8, 287–301.

Clogg, C.C. (1981): Latent structure models of mobility. Amer.Jour.Sociology, 86, 836–868.

Clogg, C.C. (1982a): Some models for the analysis of association in multiway cross–classification having ordered categories. Jour.Amer.Statist.Assoc. 77, 803–815.

Clogg, C.C. (1982b): Using association models in sociological research: Some examples. Amer.Jour.Sociology, 88, 114–134.

Clogg, C.C. and Goodman, L.A. (1984): Latent structure analysis of a set of multidimensional contingency tables. Jour.Amer.Statist.Soc. 79, 762–771.

Clogg, C.C. and Goodman, L.A. (1986): On scaling models applied to data from several groups. Psychometrika, 51, 123–135.

Connolly, M. A. and Liang, K–Y. (1988): Conditional logistic regression models for corre lated binary data. Biometrika. 7, 501–506.

Cox, D.R: (1970): The Analysis of Binary Data. London: Methuen and Co.

Cox, D.R. and Snell, E.J. (1968): A general definition of residuals. Jour.Royal Statist.Soc. B, 30, 248–275.

Cox, D.R. and Snell, E.J. (1971): On test statistics computed from residuals. Biometrika, 58, 589–594.

Cox, M.A.A., Przepiora, P. and Plackett, R.L. (1982): Multivariate contingency tables with ordinal data. Utilitas Mathematica. 21A, 29–42.

Cressie, N. and Holland, P.W. (1983): Characterizing the manifest probability of latent trait models. Psychometrika. 48, 129–142.

Cressie, N. and Read, T.R.C. (1984): Multinomial Goodness of fit tests. Jour.Royal Statist.Soc. B. 46, 440–464.

Dale, J.R. (1986): Asymptotic normality of goodness–of–fit test statistics for sparse product multinomials. Jour.Royal Statist.Soc. B. 48, 48–59.

Darroch, J.N., Lauritzen, S.L. and Speed, T.P. (1980): Markov fields and log–linear interaction models for contingency tables. Annals Statist. 8, 522–539.

Dayton, C.M. and MacReady, G.B. (1980): A scaling model with response errors and intrinsically inscalable respondents. Psychometrika. 45, 343–356.

Deming, W.E. and Stephan, F.F. (1940): On least squares adjustment of a sampled frequency table when the expected marginal totals are known. Ann.Math.Stat. 11, 427–444.

Dempster, A.P., Laird, N.M. and Rubin, D.B. (1977): Maximum likelihood from incomplete data via the EM–algorithm. Jour.Royal Stat.Soc. B, 39, 1–22.

Diaconis, P. and Efron, B. (1985): Testing the independence of two–way table: New interpretations of the chi–square statistic. Annals Statist. 13, 845–913.

Edwards, D. and Kreiner, S. (1983): The analysis of contingency tables by graphical models. Biometrika, 70, 553–565.

Edwards, D. and Havranek, T. (1985): A fast procedure for model search in multidimensional contingency tables. Biometrika. 72, 339–351.

Edwards, D. and Havranek, T. (1987): A fast model selection procedure for large families of models. Jour.Amer.Statist.Assoc. 82, 205–213.

Efron, B. (1986): Double exponential families and their use in generalized linear regression. Jour.Amer.Statist.Assoc. 81, 709–721.

Engelen, R.J.H. (1978): Semiparametric estimation in the Rasch model. Research Report 87–1, Department of Education, University of Twente.

Escofier, B. (1979): Traitement simultané de variables qualitatives et quantitatives en analyse factorielle. Cahiers de l'Analyse des Donnés, 4, 137–146.

Escoufier, Y. (1982): L'Analyse des tableaux de contingence simples et multiples. Metron, 40, 53–77.

Fienberg, S.E. (1970): Quasi independence and maximum likelihood estimation in incomplete contingency tables. Jour.Amer.Statist.Assoc. 65, 1610–1616.

Fienberg, S.E. (1972): The analysis of incomplete multi–way contingency tables. Biometrics. 28, 177–202.

Fienberg, S.E. (1980): The analysis of Cross–classified Categorical Data. 2.ed. Cambridge: The MIT Press.

Firth, D. (1989): Marginal homogeneity and the superposition of Latin squares. Biometrika. 76, 179–182.

Fischer, G.H. (1977): Some probabilistic models for the description of attitudinal and behavioral changes under the influence of mass communication. In: Kempf, W.F. and Repp, B.H. : Mathematical Models for Social Psychology. Vienna: Hans Huber.

Fischer, G.H. (1981): On the existence and uniqueness of maximum likelihood estimates in the Rasch model. Psychometrika. 46, 59–77.

Fischer, G.H. (1983): Logistic latent trait models with linear constraints. Psychometrika. 48, 3–26.

Fisher, R.A. (1935): The Design of Experiments. Edinburgh: Oliver and Boyd.

Formann, A. K. (1985): Constrained latent class models: Theory and applications. Brit.Jour.Math.Statist.Psych. 38, 87–111.

Formann, A. K. (1988): Latent class models for non–monotone dichotomous items. Psychometrika. 53, 45–62.

Fowles, E:B. (1987): Some diagnostics for binary logistic regression via smoothing. Biometrika. 74, 503–515.

Gifi, A. (1981): Non–linear Multivariate Analysis. Leiden: Department of Data Theory. University of Leiden.

Gilula, Z. (1984): On some similarities between canonical correlation models and latent class models for two–way contingency tables. Biometrika, 71, 523–530.

Gilula, Z. (1986): Grouping and association in contingency tables: An exploratory canonical correlation approach. Jour.Amer.Statist.Assoc. 81, 773–779.

Gilula, Z. and Haberman, S.J. (1986): Canonical analysis of contingency tables by maximum likelihood. Jour.Amer.Statist.Assoc. 81, 780–788.

Glas, C.A.W. (1988): The derivation of some tests for the Rasch model from the multinomial distribution. Psychometrika. 53, 525–546.

Gleser, L.J. and Moore D.S. (1985): The effect of positive dependence on chi–square tests for categorical data. Jour.Royal Statist.Soc. B. 47, 459–465.

Gokhale, D.V. (1981): The minimum discriminant information analysis of contingency tables in traffic safety studies. In: Fleischer, G.A. (ed.): Contingency Table Analysis for Road Safety Studies. Alphen aan den Rijn: Sijthoff and Nordhoff.

Goldstein, H. (1987): The choice of constraints in correspondance analysis. Psychometrika. 52, 207–215.

Goodman, L.A: (1965): On the statistical analysis of mobility tables. Amer.Jour. Sociol. 70, 564–584.

Goodman, L.A. (1968): The analysis of cross–classified data: Independence, quasi–independence and interactions in contingency tables with and without missing entries. Jour. Amer. Statist. Ass. 63, 1091–1131.

Goodman, L.A. (1971): Some multiplicative models for the analysis of cross classified data. Sixth Berkeley Symposium on Probability and Mathematical Statistics. Vol.I, 649–696

Goodman, L.A. (1972): A general model for the analysis of surveys. Amer.Jour.Sociology. 77, 57–109.

Goodman, L.A. (1973): Causal analysis of data. Amer.Jour.Sociology. 78, 173–229.

Goodman, L.A. (1974): Exploratory latent structure analysis using both identifiable and unidentifiable models. Biometrika. 61, 215–231.

Goodman, L.A. (1975): A new model for scaling response patterns: An application of the quasi–independence concept. Jour.Amer. Statist.Assoc. 70, 363–401.

Goodman, L.A. (1978): Analyzing Qualitative/Categorical Data.Log–linear Models and Latent Structure Analysis. London: Addisson and Wesley.

Goodman, L.A. (1979a): On the estimation of parameters in latent structure analysis. Psychometrika. 44, 123–128.

Goodman, L.A. (1979b): Simple methods for the analysis of association in cross–classifications having ordered categories. Jour. Amer.Statist.Assoc. 74, 537–552.

Goodman, L.A. (1981a): Association models and canonical correlation in the analysis of cross–classifications having ordered categories.Jour.Amer.Statist.Assoc. 76, 320–334.

Goodman, L.A. (1981b): Association models and the bivariate normal for contingency tables with ordered categories. Biometrika, 68, 347–355.

Goodman, L.A. (1985): The analysis of cross–classified data having ordered and/or unordered categories: Association models, correlation models and asymmetry models for contingency tables with and without missing entries. Annals Statist. 13, 10–69.

Goodman, L.A. (1986): Some useful extensions of the usual correspondance analysis approach and the usual log–linear models approach in the analysis of contingency tables. Int.Statist.Rev. 54, 243–309.

Green, (1984): Iterative reweighted least squares for maximum likelihood estimation and some robust and resistant alternatives. Jour.Royal Statist.Soc.B. 46, 149–192.

Greenacre, M. J. (1984): Theory and applications of Correspondance Analysis. Academic Press.

Greenacre, M.J. (1988): Correspondance analysis of multivariate categorical data by weighted least–squares. Biometrika. 75, 457–467.

Greenacre, M. and Hastie, T. (1987): The geometric interpretation of correspondance analysis. Jour.Amer.Statist.Assoc. 82, 437–447.

Guerro, V.M. and Johnson, R.A. (1982): Use of the Box–Cox transformation with binary response models. Biometrika, 69, 309–314.

Gustavsson, J.E. (1980): Testing and obtaining fit of data to the Rasch model Brit.Jour.Math. Statist. Psych. 33, 205–233.

Haberman, S.J. (1973a): The analysis of residuals in cross–classified tables. Biometrics, 29, 205–220.

Haberman, S.J. (1973b): Log–linear models for frequency data: Sufficient statistics and likelihood equations. Ann.Stat. 1, 617–632.

Haberman, S.J. (1974a): Log–linear model for frequency tables with ordered classification. Biometrics. 30, 589–600

Haberman, S.J. (1974b): The Analysis of Frequency Data. Chicago: The University of
Chicago Press.

Haberman, S.J. (1977a): Log–linear models and frequency tables with small cells counts.
Annals Statist. 5, 137–145.

Haberman, S.J. (1978): Analysis of Qualitative Data. Vol.I. New York: Academic Press.

Haberman, S.J. (1979): Analysis of Qualitative Data Vol.II. New York: Academic Press.

Hansen, E.J. (1978): The Distribution of Living Conditions. Main Results from the
Welfare Survey. (In Danish). Danish National Institute for Social
Research. Publication 82. Copenhagen: Teknisk Forlag.

Hansen, E.J. (1984): Social Groups in Denmark (In Danish). Study No.48. Danish
National Institute for Social Research. Copenhagen. Teknisk Forlag.

Havranek, T. (1984): A procedure for model search in multidimensional contingency
tables. Biometrics, 40, 95–100.

van der Heijden, P.G.M. and de Leeuw, J. (1985): Correspondance analysis used
complementary to log–linear analysis. Psychometrika. 50, 429–447.

van der Heijden, P.G.M. (1987): Correspondance analysis of Longitudinal Categorical
Data. Leiden: DSWO Press.

van der Heijden, P.G.M., Falguerolles, A. and de Leeuw, J. (1987): A combined approach
for contingency table analysis using correspondance analysis and log–
linear analysis. Research Report. Leiden University: Department of
Psychology.

Henry, N.W. and Lazarsfeld, P.F. (1968): Latent Structure Analysis. Boston: Houghton
Mifflin and Co.

Holm, S. (1979): A simple sequentially rejective multiple test procedure. Scand.
Journ.Stat., 6, 65–70.

Holt, D., Scott, A.J. and Ewings, P.D. (1980): Chi–squared tests with survey data.
Jour.Royal Stat.Soc. A, 143, 303–320.

Hommel, G. (1988): A stagewise rejective multiple test procedure based on a modified
Bonferroni test. Biometrika. 75, 383–386.

Imrey, P.B, Koch, G.G. and Stokes, M.E. (1981): Categorical data analysis: Some reflections on the log–linear model and logistic regression. Part I: Historical and methodological overview. Int.Statist.Review, 49, 265–283.

Imrey, P.B, Koch, G.G. and Stokes, M.E. (1982): Categorical data analysis: Some reflections on the log–linear model and logistic regression. Part II: Data analysis. Int.Statist. Review, 50, 35–63.

Jennings, D.E. (1986): Outliers and residual distributions in logistic regression. Jour.Amer.Statist.Assoc. 81, 987–990.

Johnson, W. (1985): Influence measures for logistic regresssion: Another point of view. Biometrika, 72, 59–65.

Jørgensen, B. (1984): The delta algorithm and GLIM. Int.Statist.Review, 52, 283–300.

Kay, R. and Little, S. (1987): Transformations of the explanatory variables in the logistic regression model for binary data. Biometrika. 74, 495–501.

Kelderman, H. (1984): Log–linear Rasch model tests. Psychometrika. 49, 223–245.

Kjær, A. (1978): Redundancy in the Labour Market. Literature and Concepts. (In Danish). Study No.36. The Danish National Institute of Social Research. Copenhagen: Teknisk Forlag.

Koehler, K.J. (1986): Goodness of fit test for log–linear models in sparse contingency tables. Jour.Amer.Statist.Assoc. 81, 483–493.

Kotze, T.J. v W. and Hawkins, D.M. (1984): The identification of outliers in two–way contingency tables using 2x2 subtables. Appl.Statist. 33, 215–223.

Kreiner, S. (1987): On collapsability of multidimensional contingency tables. Symposium in Applied Statistics. Copenhagen: UNI–C.

Landwehr, J.M., Pregiborn, D. and Shoemaker, A.C. (1984): Graphical methods for assessing logistic regression models. Jour.Amer.Statist.Assoc. 79, 61–83.

Lawley, D.N. (1943): On problems connected with time selection and test construction. Proc.Royal Society of Edinburgh. 61, 273–287.

Lazarsfeld, P.F. (1950): The logical and mathematical foundation of latent structure analysis. In: Stouffer et al.: Measurement and Prediction. Princeton: Princeton University Press.

Lebart, L., Morineau, A. and Warwick, K. (1984): Multivariate Descriptive Statistical Analysis. New York: J. Wiley and Sons.

Lehmann, E.L. (1959): Testing Statistical Hypotheses. N.Y.: John Wiley and Sons.

de Leeuw, J. and Verhelst, N. (1986): Maximum likelihood estimation in generalized Rasch models. Jour.Educ.Statist. 11, 183–196.

de Leeuw, J. and van der Heijden, P.G.M. (1988): Correspondance analysis of incomplete contingency tables. Psychometrika. 53, 223–233.

Lindsay, B.G. (1987): Semi–parametric estimation in the Rasch model. Research Report. Department of Statistics. Pennsylvania State University.

Lindsay, B. Clogg, C.C. and Grego, J. (1989): Semi–parametric estimation in the Rasch model and related exponential response models, including a simple latent class model for item analysis. Research Report. Department of Statistics. Pennsylvania State University.

Lord, F.M. (1952): A Theory of Test Scores. Psychometric Monograph. No.7.

Lord, F.M. (1967): An analysis of the Verbal Scholastic Aptitude Test using Birnbaum's three–parameter logistic model. Research Bulletin 67–34. Princeton: Educational Testing Service.

Lord, F.M. and Novick, M.R. (1968): Statistical Theories of Mental Test Scores. Reading: Addison and Wesley.

Louis, T.A: (1982): Finding the observed information matrix when using the EM–algorithm. Jour.Royal Statist.Soc. B. 44, 226–233.

Manski, C.F. (1981): Econometric models of probabilistic choice. In: Manski, C.F. and McFadden,D.: Structural Analysis of Discrete Data. Cambridge: MIT Press.

Masters, G.N. (1982): A Rasch model for partial credit scoring. Psychometrika. 47, 149–174.

Masters, G.N. (1985): A comparison of latent trait and latent class analyses of Likert–type data. Psychometrika. 50, 69–82.

Masters, G.N. and Wright, B.D: (1984): The essential process in a family of measurement models. Psychometrika. 49, 529–544.

McCullagh, P. (1980): Regression models for ordinal data. Jour.Royal Stat.Soc. B, 42, 109–142.

McCullagh, P. (1982): Some applications of quasisymmetry. Biometrika, 69, 303–308.

McHugh, R.B. (1956): Efficient estimation and local identification in latent class analysis. Psychometrika. 21, 331–347.

McNemar, H. (1947): Note on the sampling error of the differences between correlated proportions or percentages. Psychometrika. 12, 153.157

Mislevy, R.J. (1984): Estimating latent distributions. Psychometrika. 49, 359–381.

Mislevy, R.J. (1987): Exploiting auxiliary infernation about examinees in the estimation of item parameters. Appl. Psych.Meas. 11, 81–91.

Molenaar, I.W. (1983): Some improved diagnostics for failure of the Rasch model. Psychometrika. 48, 49–72.

Muthen, B. (1978): Contributions to factor analysis of dichotomous variables. Psychometrika. 43, 551–560.

Muthen, B. (1979): A structural probit model with latent variables. Jour.Amer. Statist.Assoc. 74, 807–811.

Nelder, and McCullogh, (1983): Generalized linear models. London: Chapman and Hall.

Nelder, and Wedderburn (1972): Generalized linear models. Jour.Royal Statist.Soc. A, 135, 370–384.

Nishisato, S. (1980): Analysis of Categorical Data. Dual Scaling and its Applications. Toronto: University of Toronto Press.

Nordberg, L. (1980): Asymptotic normality of maximum likelihood estimates based on independent, unequally distributed observations in exponential family models. Scand. Jour.Statist. 7, 27–32.

Obel, E.B. (1975): A comparative study of patients with cancer of the ovary who have survived more or less than 10 years. Acta Obst. Gynecol.Scand. 55, 429–439.

Olsen, H. and Hansen, G. (1977): Retirement from work. Publication No.79. (In Danish). Danish National Institute for Social Research. Copenhagen: Teknisk Forlag.

Olsen, H. (1984): Early retirement or partial pension. (In Danish). Communication No.42. Copenhagen: The Danish National Institute for Social research.

Petersen, E. (1968): Job satisfaction in Denmark. (In Danish). Copenhagen: Mental–hygiejnisk Forlag.

Pierce, D.A. and Schafer, D.W. (1986): Residuals in generalized linear models. Jour. Amer.Statist. Assoc. 81, 977–986.

Plackett, R.L. (1981): The Analysis of Categorical Data. 2.ed. London: Griffin.

Poulsen, C.S. (1981): Latent class analysis of consumer complaining behaviour. In: Hoskulson et al. (ed.) Symposium in Applied Statistics. Copenhagen: NEUCC.

Pregiborn, D. (1981): Logistic regression diagnostics. Ann.Statist. 9, 705–724.

Qu, Y.S., William, G.W., Beck, G.J. and Goormastic, M. (1987): A generalized model of logistic regression for correlated data. Comm.Statist.A, 16, 3447–76.

Rao, C.R. (1973): Linear Statistical Inference and its Applications. 2.ed. New York: John Wiley and Sons.

Rao, J.N.K. and Scott, J. (1981): The analysis of categorical data from complex sample surveys: Chi–squared tests for goodness of fit and independence in two–way tables. Jour.Amer.Statist.Assoc. 76, 221–230.

Rasch, G. (1960): Reprinted as Rasch (1980).

Rasch, G. (1961): On general laws and the meaning of measurement in psychology. Proc. Fourth Berk.Symp.Math.Stat. and Prob. 5, 321–333.

Rasch, G. (1966): The Theory of Statistics. Lecture notes by Ulf Christiansen. (In Danish) Copenhagen: Department of Statistics. University of Copen–hagen.

Rasch, G. (1980): Probabilistic Models for Some Intelligence and Attainment Test. Chicago: The University of Chicago Press.

Read, C.B. (1978): Tests of symmetry in three–way contingency tables. Psychometrika. 43, 409–420.

Rigdon, S.E. and Tsutakawa, R.K. (1983): Parameter estimation in latent trait models. Psychometrika. 48, 567–574.

Roberts, G., Rao, J.N.K. and Kumar, S. (1987): Logistic Regression analysis of sampling survey data. Biometrika, 74, 1–12.

Rost, J. (1988): Rating scale analysis with latent class models. Psychometrika. 53, 327–348.

Samejima, F. (1969): Estimation of latent ability using a response pattern of graded scores. Psychometrika. Monograph Supplement 17.

Sanathanan, L. and Blumenthal, S. (1978): The logistic model and estimation of latent structure. Jour.Amer.Statist.Assoc. 73, 794–799.

Schaffer, J.P. (1986): Modified sequentially rejective multiple test procedures. Jour.Amer.Statist.Assoc. 81, 826–831.

Simon, G. (1974): Alternative analyses for the singly–ordered contingency table. Jour.Amer.Statist.Assoc. 69, 971–976.

Sundberg, R. (1974): Maximum likelihood theory for incomplete data from an exponential family. Scand.Jour.Stat. 2, 1–49.

Svalastoga, K. (1959): Prestige, Class and Mobility. London: William Heinemann.

Tavare, S. and Altham, P.M.E. (1983): Serial dependence of observations leading to contingency tables, and corrections to chi–squared statistics. Biometrika. 70, 139–144.

Tenenhaus, M. and Young, F.W. (1985): An analysis and synthesis of multiple correspondance analysis, optimal scaling, dual scaling, homogeneity analysis and other methods for quantifying categorical data. Psychometrika. 50, 91–119.

Thissen, D. (1982): Marginal maximum likelihood estimation for the one–parameter logistic model. Psychometrika. 47, 175–186.

Thissen, D. and Steinberg, L. (1986): A taxonomy of item response models. Psychometrika. 51, 567–577.

Tjur, T. (1982): A connection between Rasch's item analysis model and a mutiplicative Poisson model. Scand.Jour.Stat. 9, 23–30.

Wermuth, N. (1976): Model search among multiplicative models. Biometrics. 32, 253–263.

Wermuth, N. (1987): Parametric collapsability and the lack of moderating effects in contingency tables with a dichotomous response variable. Jour.Royal Statist.Soc. B, 49, 353–364.

Wermuth, N. and Lauritzen, S.L. (1983): Graphical and recursive model for contingency tables. Biometrika, 70, 537–552.

White, A.A., Landis, J.R. and Cooper, M.M. (1982): A note on the equivalence of several marginal homogeneity test criteria for categorical data. Int.Statist.Review, 50, 27–34.

Williams, E.J. (1952): Use of scores for the analysis of association in contingency tables. Biometrika, 39, 274–289.

Wollenberg, A.L. (1982): Two new test statistics for the Rasch model. Psychometrika. 47, 123–140.

Wu, C.F.J. (1983): On the convergence properties of the EM–algorithm. Annals of Statist. 11, 95–103.

Author Index

Subject Index

Examples With Data

E.B. Andersen,
N.-E. Jensen,
N. Kousgaard,
University of
Copenhagen,
Denmark

Statistics for Economics, Business Administration and the Social Sciences

1987. XI, 439 pp. 122 figs. Softcover DM 55,-
ISBN 3-540-17720-5

Contents: Introduction.- Descriptive Statistics.- Probability Theory.- Probability distributions on the real line and random variables.- Mean values and variances.- Special discrete distributions.- Special continuous distributions.- Multivariate distributions.- The distribution of sample functions and limit theorems.- Estimation.- Confidence intervals.- Testing statistical hypotheses.- Models and tests related to the normal distribution.- Simple linear regression.- Multiple linear regression.- Heteroscedasticity and autocorrelation.- Survey sampling.- Applications of the multinomial distribution.- Analysis of contingency tables.- Appendix table.- Index.- Index of examples with real data.

The aim of this book is to enable the student to understand the reasoning underlying a statistical analysis and to apply statistical methods to problems likely to be met within the fields of economics, public administration and business administration. The topics covered by the book are:

- methods for exploratory data analysis

- probability theory and standard statistical distributions

- statistical inference theory

- and three main areas of application: regression analysis, survey sampling and contingency tables.

Springer-Verlag
Berlin Heidelberg
New York London
Paris Tokyo
Hong Kong

The treatment of exploratory data analysis, regression analysis and the analysis of contingency tables are based on the most recent theoretical developments in these areas. Most of the examples have never been presented before in English textbooks.

06/23607/SF